SPACE TELESCOPE SCIENCE INSTITUTE

SYMPOSIUM SERIES: 7
*Series Editor* S. Michael Fall, Space Telescope Science Institute

# EXTRAGALACTIC BACKGROUND RADIATION

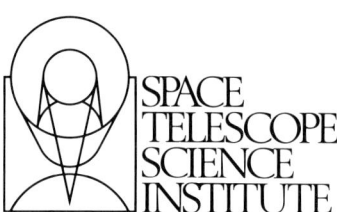

Other titles available in the Space Telescope Science Institute Symposium Series

1   Stellar Populations
    Edited by C. A. Norman, A. Renzini and M. Tosi  1987  0 521 33380 6
2   Quaser Absorption Lines
    Edited by C. Blades, C. A. Norman and D. Turnshek  1988  0 521 34561 8
3   The Formation and Evolution of Planetary Systems
    Edited by H. A. Weaver and L. Danly  1989  0 521 36633 X
4   Clusters of Galaxies
    Edited by W. R. Oegerle, M. J. Fitchet and L. Danly  1990  0 521 38462 1
5   Massive Stars in Starbursts
    Edited by C. Leitherer, N. R. Walborn, T. M. Heckman and C. A. Norman
    1991  0 521 404657
6   Astrophysical Jets
    Edited by D. Burgarella, M. Livio and C. P. O'Dea
    1993  0 521 44221 4

# EXTRAGALACTIC BACKGROUND RADIATION
## A Meeting in Honor of Riccardo Giacconi

Proceedings of the Extragalactic Background Radiation Meeting
Baltimore 1993 May 18–20

*Edited by*

DANIELA CALZETTI
*Space Telescope Science Institute*

MARIO LIVIO
*Space Telescope Science Institute*

PIERO MADAU
*Space Telescope Science Institute*

Published for the
Space Telescope Science Institute

CAMBRIDGE
UNIVERSITY PRESS

Published by the Press Syndicate of the University of Cambridge
The Pitt Building, Trumpington Street, Cambridge CB2 1RP
40 West 20th Street, New York, NY 10011-4211, USA
10 Stamford Road, Oakleigh, Melbourne 3166, Australia

© Cambridge University Press 1995
Individual papers in this edition are © Space Telescope Science Institute 1995

First published 1995

Printed in Great Britain at the University Press, Cambridge

*A catalogue record of this book is available from the British Library*

*Library of Congress cataloguing in publication data available*

ISBN 0 521 49558 X hardback

# CONTENTS

*Preface*     xi

*Participants*     xiii

*Chapter 1*

**Introduction**     1
P. J. E. PEEBLES
Introduction     1
Large-Scale Structure     1
Cosmogony     5
References     9
Discussion     10

*Chapter 2*

**Extragalactic Gamma-Ray Background**     15
N. GEHRELS, C. CHEUNG
Introduction     15
Low-Energy Measurements (0.1–10 MeV)     17
Interpretation of Low-Energy GRB Spectrum     18
High-Energy Measurements of GRB (>10 MeV)     20
Interpretation of High Energy GRB Spectrum     21
*COMPTON* Results and Implications for GRB     24
Other Topics     28
References     29
Discussion     32

*Chapter 3*

**The X-Ray Background (Observations)**     37
G. ZAMORANI
Introduction     37
Early History     38
ROSAT Deep Surveys Data     39
AGN Spectra and Fits to the XRB Spectrum     44
Conclusions     46
References     46
Discussion     48

## Chapter 4

**Extragalactic Ultraviolet Background Radiation** — 51
R. C. HENRY, J. MURTHY

| | |
|---|---|
| Introduction | 51 |
| The Ultraviolet Background at High Latitude | 52 |
| Observations of Diffuse Background Generally | 56 |
| Line Emission from the Interstellar Medium | 56 |
| Starlight Scattering from Interstellar Dust | 57 |
| Voyager Observations of the Diffuse Background | 65 |
| Discussion | 67 |
| Conclusion | 70 |
| References | 70 |
| Discussion | 72 |

## Chapter 5

**Ultraviolet Background (Theory)** — 75
P. JAKOBSEN

| | |
|---|---|
| Introduction | 75 |
| Diffuse UV Emission from the Intergalactic Medium | 76 |
| The Integrated UV Light of Galaxies and Quasars | 87 |
| Radiative Decay of Exotic Particles | 90 |
| Summary and Conclusions | 92 |
| Appendix: Diffuse Background from Extragalactic Sources | 93 |
| References | 94 |
| Discussion | 96 |

## Chapter 6

**The Optical Extragalactic Background Radiation** — 103
J. A. TYSON

| | |
|---|---|
| Introduction | 103 |
| Faint Surface Brightness Surveys | 106 |
| The Nature of the Faint Blue Galaxies | 110 |
| The Optical EBL from Discrete Objects | 121 |
| Summary | 125 |
| References | 126 |
| Discussion | 128 |

## Chapter 7

**Infrared Background (Observations)** — 135
M. G. HAUSER

| | |
|---|---|
| Introduction | 135 |
| The Search for the Cosmic Infrared Background | 136 |
| The Observational Challenge | 137 |
| The COBE Diffuse Infrared Background Experiment | 137 |
| Current Limits on the Cosmic Infrared Background Radiation | 138 |
| Conclusion | 140 |
| References | 141 |
| Discussion | 141 |

## Chapter 8

**The Infrared Background (Theory)** — 145
C. J. LONSDALE
Introduction — 145
Integrated Infrared Emission of Galaxies — 146
Primeval Galaxies and the Far-Infrared Background — 155
The Future in Space: DIRBE, ISO and SIRTF — 161
Conclusions — 163
References — 164
Discussion — 166

## Chapter 9

**Microwave Background Radiation (Observations)** — 169
J. C. MATHER
Introduction — 169
COBE Mission — 171
FIRAS Instrument and Measurements — 173
FIRAS Interpretation — 177
DMR Instrument and Observations — 180
DMR Results — 182
COBE Data Products — 186
Summary — 186
References — 187
Discussion — 188

## Chapter 10

**Detection of Degree Scale Anisotropy** — 191
P. M. LUBIN
Introduction and Current Status — 191
CBR Anisotropy Measurements — 193
Instrumental Considerations — 193
History of the ACME Experiments — 195
The MAX Experiment — 196
Results — 198
Goals for the Future — 199
Atmosphere — 200
Galactic Emission — 200
Extra-Galactic Sources — 200
Sidelobe Issues (Off Axis Response) — 201
Detector Limitations—Present and Fundamental — 201
Spectrum Measurements — 202
Polarization — 202
To Space — 203
References — 203
Discussion — 204

## Chapter 11

**Cosmic Microwave Background Anisotropies and Structure Formation in the Universe** — 209
N. VITTORIO

| | |
|---|---|
| Introduction | 209 |
| CMB Anisotropy on Large Angular Scales | 210 |
| LSS Models | 212 |
| CMB Anisotropy on Small and Intermediate Angular Scales | 213 |
| X-Ray Clusters | 216 |
| Conclusions | 217 |
| References | 218 |
| Discussion | 219 |

## Chapter 12

**The Radio Background Emission—The Long and Short of It** — 223
M. S. LONGAIR

| | |
|---|---|
| Salutations | 223 |
| The Radio Background Radiation | 224 |
| The Background Radiation and Galaxy Formation Revisited | 227 |
| Submillimetre Cosmology | 230 |
| The Background Radiation and Galaxy Formation | 232 |
| References | 234 |
| Discussion | 235 |

## Chapter 13

**The Radio Background: Radio-Loud Galaxies at High and Low Redshifts** — 237
J. A. PEACOCK

| | |
|---|---|
| Introduction | 237 |
| The Smooth Radio Background | 238 |
| What Makes a Radio-Loud Galaxy? | 238 |
| Luminosity Functions | 241 |
| HI Searches for High-Redshift Galaxies | 246 |
| Stellar Populations at High Redshift | 247 |
| Conclusions | 253 |
| References | 254 |
| Discussion | 256 |

## Chapter 14

**Conference Summary** — 259
M. J. REES

## POSTER PAPERS

The Largest Structures: Limits From the
Radio-Source Background  265
C. R. BENN, J. V. WALL

Biasses in the Estimation of the UV Background Strength  269
B. ESPEY

The Case for an Extragalactic Origin for the High Galactic
Latitude Diffuse Ultraviolet Background  271
R. C. HENRY, J. MURTHY

On the Absorption of UV Radiation in Ly$\alpha$ Clouds  275
V. KHERSONSKY, D. TURNSHEK

On the Contribution of Star Forming Galaxies to the
UV Background Radiation  279
V. KHERSONSKY, D. TURNSHEK

Hitchhiker, Dwarfs and the EBL  285
I. MORGAN, S. P. DRIVER

Antiproton Lifetime Limit from Observation of Nucleon Decay
in Clusters of Galaxies  289
D. J. O'CONNOR

Secondary Fluctuations in the Cosmic Microwave Background  291
R. B. PARTRIDGE

# PREFACE

The principal goal of this symposium was to discuss recent developments in the study of the extragalactic background radiation, covering the entire electromagnetic spectrum from gamma rays to radio. At the same time, we thought that it would be very appropriate to hold a meeting in honor of Riccardo Giacconi, who not only discovered the x-ray background more than 30 years ago, but who continues to contribute significantly to the research in this field.

From a more personal point of view, it has been our honor to hold the meeting at the Space Telescope Science Institute, which Dr. Giacconi directed from 1981 until 1993.

An enormous effort was invested in recording and editing the discussions which followed each of the invited talks, so that these could be included as a valuable resource in the proceedings.

We hope that this comprehensive summary of all the topics associated with the extragalactic background radiation will stimulate and facilitate future research.

Daniela Calzetti
Mario Livio
Piero Madau

# PARTICIPANTS

Ron Allen
John N. Bahcall
Anthony J. Banday
Kevin Black
Elihu Boldt
David Bowen
Margaret Burbidge
Richard Burg
Daniela Calzetti
Claude R. Canizares
Alfonso G. Cavaliere
Cynthia Cheung
Thomas L. Cline
Michael Dahlem
Ruth A. Daly
Laura Danly
Brian Espey
Giuseppina Fabbiano
Michael Fall
Eric Feigelson
Carl Fichtel
Neil Gehrels
Gabriele Ghisellini
Riccardo Giacconi
Emanuele Giallongo
Richard Griffiths
Hashima Hasan
Michael Hauser
Timothy Heckman
Richard C. Henry
Peter Jakobsen
Ed Kaita
Valery Khersonsky

Anne Kinney
Varsha P. Kulkarni
Vladimir Kurt
David Leisawitz
Darryl Leiter
Mario Livio
Malcolm Longair
Carol Lonsdale
James Lowenthal
Phil Lubin
Piero Madau
Barry Madore
Stephen Maran
John Mather
Avery Meiksin
Kenneth Mitchell
Warren Moos
Ian Morgan
Windsor Morgan
Jeremy Mould
Colin A. Norman
Daniel O'Connor
Giorgio G. C. Palumbo
Francesco Paresce
R. Bruce Partridge
John A. Peacock
James Peebles
Larry Petro
Ryszard Pisarski
Marc Postman
Martin Rees
Katherine C. Roth
Bernard Sadoulet

Rex Saffer
David Schlegel
Maarten Schmidt
Ethan Schreier
Daniel A. Schwartz
Salvatore Serio
Rick Shafer
Hugues Sicotte
David Soderblom
Caroline K. Stahle
Floyd Stecker
Peter Stockman
Rashid Sunyaev
Alex Szalay
Miyaji Takamitsu
David J. Thompson
Luigi Toffolatti
Sachiko Tsuruta
J. Anthony Tyson
C. Megan Urry
Piet van der Kruit
Nicola Vittorio
Michael Vogeley
Peter von Gronefeld
J. V. Wall
Boqi Wang
Simon White
Bruce Woodgate
Bogdan Wszolek
Giovanni Zamorani
Andrzej Zdziarski
Piotr Zycki

# INTRODUCTION

P. J. E. Peebles
Joseph Henry Laboratories
Princeton University
Princeton, NJ 08544
USA

## 1. INTRODUCTION

The study of the extragalactic background radiation has a long and honorable history. Landmarks include the discussion by de Vaucouleurs (1949) of the extragalactic contribution to the mean optical surface brightness of the sky (de Vaucouleurs' number, $S_{10} \sim 0.5$ tenth magnitude stars per square degree, is quite close to modern estimates); the discovery by Giacconi *et al.* (1962) of the X-ray background radiation; and the discovery by Penzias and Wilson (1965) of the microwave background radiation. It is only in the last decade that the subject has reached the very rich level you will find in the contributions in this volume, however. My comments on the state of the subject are organized by physical issues, and I use order-of-magnitude estimates in an attempt to illustrate the significance of the observations of the background fields for these issues. In some cases we do not know how to go much beyond the orders of magnitude, while in others detailed analyses are feasible and significant, as you will see in the other papers in this volume.

## 2. LARGE-SCALE STRUCTURE

### 2.1. The Cosmological Principle

It is meaningful to talk of cosmic or universal radiation fields, rather than the radiation received from this or that system in a heirarchy of structures, because the evidence is that the universe we can see back to high redshift is close to homogeneous and isotropic. I discuss here some of the evidence and the measures of large-scale departures from homogeneity.

A useful foil to the standard homogeneous world picture models the large-scale distribution of matter as a scale-invariant fractal with dimension $D$. If $D < 2$ the mean column density of material along a line of sight converges, meaning there is no Olbers paradox of unbounded sky brightness even if the universe were static and stars lived forever. This case is easy to rule out, however, because the angular distribution of the column density of material integrated to distance $R$ has the same fractal dimension $D$, independent of $R$ (as one must expect, for in a scale-invariant fractal there is no characteristic length to set the scale for $R$). That is, the angular fluctuations in counts of objects across the sky are statistically independent of the depth to which the objects are counted, which is quite contrary to the observations. (This prediction and its observational tests are discussed in sections 3 and 7 of *Principles of Physical Cosmology*, Peebles 1993, hereafter PC).

If $2 < D < 3$ the line-of-sight integral for the background radiation surface brightness has to be cut off at some limiting distance $R_{\text{lim}}$, by the finite lifetime of the sources or by the cosmological redshift. We know an appreciable fraction of the X-ray background (the XRB) is produced by sources with redshifts on the order of unity, that is, that in this case $R_{\text{lim}}$ is on the order of the Hubble distance $cH_o^{-1} = 3000h^{-1}$ Mpc (where $H_o = 100h$ km s$^{-1}$ Mpc$^{-1}$). It is an easy exercise to check that in a scale-invariant fractal model with $2 < D < 3$, and with sources turned off at distances greater than $R_{\text{lim}}$, the quadrupole anisotropy of the integrated background is

$$\frac{\delta f}{f} \sim 3 - D, \tag{1}$$

again independent of $R_{\text{lim}}$. The large-scale anisotropy of the XRB is not more than a few tenths of one percent, meaning that the fractal dimension of the large-scale space distribution of X-ray sources differs from $D = 3$ by only a few parts in 1000.

The bound on $D$ is removed if we are close to the center of a universe spherically symmetric about one position, but since our galaxy and its near neighbors appear to be quite ordinary this does not seem to be a very sensible model.

The large-scale isotropy of the 3 K thermal cosmic background radiation (the CBR) gives an even better limit on large-scale mass fluctuations, if we are willing to assume general relativity theory is a useful approximation to physics on the scale of the Hubble length. Here a useful foil assumes mass cluster with galaxies on scales $\sim 10h^{-1}$ Mpc, and that the mass autocorrelation function $\xi(r) = \langle \rho(\vec{r}+\vec{s})\rho(\vec{s})\rangle/\langle \rho\rangle^2 - 1$ vanishes at larger separations. Then the large-scale rms density fluctuations are characterized by the integral

$$J_3 = \int r^2 dr \xi(r)$$
$$= 1000 \pm 200 \ h^{-3} \ \text{Mpc}^3. \tag{2}$$

The second line is based on the galaxy two-point correlation function. In this model the rms fluctuation in the mass in a randomly placed sphere of radius $r$ large compared to the support of $\xi$ is

$$\frac{\delta M}{M} = \left[\frac{3 J_3}{r^3}\right]^{1/2} = 0.2 \quad \text{for} \quad r = 50h^{-1} \ \text{Mpc}, \tag{3}$$

and the rms value of the velocity averaged over the material within the sphere is (PC §§13 and 21)

$$\bar{v} = \left[\frac{6}{5}\frac{H_o^2 \Omega^{1.2} J_3}{r}\right]^{1/2}$$
$$= 500 \ \Omega^{0.6} \ \text{km s}^{-1} \quad \text{for} \quad r = 50h^{-1} \ \text{Mpc}. \tag{4}$$

The cosmological density parameter $\Omega$ is the ratio of the mean mass density to the critical Einstein-de Sitter value. The indicated scaling with $\Omega$ is a good approximation if the cosmological constant $\Lambda$ vanishes or if $\Lambda$ is present and space curvature is negligible. The numbers in equations (3) and (4) are comparable to the direct estimates from galaxy counts and galaxy peculiar motions.

In this model for the mass distribution, and with an Einstein-de Sitter cosmological model, the CBR quadrupole anisotropy produced by the gravitational potential fluctuations at the Hubble distance (the Sachs-Wolfe effect) is

$$a_2 = \left[\frac{4\pi J_3 H_o^3}{105 c^3}\right]^{1/2} = 7 \times 10^{-5}, \qquad (5)$$

an order of magnitude above the COBE quadrupole anisotropy Mather and Vittorio discuss in these Proceedings.

The lesson is that our model gives a not unreasonable picture for the mass fluctuations on scales $\lesssim 100$ Mpc, but it significantly overestimates fluctuations on larger scales. An equivalent way to put it is that the power spectrum of mass fluctuations has to decrease with increasing scale at wavelengths $\gtrsim 100$ Mpc. This was predicted by the inflation theory incorporated in the adiabatic Cold Dark Matter cosmogony (Frenk 1991 and references therein), an impressive success. One should bear in mind, however, that the wanted decrease in the mass fluctuation spectrum at large separations appears in other scenarios, such as cosmic strings and textures (Turok 1991).

Equation (5) does assume general relativity theory, and it would be very interesting to have a check on this constraint from a direct measurement of the fluctuations in the distribution of sources on the scale of the Hubble length. The constraint says the fluctuations are less than one part in $10^4$, however, which will not be easy to get at.

A related issue is the interpretation of the CBR dipole anisotropy. The COBE measurements described by Mather show that the CBR anisotropy is very close to a pure dipole in thermodynamic temperature. This is consistent with the usual assumption that the dipole is caused by our peculiar motion relative to the general Hubble expansion of the universe, that is, that the rest frame defined by the CBR agrees with the rest frame defined by the mean motion of the matter within the Hubble length. Detection of the departure from a pure dipole due to the second-order Doppler shift would be an important check, but I gather not yet within reach.

A detection of the mass fluctuations whose gravitational pull drive our motion relative to the CBR would be a key test of the velocity interpretation. One can estimate the peculiar gravitational acceleration caused by the galaxies taken to be tracers of the mass within distance $R$. If the direction and magnitude of the acceleration are unchanged as $R$ is increased, and agree with our motion relative to the CBR for a sensible value of the mass per tracer galaxy, one has reason to believe the mass fluctuations whose gravity produced our peculiar motion are detected with $R$. In a recent application, Strauss et al. (1992) find that the direction of the gravitational acceleration of IRAS galaxies taken to be mass tracers converges to within $20°$ of the CBR dipole direction at $H_o R \sim 3000$ km s$^{-1}$, and stays within that angular distance as one adds IRAS galaxies out to $H_o R \sim 20,000$ km s$^{-1}$. This suggests the source of the motion of our Local Group of galaxies is within 3000 km s$^{-1}$ distance. If that were so, however, one would expect to find an appreciably smaller mean peculiar velocity in the average over the material within a sphere of radius larger than 3000 km s$^{-1}$, contrary to the present evidence. The mean peculiar velocity relative to the CBR for the galaxies within 6000 km s$^{-1}$ is about 400 km s$^{-1}$, directed $\sim 40°$ away from the motion of the Local Group relative to the CBR (Faber and Burstein 1988; Mould et al. 1993). The Postman-Lauer (1993) mean peculiar flow for Abell clusters at $R < 15,000$ km s$^{-1}$ has a similar magnitude but swings to ∼ 90° from the motion of the Local Group relative to the CBR. In short, our picture for the large-scale peculiar velocity field and the mass density fluctuations that may have caused it remains confused.

It would be very helpful to have a measurement of the XRB dipole. If the CBR dipole were the result of our peculiar motion caused by the gravitational pull of mass fluctuations well within the Hubble length then we would expect to see a parallel dipole anisotropy in the XRB. If on the other hand the CBR dipole were due to a very smooth gradient in the primeval entropy per unit mass the X-ray dipole from sources within the Hubble length would not be expected to stand out above the other low order multipole moments of the XRB, or to be aligned with the CBR dipole. I gather the experts are not yet willing to make a definite pronouncement on this important test.

## 2.2 The Mean Optical Luminosity Density

My conclusion is that we have a good case for Einstein's cosmological principle, that the universe is close to homogeneous and isotropic in the large-scale average.*
This means we can consider the mean mass density of the universe, and the mean density of radiation as a function of wavelength. Of particular interest is the quantity de Vaucouleurs (1949) considered, the density of starlight, as a check on our inventory of luminous objects in the universe (Peebles and Partridge 1967). This is motivated by the notorious mass problem in cosmology, that the starlight in the known galaxies, assigned a mass-to-light ratio characteristic of familiar star populations, yields a mass insufficient to account for the motions of galaxies within groups and clusters, and two orders of magnitude below the Einstein-de Sitter density usually considered to be the most reasonable possibility. The commonly discussed interpretation is that the mass of the universe is dominated by nonbaryonic matter, perhaps massive neutrinos or some other weakly interacting particle, maybe something even more exotic. But it behooves us to consider the possibility that our sums have missed luminous objects too compact to be distinguished from stars, or too low in surface density to be seen against the light of the night sky. We have a test, from the comparison of the mean surface brightness contributed by the known galaxies and the measured extragalactic sky brightness (after correction for airglow, the zodiacal light, and the diffuse light of the Milky Way).

Tyson's (1990) value for the mean surface brightness of the sky at 4500 Å from the sum of observed galaxies is

$$\nu i_\nu = 3 \times 10^{-6} \text{ erg cm}^{-2} \text{ s}^{-1} \text{ ster}^{-1}. \tag{6}$$

This is de Vaucouleurs' (1949) computation; it is impressive that the result has not changed by more than a factor of about two. The equivalent mean luminosity density is

$$j = 1 \times 10^8 h L_\odot \text{ Mpc}^{-3}. \tag{7}$$

Mattila's (1990) survey indicates the measured extragalactic sky brightness satisfies

$$\nu i_\nu \lesssim 3 \times 10^{-5} \text{ erg cm}^{-2} \text{ s}^{-1} \text{ ster}^{-2}. \tag{8}$$

---

* A lesson from the inflation cosmology is that the universe at great distances could be very different from what we see. The more careful statement thus is that mass density variations across the Hubble length have to be less than about one part in $10^4$. The issue seldom debated is whether Einstein was right for the right reason, and the universe really is close to uniform everywhere.

The difference between this and the surface brightness contributed by ordinary galaxies (eq. [6]) leaves room for new classes of objects with luminosity density an order of magnitude greater than that of the known galaxies. With conventional mass-to-light ratios for star populations this leaves the bound on the mass density of luminous material an order of magnitude below the Einstein de-Sitter value.

The relation between equations (6) and (7) is not sensitive to the cosmological model, because the cosmological redshift suppresses the contribution from ordinary galaxies at the Hubble distance. But we see that there is considerable room for galaxy evolution that could have produced ultraviolet emission which has been redshifted into the visible band. Advances in the art of measuring the extragalactic optical and near infrared background, as discussed here by Hauser, thus will be followed with great interest.

## 3. COSMOGONY

As we have just noticed, the radiation backgrounds contain evidence of what has been happening in the universe now and in the distant past. I discuss here the histories of heavy element formation and the ionization of diffusely distributed matter. Both depend on opinions on the epoch of first star formation in galaxies or pre-galaxy star clusters.

### 3.1. The Epochs of Structure Formation

The discovery of the CBR, and the demonstration that its spectrum is close to a single Planck function, had two important consequences. First, it gave considerable credibility to the evolving relativistic cosmology, because all sides I have encountered agree that there is no way this radiation could have been thermalized in the universe as it now: the universe has to have expanded from a denser and hotter state. Second, the presence of this radiation, if truly primeval, yields a key constraint on structure formation in the early universe, as follows.

If the CBR is primeval the very early universe was dominated by relativistic material with Jeans length comparable to the Hubble length. That means any departure from homogeneity large enough to break away from the general expansion and form a gravitationally bound system would have had gravitational potential energy large enough for relativistic collapse to a black hole. Unless such black holes are small enough to evaporate by Hawking radiation, their mean energy density scales as $a(t)^{-3}$, where $a(t)$ is the expansion factor for the expanding universe, while the energy density of the relativistic material scales as $a(t)^{-4}$. Thus structure formation in the very early universe had to have been rare, for otherwise black holes would have made an unacceptably large contribution to the present mean mass density.

In the standard cosmological model, the universe at redshifts less than $z_{eq} \sim 2 \times 10^4 \Omega h^2$ is dominated by nonrelativistic matter. In this matter-dominated phase gravity causes the growth of small-scale fluctuations in the distribution of any nonrelativistic nonbaryonic matter that is unaffected by radiation drag or its own pressure. At redshift $z_{dec} \sim 1500$ the CBR becomes cool enough to allow the primeval baryonic plasma to combine and decouple from radiation drag, and this matter can join the growing clustering of any nonbaryonic matter. Thus we conclude that in the standard hot evolving cosmology nothing much in the way of structure formation can have happened prior to redshift $z \sim 1000$ to $10{,}000$.

What was happening at redshift $z = 1000$? In the Cold Dark Matter model (Frenk 1991) the primeval density fluctuations are adiabatic (fixed entropy per conserved particle number) and Gaussian, with the scale-invariant Zel'dovich power spectrum for which the rms value of the mass density fluctuations appearing at the Hubble length is independent of epoch. In this model the normalization to the density fluctuations appearing at recent epochs implies that the density fluctuations at $z = 1000$ are small on all scales: baryon structure formation commences at $z \sim 30$ in clouds at the baryon Jeans mass, about that of a globular star cluster (Peebles 1983). If the fluctuation spectrum were bent to add power on smaller scales it would make structure formation commence earlier. Under isocurvature initial conditions (uniform mass density, entropy per baryon number a random function of position) structure formation could have commenced still earlier, at $z \sim 1000$.

If star formation at $z \sim 1000$ reionized an appreciable fraction of the baryons it would recouple the plasma to the CBR, tending to suppress further star formation.* I am not aware of any detailed analysis of this self-limiting star formation effect. Perhaps it is a significant coincidence that when radiation drag on diffuse ionized matter becomes unimportant, at $z \sim 100$, the mean mass density is equivalent to about $10\Omega h^2$ protons cm$^{-3}$, which for reasonable values of $\Omega$ and $h$ is on the order of the characteristic density within the luminous part of normal giant galaxies.

The material in a galaxy could not have been assembled as a gravitationally bound system until the mean mass density within the object is greater than the cosmological mean. That says the bright parts of normal giant galaxies could not have been assembled before $z \sim 100$. A more precise and possibly more accurate bound comes from the spherical model. I have trouble believing spherical symmetry is a useful approximation after a protogalaxy has broken away from the general expansion and started collapsing, so I am inclined to put the epoch of assembly at the epoch of maximum expansion in the spherical model (following Partridge and Peebles 1967). This would say the material within $r \sim 10h^{-1}$ kpc of an $L_*$ galaxy was assembled at redshift

$$1 + z \sim 30\Omega^{-1/3}, \qquad (9)$$

and the material within the Abell radius of an Abell cluster was assembled at

$$1 + z \sim 2\Omega^{-1/3}. \qquad (10)$$

There are clusters of galaxies at redshift $z \sim 1$ with the velocity dispersion and central mass concentration characteristic of Abell clusters, so equation (10) does not seem unreasonable. Equation (9) is not inconsistent with the observation of high column density atomic hydrogen gas clouds (the damped Ly$\alpha$ systems) detected as broad absorption features in quasar spectra at $z \sim 3$ (Wolfe 1989). These clouds have the surface densities and comoving number density characteristic of galaxies. Our political leaders have taught us that what quacks like a duck and waddles like a duck very likely is a duck. On the same principle, I think it is a good bet that the damped Ly$\alpha$ clouds are members of the long-sought population of young galaxies. To be debated is whether they are typical of the galaxies at that epoch, and whether they are pieces of protogalaxies to be assembled or maybe the first parts of protodiscs collecting around

---

* A second effect, that the radiation drag on the free electrons would perturb the CBR spectrum, is discussed in the next section.

previously assembled spheroids. Meanwhile, a reasonable guess is that the redshift of assembly of the mass in the central parts of galaxies is somewhere between equation (9) and the epoch of the damped Ly$\alpha$ systems.

## 3.2. The History of Element Formation

One has to define what is meant by the epoch of galaxy formation, because different parts of galaxies may have been assembled at quite different times. Thus equation (9) indicates that the inner parts of giant galaxies might have been assembled well before their extended massive dark halos could have been attached. One similarly has to specify what is meant by the epoch of heavy element production, because element production continues to the present day. The present rate in galaxies like the Milky Way is too low to have produced the heavy elements in the time available, however, so there had to have been an epoch when the element production rate per unit mass in the typical progenitor of a normal giant galaxy was considerably larger than it is now. The typical luminosity per unit of mass likely was considerably larger then as well.

We have noted that the damped L$\alpha$ systems look like young galaxies. Their heavy element abundances are estimated to be about 10% of solar (Hunstead, Pettini, and Fletcher 1990). This suggests the bulk of element formation in the discs of spirals is at $z_e \lesssim 3$.

We have a check from the residual optical/IR background from the starlight that accompanied element formation:

$$\nu_o i_{\nu_o} \sim \frac{c}{4\pi} \frac{\rho_* Z \epsilon c^2}{1+z_e} \tag{11}$$
$$= \frac{8 \times 10^{-3} \Omega_* h^2}{(1+z_e)} \text{ erg cm}^{-2} \text{ s}^{-1} \text{ ster}^{-1}.$$

This assumes the mass fraction $Z = 0.03$ of baryonic matter with mean mass density $\rho_* = \Omega_* \rho_{crit}$ has burned from hydrogen to heavier elements, releasing a fraction $\epsilon = 0.007$ as starlight. The integrated local background radiation has a minimum near 3 microns, with

$$\nu_o i_{\nu_o} \sim 3 \times 10^{-5} \text{ erg cm}^{-2} \text{ s}^{-1} \text{ ster}^{-1}. \tag{12}$$

If $1 + z_e \sim 3$, and an appreciable part of the starlight is in the infrared, this would bound the mass density of the heavy element producing material at $\Omega_* \lesssim 0.01 h^{-2}$, comparable to the baryon density in the standard model for Big Bang nucleosynthesis. Hauser explains the prospects for improving this constraint.

## 3.3. The Ionization History of Diffuse Matter

Quasar absorption spectra have yielded a remarkably detailed picture of the state of the universe at redshift $z \sim 3$ (Blades, Turnshek, and Norman 1988). At this epoch the Ly$\alpha$ forest clouds fill on the order of 1% to 10% of space, about as much as possible for irregularly shaped clouds. A cloud at the detection limit contains just a few hundred solar masses of neutral atomic hydrogen. The clouds are optically thin to ionizing radiation, however, and the radiation from quasars is sufficient to keep the cloud material highly ionized. At the lowest detected column densities the plasma mass of a cloud is estimated to be about $10^7$ M$_\odot$, that of a present-day dwarf galaxy. The net baryonic mass in these clouds is comparable to what is present now in the galaxies. That is, we can conclude that at $z \sim 3$ an appreciable fraction of the baryonic matter in our universe was in diffuse ionized clouds, along with a comparable fraction in the neutral damped Ly$\alpha$ systems.

What was the universe like a factor of two further back in expansion, at $z \sim 7$? The rate of intersection of high column density neutral clouds along a line of sight increases with increasing redshift up to the largest presently known, $z \sim 5$ (Schneider, Schmidt, and Gunn 1991; Lanzetta 1991). This increase in opacity to ionizing radiation suggests that at $z \sim 7$ the L$\alpha$ forest clouds were considerably less strongly ionized, if they had formed.

The possible ionization history at diffuse baryonic matter back to its early tight coupling to the CBR is constrained by the effect on the spectrum of the CBR by scattering by moving electrons (Zel'dovich and Sunyaev 1969). This is usefully measured by the Sunyaev-Zel'dovich parameter

$$y = \int \frac{kT_e}{m_e c^2} \sigma_t n_e c \, dt. \tag{13}$$

The plasma has mean pressure $n_e k T_e$, and $\sigma_t$ is the Thomson cross section. The analysis by Field and Perrenod (1977) of models for the origin of the XRB by thermal bremsstrahlung emission from a hot intergalactic plasma predicted values of $y$ considerably in excess of the bounds from the beautiful COBE measurements described by Mather in these Proceedings. There does not seem to be any reasonable way to produce the XRB in hot diffuse plasma either before or after the epoch of the Ly$\alpha$ forest; the XRB has to have come from compact sources. Gehrels discusses the fascinating enigma of the XRB sources.

Could a significant part of the baryonic matter have been in diffuse plasma from $z \sim 1000$ to incorporation in stars at a much lower redshift? The analysis by Bartlett and Stebbins (1991) indicated that relatively cool plasma, as would be produced by photoionization by starlight, need not violate the preliminary COBE bound on $y$ (or the bound on the bremsstrahlung contribution to the 30 cm radio background). The significance of the latest COBE limit on $y$ is under discussion.

Bulk peculiar motions of the plasma act as an effective temperature in the integral in equation (13). Thus plasma in protogalaxies with one-dimensional rms peculiar motion $\sigma$ at redshift $z$ contributes

$$\begin{aligned} y &\sim \sigma_t n_e c t (\sigma/c)^2 \\ &\sim 10^{-8} \sigma_{100}^2 h \Omega_p \Omega^{-1/2} (1+z)^{3/2}. \end{aligned} \tag{14}$$

Here $\Omega$ is the cosmological density parameter, $\Omega_p$ is the density parameter in optically thin plasma in the protogalaxies, and the velocity dispersion is measured in units of 100 km s$^{-1}$. The conclusion is that it is not difficult to choose parameters that reconcile the COBE bound on $y$ with the assumption that the bulk of the baryons are in plasma in protogalaxies at the redshift $z \sim 30$ in equation (9).

The mean optical depth for scattering by electrons is unity at redshift

$$1 + z_s = 8 \, \Omega^{1/3} (h \Omega_p)^{-2/3}. \tag{15}$$

If structure formed at high redshift in a low density cosmology, with $\Omega \sim \Omega_p \sim 0.1$, then the CBR would have been last scattered at $z_s \sim 25$. This would have smoothed primeval fluctuations in the CBR on angular scales $\lesssim \theta_s \sim 3°$. The character of the angular fluctuations in the CBR produced by the scattering plasma depends on what was happening at $z_s$. One possibility is that the plasma was in protogalaxies with typical angular size $\theta_g$. Then the contribution to the CBR angular autocorrelation function at separation $\theta \gtrsim \theta_g$ is set by the peculiar velocity autocorrelation function at the mean distance between intersection of clouds along the line of sight, which can be quite small. If this is what happened, untangling the mean of the measurements of the CBR anisotropy may be a considerable challenge.

## 3.4. Concluding Remarks

Research in the extragalactic background radiation fields has come a long way from the early steps I mentioned in my introductory remarks, but the generally accepted pieces to the puzzle of what it all means still do not fit into any very compelling pattern, which is to say that cosmogony is not yet a very mature branch of physical cosmology. It is one of the most active, however, and one is impressed to note the progress in the last decade in establishing a detailed observational picture of the intergalactic medium back to $z \sim 5$ and the very significant constraints on what was happening at higher redshifts. It will be fascinating to see what turns up next.

This work was supported in part by the US National Science Foundation.

## REFERENCES

Bartlett, J. G. and Stebbins, A. 1991, *Ap. J.*, **371**, 8.
Blades, J. C., Turnshek, D. A., and Norman, C. A. 1988, *QSO Absorption Lines— Probing the Universe* (Cambridge: Cambridge University Press).
de Vaucouleurs, G. 1949, *Annales d'Astrophysique*, **12**, 162.
Faber, S. M. and Burstein, D. 1988, in *Large-Scale Motions in the Universe*, ed. V. C. Rubin and G. V. Coyne (Princeton: Princeton University Press), p. 115.
Field, G. B. and Perrenod, S. C. 1977, *Ap. J.*, **215**, 717.
Frenk, C. S. 1991, *Physica Scripta*, **T36**, 70.
Giacconi, R., Gursky, H., Paolini, F., and Rossi, B. 1962, *Phys. Rev. Letters*, **9**, 439.
Hunstead, R. W., Pettini, M., and Fletcher, A. B. 1990, **Ap. J.**, **356**, 23.
Lanzetta, K. M. 1991, *Ap. J.*, **375**, 1.
Mattila, K. 1990, in *The Galactic and Extragalactic Background Radiation*, ed. S. Bowyer and C. Leinert (Dordrecht: Kluwer), p. 257.
Mould, J. R. et al. 1993, *Ap. J.*, **409**, 14.
Partridge, R. B. and Peebles, P. J. E. 1967, *Ap. J.*, **147**, 868.
Peebles, P. J. E. 1983, *Ap. J.*, **277**, 470.
Peebles, P. J. E. 1993, *Principles of Physical Cosmology* (Princeton: Princeton University Press).
Peebles, P. J. E. and Partridge, R. B. 1967, *Ap. J.*, **148**, 713.
Penzias, A. A. and Wilson, R. W. 1965, *Ap. J.*, **142**, 1149.
Postman, M. and Lauer, M. 1993, in *Cosmic Velocity Fields*, Paris, July 1993.
Schneider, D. P., Schmidt, M., and Gunn, J. E. 1991, *A. J.*, **102**, 837.
Strauss, M. A. et al. 1992, *Ap. J.*, **397**, 395.
Turok, N. 1991, *Physica Scripta*, **T36**, 135.

Tyson, J. A. 1990, in *The Galactic and Extragalactic Background Radiation*, ed. S. Bowyer and C. Leinert (Dordrecht: Kluwer), p. 245.

Wolfe, A. M. 1989, in *The Epoch of Galaxy Formation*, ed. C. S. Frenk *et al.* (Dordrecht: Kluwer), p. 101.

Zel'dovich, Ya. B. and Sunyaev, R. A. 1969, *Ap. Space Sci.*, **4**, 301.

## DISCUSSION

**V. Khersonsky**: Your expectation of galaxy formation epoch is mainly based on the consideration of gravitational instability, right? So what do you think about the role of turbulent processes, which may shift completely the timescales?

**Peebles**: You are quite right in pointing out that turbulence must be important, for after all one sees turbulence in the material within a galaxy. My opinion on the role of turbulence in initiating the formation of structures is based on the theoretical expectation within a relativistic cosmology that the universe is gravitationally unstable. And therefore, that if there were primeval turbulence present at redshifts greater than $z_{eq}$, and which could have been present way back, then at decoupling the turbulence would have become supersonic. Supersonic motion immediately initiates shock waves leading to a burst of structure formation at decoupling, which is at a density too high to allow for ordinary galaxies. So my bet is the prime mover in the formation of galaxies and clusters of galaxies had to have been gravity. But once gravity has started the non-linear development of a mass concentration, all of the vexed complexity of non-linear dynamics comes into play. A standard way to analyze the final stages of collapse uses N-body computations. That approach is important, but I would love to see better theoretical understanding of the complex non-linear dynamics accompanying structure formation.

**S. White**: So, to carry on from the same question, you made the remark that spherical models for gravitational collapse would allow you to calculate the time when, for example, a rich cluster of galaxies was assembled, and you noted that you did not believe in the spherical model or you thought at least it needed more investigation. I'd just like to say that I think there has been considerable more investigation and that it is clear that the formation times from the spherical model are in error; but I think it is also clear that they are in error in the opposite sense than you said. So, for example, if you take a present rich Abell cluster and you ask at what redshift was the material currently inside the Abell radius assembled, meaning that it had more than half its present mass, then the median for a flat universe is something like a redshift of 0.2, while for a universe with a density parameter of 0.2 it is something like 0.7. And I think that is understood analytically as well as through N-body simulations and it means that, in terms of assembling the material into one lump, it actually occurs more recently than the spherical model suggests, not earlier.

**Peebles**: You have touched, Simon, two key issues. First, when were the mass concentrations in galaxies assembled? Could it be as low as a redshift of 0.2? I have a bottle of fine Scotch whiskey riding on this issue with Carlos Frank. I would be amazed if it turns out that the mass within the Abell radius in an Abell cluster typically is assembled at a redshift as low as 0.2.

Second is the issue of the nonlinear dynamics of cluster formation. I expect non-radial motion generally inhibits the collapse of a mass concentration, which is to say that the collapse factor after maximum expansion surely is less than the factor of two one usually adopts in a spherical model. That is what led me to adopt a relatively small density contrast at the epoch of virialization of a newly formed clump.

Let me remind you of one other point, which you can see illustrated in figure 25.1 in my book, The Large-Scale Structure of the Universe. Material falling radially onto an isolated mass concentration is strongly sheared by the tidal field of the concentration, but the tidal field has very little effect on the mass density of the accreting material until first orbit crossing, where the mass density rises by about an order of magnitude. I have not encountered evidence for shoulders of this sort in the density runs around great clusters of galaxies. You do not see this shoulder effect in N-body computations, but I wonder if that is because you have the clusters just forming now, so the shoulder has not yet formed.

**S. White**: Just let me reply to that. I think we do actually see that structure, but the reason we do not see the sharp edge is because the clusters are non-spherical. And, in fact, they do not form mainly by infall, they increase their mass mostly by merging with other lumps of similar or only somewhat lower mass. At least in a hierarchical clustering universe, the growth in mass of objects is not primarily by spherical accretions, as that model assumes.

**Peebles**: We are agreeing that it is puzzling that one would put much faith in spherical accretion.

**J. Mould**: You are going to get a rise out of somebody with your great rise model or cartoon, so it may as well be me in the absence of the Seven Samurai. If you take something like that and project it back to the surface of last scattering, does not it lead to very observable consequences on the anisotropy of the Microwave Background?

**Peebles**: I would be just amazed if the "great rise" were a useful approximation to reality. Whether or not it is easily ruled out by such constraints as the anisotropy of the Cosmic Background Radiation depends on your choice of cosmology. Recall that in an open cosmological model with $\Lambda = 0$ and density parameter $\Omega = 0.1$ the angular size distance back to last scattering is $x = 2/H_o\Omega = 60,000 h^{-1}$ Mpc, so that the angular scale $\sim 10°$ of the COBE anisotropy detection subtends a comoving length of about $10^4 h^{-1}$ Mpc, much larger than the great rise. That is, there does not seem to be a direct relation between the two phenomena. But the shorter answer is, I do not know whether one could count contrive a "great rise" cosmogony consistent with all the observational constraints.

**J. Peacock**: To continue this topic of the great rise, you have sketched it as though you wanted density contours but, of course, the relevant thing is more the velocity field, which is much more sensitive to the large wavelength perturbations. So you would not necessarily expect to see a big density structure. I think something like the great rise is not necessarily so implausible because if you think about the small scale correlations as you have sketched them, you have in effect a spectral index of $-1$ and a bit, which therefore will lead to a velocity field which is dominated by progressively larger velocity perturbations. And that continues until the power spectrum breaks and turns over at a wavelength of about 200 Mpc. So it is not surprising, given what we know about the density field's statistics, to see that there is a velocity field of a very large contribution is on many hundred Mpc scales.

**Peebles**: This is a fascinating remark. The next issue is, would you expect a random Gaussian process with the required power spectrum to produce a velocity field that varies so smoothly on such a large scale?

**J. Peacock**: I have actually published such a realization, though not for the case of the Berkeley IRAS results you were talking about, but the QDOT results. It is perfectly possible to attain an integrated dipole signal out to the radius they use, which is 150 Mpc, which agrees within a few degrees of the direction of the CBR dipole, but predicts be velocity which is wrong by fully a factor of 2, although it converges in direction very quickly, as expected on a Gaussian model.

**Peebles**: Gaussian fluctuations do not usually make spheres, but I'll look this up. Thank you.

**S. White**: How about a question about the diffuse backgrounds? I mean you made the point which I think is an important one, that if you take the amount of light that you would expect to be produced by star formation, it produces a bound of energy that we should be able to see. You normalized to something like 3% of the critical density in the form of stars. That is actually a little high because if you take the luminosity density of stars inside observed galaxies on the typical mass to light ratios that are found dynamically for them, with present numbers you get something which is closer to 0.5%. If you normalize your number down by that factor of six, what do we expect for the stars we have already counted, then in fact, the $\nu I(\nu)$ you get is quite comparable to what we have already detected in the form of faint galaxy counts. You just integrate the counts of the blue galaxies to the faintest levels and you get roughly that value. So I think you can at least try to make an argument we have actually seen most of the star formation already.

**Peebles**: A big ambiguity is the heavy element abundance in the baryonic material that might be outside the bright parts of galaxies. If the massive dark halos of galaxies are baryonic, and if these baryons are sequestered in stars and star remnants, then it seems conceivable that the starlight production per unit mass in the now dark halos was comparable to the production in the central parts of the galaxies.

As you indicate, the flat $f(\nu)$ objects are observed to be making heavy elements at a prodigious rate, but I wonder if these could be ordinary $L^*$ galaxies, because they have the wrong spectrum for objects that are identified at redshifts less than unity as giant galaxies. Isn't it a good bet that one is seeing element formation in dwarf galaxies?

**S. White**: Well, roughly half the flux comes from things which are still too faint for us to know what the redshift distribution is, so there is not an extrapolation.

**Peebles**: The extrapolation I have in mind is the picture that the elements being produced in the flat $f(\nu)$ galaxies are ending up in normal $L^*$ galaxies.

**C. Norman**: You computed a very large mass for the amount of mass in the Lyman $\alpha$ forest and I just wonder how you got such a large amplification factor over the observed parameters. It is certainly due to ionization balance, but I am surprised one could compute it with any certainty.

**Peebles**: From the proximity effect one has an estimate of the ionizing radiation density. From the column densities of these objects, and an estimate of the characteristic size, which I took to be 10 kiloparsecs, one has an estimate of the density of neutral material within these optically thin clouds. That plus the ionizing radiation density indicates the neutral fraction is one part in $10^5$. The density parameter contributed by the plasma in the clouds thus is about $10^5$ times what is observed in neutral hydrogen. The result is a net baryon density $\Omega \sim 0.01$. I too am surprised that there is so much baryonic matter in the Lyman $\alpha$ forest clouds, but people seem to agree that the number is secure to a factor of three.

**Q.**: I might just add a comment about a new observational piece of evidence. Unfortunately, the magnitude of which I cannot quite recall, but it was a paper that was just presented last week at Cambridge by Martin Elvis and his colleagues. The paper is about the X-ray absorption seen in spectral quasars at redshift of 3, which indicates that there is a significant column to some but not all quasars out to that redshift, possibly associated with the damped Lyman $\alpha$ systems along the line of sight.

**Peebles**: I see, not intrinsic to the quasar host.

**Q.**: Possibly. One certainly cannot rule that out. If it is intrinsic, then it sets a quite stronger limit on what else is along the line of sight.

**Peebles**: Right.

# EXTRAGALACTIC GAMMA-RAY BACKGROUND

Neil Gehrels and Cynthia Cheung
NASA/Goddard Space Flight Center
Greenbelt, MD 20771

**Abstract.** We review the observational data of the extragalactic gamma-ray background obtained before the launch of the *Compton* Gamma Ray Observatory and the theoretical interpretations. We also summarize the latest AGN observations by *Compton* which pertain to the origin of the extragalactic gamma-ray background and their implications.

## 1. INTRODUCTION

Historically, the measurement of the extragalactic gamma-ray background (GRB) was beset by great experimental difficulties. The high charged particle intensities in the balloon and space environment produce an intense background in gamma-ray instruments. Most early instruments had large fields of view and poor spatial resolution, which added complexities to the extraction of the diffuse extragalactic component. Data were obtained by many balloon- and satellite-borne instruments the last 30 years (Table 1), with a gradual improvement in the detector sensitivity achieved by larger detector sizes and better background-reduction techniques.

Though some of the earlier results gave only upper limits, we have now measured with reasonable accuracy the general form and magnitude of the GRB spectrum (Fig. 1). The low-energy portion (10 keV to 60 keV) of the GRB is characterized by a bremsstrahlung spectral form that can be approximated by a power-law segment of energy index $\sim 0.4$. The energy spectrum transitions to a power law of index $\sim 1.6$ above 60 keV, flattens to an index of $\sim 0.7$ around 1 MeV, then steepens again above several MeV to an index of $\sim 1.7$ that applies to energies well above 100 MeV. When plotted in intensity per logarithmic energy interval $EI_E$ (Fig. 2), the GRB spectrum exhibits two peaks, one at $\sim 30$ keV and the other at a few MeV (the MeV bump) (Zycki 1993). There is still the lack of good measurements around the 10 to 30 MeV range due to the extraordinary high instrumental and atmospheric background in that regime. Results from the GRB observations by the *Compton* Gamma Ray Observatory (*Compton*) now in orbit will eventually fill in this range, and also vastly improve our knowledge at higher energies.

The GRB is of particular interest to cosmology because of the transparency of the Universe to gamma rays back to redshifts of 100 or more. A determination of the isotropy of the GRB can be a sensitive test of its cosmological origin and a de-

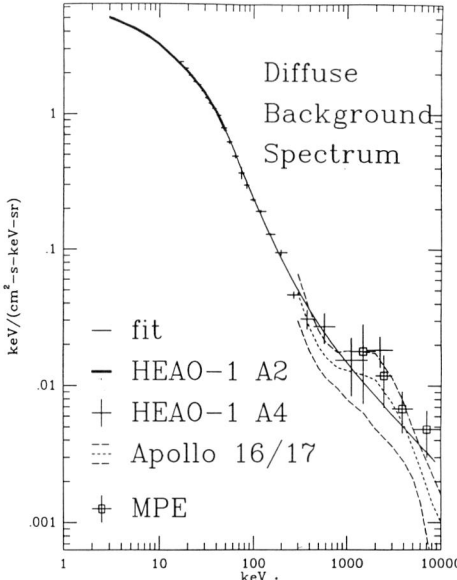

**Figure 1.** *Energy spectrum of the diffuse extragalactic background (Gruber 1992).*

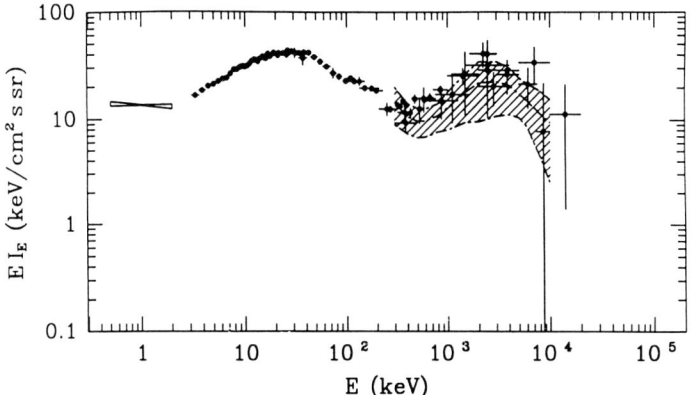

**Figure 2.** *The extragalactic background spectrum in intensity per logarithmic energy interval (adapted from Zycki et al. 1993). Circles correspond to the data points compiled by Gruber (1992) in Fig. 1. The elongated 0.5–2 keV error contour gives the* ROSAT *results (Hasinger 1992). The dashed region is the systematic error contour for the* Apollo *results, with the middle dot-dashed curve the best estimate of the spectrum. The 35–100 MeV error contour is the* SAS-2 *result from Fichtel et al. (1978).*

tailed study of the complete GRB spectrum and its spatial fluctuations may provide constraints to different cosmological models. There are two possibilities for the origin of the extragalactic GRB: (1) it is truly diffuse; and (2) it is the integrated emission of various distant unresolved gamma-ray sources. It may also be a combination of diffuse and point sources and may have different origins in different portions of the gamma-ray band. From below 10 MeV to 100 MeV, particle-antiparticle annihilation, bremsstrahlung and inverse Compton interactions between cosmic ray particles and lower-energy photons are the most likely gamma-ray production mechanisms. Above 100 MeV, the dominant process is $\pi^\circ$ decay from nucleon interactions.

In this paper we summarize the state of our knowledge on the extragalactic GRB. We present the observations and their interpretations in two parts: (1) for energies from 0.1 to 10 MeV, and (2) for energies greater than 10 MeV. We then present the recent results from AGN observations by *Compton* and discuss their relevance to the GRB.

Table 1.
**Observations of the Diffuse Gamma-Ray Background**
(adapted from Gruber (1992))

| Energy Band | Detector | Platform | Reference |
|---|---|---|---|
| 0.1–3 MeV | scintillator | Ranger 3 | Arnold et al. (1962) |
| 1–6 MeV | scintillator | ERS18 | Vette et al. (1969) |
| 1–5 MeV | scintillator | balloon | Vedrenne et al. (1971) |
| 0.1–8.5 MeV | scintillator | balloon | Daniel et al. (1972) |
| 0.03–4.1 MeV | scintillator | Kosmos | Mazets et al. (1975) |
| 0.1–4 MeV | scintillator | balloon | Fukuda et al. (1975) |
| 0.3–10 MeV | scintillator | Apollo | Trombka et al. (1977) |
| 2–10 MeV | Compton telescope | balloon | White et al. (1977) |
| 0.3–6 MeV | scintillator | balloon | Mandrou et al. (1979) |
| 3–60 keV | proportional counter | HEAO-1 A2 | Marshall et al. (1980) |
| 1.1–10 MeV | scintillator | balloon | Schönfelder et al. (1980) |
| 0.013–4.0 MeV | scintillator | HEAO-1 A4 | Gruber et al. (1985) |
| > 50 MeV | scintillator/Cerenkov | OSO-3 | Kraushaar et al. (1972) |
| 30–50 MeV | spark chamber | balloon | Pinkau et al. (1973) |
| 35–200 MeV | spark chamber | SAS-2 | Fichtel et al. (1978) |
| 4–100 MeV | spark chamber | balloon | Lavine et al. (1982) |

## 2. LOW-ENERGY MEASUREMENTS (0.1–10 MeV)

### 2.1 Spectral Characteristics

The first measurement of the GRB was obtained by a CsI scintillation detector onboard the *Ranger 3* Moon probe in 1962 in the energy range of 0.1 to 3 MeV (Arnold et al. 1962). The lunar missions *Apollo 15, 16 & 17* flown in the early 1970's carried identical NaI scintillation spectrometers and obtained data in the 0.3 to 10 MeV energy range (Trombka et al. 1977). The measured spectrum was well-fitted by an electron bremsstrahlung model, after subtracting the unwanted background components caused by charged particle activation of the detector and the spacecraft. The results agreed with other data points (Table 1) within experimental uncertainties and connected smoothly to high-energy gamma-ray measurements by *SAS-2* in 1972 (Fichtel et al. 1975) (Fig. 2). There was some indication of a flatter slope in the 1 to 5 MeV region (Fig. 1).

HEAO-1 was launched in 1977 with one of its primary objectives to measure the spectrum of the cosmic diffuse background. The A-2 and A-4 experiments together spanned the X-ray to low-energy gamma ray regime from 3 keV to 4.0 MeV. The two experiments used different techniques, yet obtained two spectra that joined smoothly and fitted remarkably well to a single thermal bremsstrahlung model of kT $\sim$ 40 keV in the transitional energy range of 60 to 100 keV (Fig. 1) (Marshall et al. 1980; Gruber et al. 1985). Besides measuring the GRB, the A-4 experiment also detected about 50 X-ray sources in the Galactic plane (Levine et al. 1984), the diffuse Galactic ridge that extends approximately $\pm 50°$ in longitude and $\pm 5°$ in latitude (e.g., Gehrels & Tueller 1993 and references therein), and around 20 active galactic nuclei (AGNs). The AGNs have simple power-law spectra with energy indices clustered around the value of $\sim 0.7$ (Mushotzky 1982, 1984; Rothschild et al. 1983).

There was some evidence in the A-4 data of spectral flattening around a few MeV (Fig. 1). Data obtained by balloon-borne Compton telescopes (Schönfelder et al. 1980; White et al. 1977) in the energy range of 1 to 20 MeV were consistent with the existence of the MeV bump. Beyond 5 MeV, the GRB spectrum steepens to a power law of energy index $\sim 2$ (Schönfelder et al. 1980), in excellent agreement with the extrapolation of SAS-2 measurements (Fichtel et al. 1978) towards lower energies.

Gruber (1992) has performed a best fit after reviewing the available spectral data published prior to 1990 (Fig. 1). He found that below 60 keV, the diffuse background energy flux was fitted well by a bremsstrahlung spectrum. But above 60 keV, he needed to add to the bremsstrahlung extension a power law component with an index of $\sim 0.7$, which characterizes the energy spectra of AGN measured by HEAO-1. His empirical functional fit to the energy flux in units of keV per $cm^2$-s-keV-sr is:

$$7.877 E^{-0.29} \exp\left(\frac{-E}{41.13 keV}\right) \quad 3\ keV < E < 60\ keV$$

$$1652 \times E^{-2.00} + 1.75 \times E^{-0.70} \quad 60\ keV < E < 6\ MeV$$

## 2.2 Spatial Characteristics

Comprehensive studies have been performed on the spatial structure of the low-energy GRB using HEAO-1 data (Boldt 1987). The HEAO-1 detectors had nominal fields-of-view of $3° \times 3°$, $3° \times 1.5°$, and $3° \times 6°$. Shafer (1983), using the A-2 HED data from 2.5 to 60 keV, derived a dipole amplitude of $dI/I_0 = (0.5 \pm 0.2)\%$, with a peculiar velocity $v = 475 \pm 165$ km/sec and apex at $l = 282°$, $b = 30°$. Gruber (1992), using the A-4 MED data from 95 to 165 keV, found a similar dipole anisotropy with $dI/I_0 = (2.2 \pm 0.7)\%$, $v = 1450 \pm 440$ km/sec and apex at $l = 304°$, $b = 26°$. These results are consistent, within the experimental uncertainties, with each other and also with the recent microwave result from COBE (Smoot et al. 1992; Kogut et al. 1993), which gives a peculiar velocity $v = 627 \pm 22$ km/s, and apex at $l = 264°$, $b = 48°$. Gruber noted that the large peculiar velocity measured in A-4 could be due to a local enhancement of the diffuse background.

## 3. INTERPRETATION OF LOW-ENERGY GRB SPECTRUM

The empirical bremsstrahlung shape of the low-energy GRB spectrum prompted early models to attribute its origin to thermal bremsstrahlung by a hot diffuse intergalactic gas at a temperature of $\sim 4 \times 10^8$ K (Marshall et al. 1980; Daly 1988; Brown & Stecker 1979; Olive & Silk 1985). These models were essentially ruled out after the recent observations by *COBE* (Mather et al. 1990). The lack of significant deviations from a blackbody spectrum in the microwave background implies a much smaller Sunyaev-Zel'dovich $y$-parameter than predicted, thereby excluding the possibility that the Universe was filled with hot intergalactic gas in the past (Terasawa 1991; Rogers & Field 1991).

Other models attempted to explain the low-energy GRB spectrum by the integrated flux of extragalactic sources. These were confronted by the well-known spectral paradox: AGNs such as Seyferts and quasars which are the most likely contributors produce continuum spectra that are markedly different from that of the GRB. The average AGN spectrum is characterized by a power law of energy index $\alpha \sim 0.7$, whereas the GRB spectrum has an index of $\alpha \sim 0.4$ below $\sim 60$ keV. To calculate properly the contribution of cosmological sources to the GRB, one needs to take in account the evolutionary effects of the AGN luminosity function. Avni (1978) found that the contribution from AGN depends on the form and amount of density evolution, on the deceleration parameter, and on the AGN formation epoch. The observed integrated flux is given by

$$I_N(E) = \frac{1}{4\pi} \int_0^{Z_F} dz \frac{dV(z)}{dz} \frac{(1+z)B_N(E(1+z),z)}{4\pi d_L^2(z)}$$

$$\frac{dV(z)}{dz} = \frac{4\pi c}{H_0} \frac{1}{(1+z)^3} \frac{1}{\sqrt{1+2q_0 z}} d_L^2(z)$$

where $I_N(E)$ is the flux in keV per sec-cm$^2$-keV, $B_N(E,Z)$ is the emissivity in erg per sec-Mpc$^3$-keV, and $Z_F$ is the epoch of first formation. $V(z)$ is the comoving volume to redshift $z$, $d_L(z)$ is the luminosity-distance to redshift $z$ (Weinberg 1972), and $q_0$ is the deceleration parameter. For a pure density evolution where the source density $\rho(z)$ can be parametrized by $(1+z)^k$, the emissivity is given by $B_N(E(1+z),z) = (1+z)^{k+a} B_N(E, z=0)$, and $\alpha$ is the energy spectral index.

Rothschild et al. (1983) and Bassani et al. (1985) integrated the X-ray luminosity function of the *HEAO-1* AGNs as derived by Piccinotti et al. (1982) to compute the relative contributions of different classes of AGNs. The results are shown in Fig. 3. They found that Seyferts are the most important contributors to the GRB. Their contribution increases from energies beyond $\sim 40$ keV until at $\sim 160$ keV, they can account for all of the observed background. The observed spectral shape and intensity of GRB are used to set limits on the luminosity cutoff and spectrum break energy of AGNs (Rothschild et al. 1983; Zycki et al. 1993) to prevent an overproduction of diffuse background above 100 keV. Indeed recent AGN observations by *Compton* and *GRANAT* are indicating that Seyferts do typically have low-energy spectral breaks (see Section 6).

After subtracting the Seyfert contribution, a significant residual GRB remains below 100 keV that needs to be explained. Various authors have postulated the existence of different AGN populations in earlier epochs to explain the residual spectrum: for example, Eddington-limited thermal-type sources (Boldt 1987), AGN precursors with high compactness (Leiter & Boldt 1991), and AGNs with harder power-law spectra (Gruber 1992). The recent ROSAT deep survey detected an unexpectedly high number of faint QSOs at $1 < z < 2$ (Shanks et al. 1991; Griffiths et al. 1993). When the new QSO X-ray luminosity function with its derived evolution is integrated to a maximum

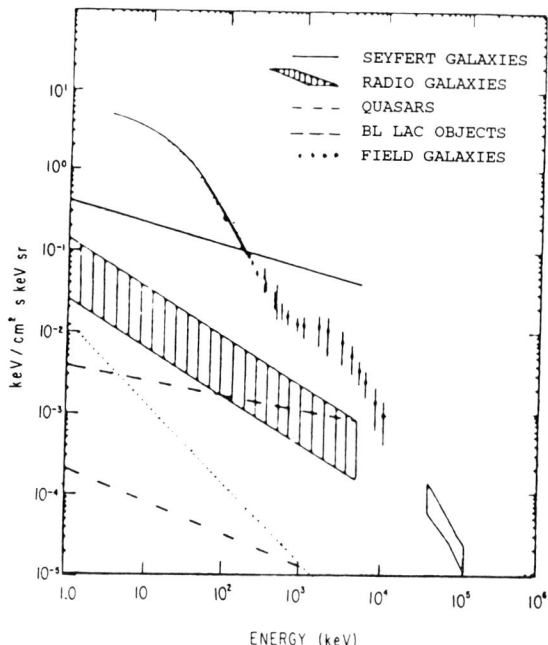

**Figure 3.** *Contributions of the different classes of AGNs to the GRB (Bassani et al. 1985).*

redshift of $z_{max} \sim 4$ (Boyle et al. 1993), it is found that QSOs can account for 30 to 90% of the diffuse background at $\sim 2$ keV.

## 4. HIGH-ENERGY MEASUREMENTS OF GRB (>10 MeV)

The first all-sky survey of cosmic gamma rays of energies above 50 MeV was carried out by *OSO-3* in 1967–1968. The directional scintillator/Cerenkov detector recorded 621 "sky events" in 16 months of operations (Fig. 4). The observations can be attributed to three components: (1) a galactic component, concentrated along the galactic plane and well-correlated with the column density as deduced from 21-cm radio measurements; (2) a galactic center component; and (3) an isotropic, extragalactic component with a steep power-law spectrum. Subsequent measurements made using an entirely different technique—spark chamber detectors—confirmed the general picture. These included the *SAS-2* (Fichtel et al. 1975) mission and several balloon flights (Table 1). *SAS-2* measured a diffuse GRB component with a very steep differential power law of energy index $\alpha \sim 1.7$ (photon index $= \alpha + 1 \sim 2.7$) between 35 MeV and 200 MeV (Fichtel et al. 1978) (Fig. 5). The extrapolated intensity of this component to lower energies agreed well with measurements at 10 MeV. Above several hundred MeV, the flux of the extragalactic GRB falls below the galactic high latitude background and the determination of its value is highly dependent on data analysis techniques.

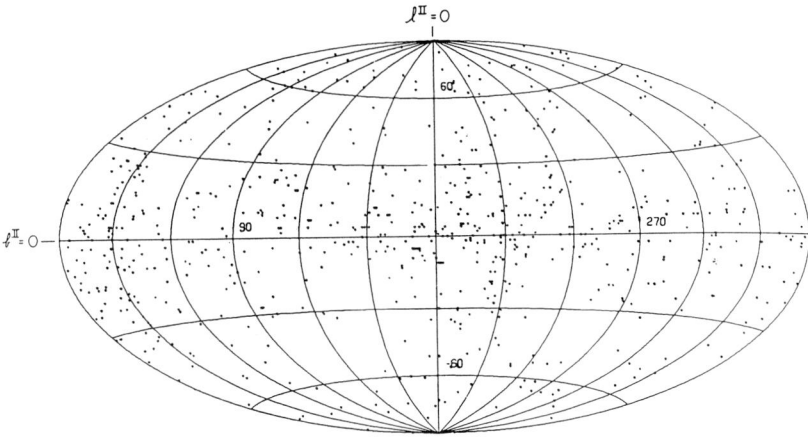

**Figure 4.** *OSO-3 sky events plotted in galactic coordinates (Kraushaar, Clark & Garmire 1972).*

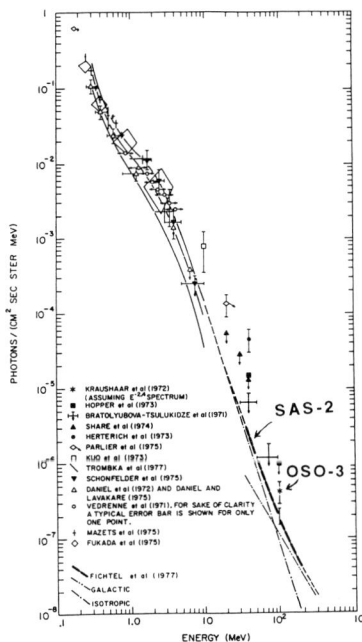

**Figure 5.** *Differential photon spectrum of the SAS-2 diffuse gamma-ray background (Fichtel et al. 1978).*

## 5. INTERPRETATION OF HIGH-ENERGY GRB SPECTRUM

Before the launch of *Compton*, only one extragalactic source, 3C 273, had been detected above 10 MeV (Bignami et al. 1979). So theorists tended to explain the origin of the high-energy GRB spectrum by diffuse mechanisms rather than by a superposition of unresolved discrete extragalactic sources. Above 100 MeV, this now has been changed by *Compton* observations of blazars (see Section 6), but in the 1–100 MeV range diffuse models are still the best interpretation. Below we describe two theoretical models that postulate a cosmological origin for the 1–100 MeV GRB.

## 5.1 GRB from Proton-Antiproton Annihilation

Stecker et al. (1971) proposed a diffuse emission model in which the GRB arises in a baryon symmetric Big Bang cosmology from matter-antimatter annihilations. In a grand-unified-theory model with spontaneous CP violation (Stecker 1985), it is possible for the Universe to have evolved very large regions of pure matter and pure antimatter containing masses the size of galaxy clusters or superclusters in its early history. These regions are essentially the fossils of the vanished CP domains. In a baryon symmetric cosmology, annihilations occur at the boundaries between these regions at all redshifts to produce an extragalactic GRB. Puget (1973) has computed the annihilation rate as a function of redshift. The annihilations produce $\pi^\circ$-mesons with gamma-ray producing decay modes. ($\pi^\circ$ decay is also the principal mechanism for producing the galactic high-energy diffuse gamma radiation.) Figure 6 shows a typical rest-frame spectrum produced by proton-antiproton annihilation with $\pi^\circ$ decay (Stecker 1971), with maximum intensity at $m_\pi c^2/2 \sim 70$ MeV. The spectrum is nearly flat near the maximum, with a minimum energy of $\sim 5$ MeV and a maximum cutoff at $\sim 1$ GeV. To arrive at the predicted GRB spectrum observed at the current epoch, it is necessary to solve a cosmological photon transport equation and take into account pair production and Compton scattering at high redshifts which may cause energy loss (Stecker et al. 1971). The resultant spectrum (Stecker 1989) matches the observed steep slope of the extragalactic GRB above $\sim 1$ MeV very well. Furthermore, it can account for the observed MeV bump (Fig. 7). Absorption due to Compton scattering and pair production causes the spectrum to bend over below $\sim 1$ MeV.

## 5.2 GRB from Primordial Blackholes

Page & Hawking (1976) postulated a population of primordial black holes (PBH) created in the early Universe that could explain the GRB. The PBHs cannot be created in the present epoch since the necessary compressional forces do not exist. They undergo quantum mechanical decay by radiating gravitons, neutrinos, electrons, positrons, and gamma rays. PBHs with initial masses less than a critical mass of $5 \times 10^4$ g would have completely evaporated by now. PBHs of slightly greater initial mass would be radiating energy at the rate of $2.5 \times 10^{17}$ ergs/s. As they reach the end of their life, they evaporate by ejecting all remaining rest mass in a very short time. The particles emitted in this final release would decay rapidly ($t \ll 1$ sec), giving a short burst of gamma-rays between 100 MeV and 1 GeV. The EGRET instrument onboard *Compton* may be able to detect PBH bursts with this particular timing signature, but has not seen any such events to date. A uniform distribution of PBHs of initial mass $\geq 5 \times 10^4$ g, when integrated over the cosmological time scale, would give a GRB photon spectrum of power law index $\sim 3$ above 120 MeV. This matches the observed GRB spectral slope in the 100 MeV range rather well. Below 120 MeV, the spectrum may flatten depending on the space density of PBHs.

Extragalactic Gamma-Ray Background 23

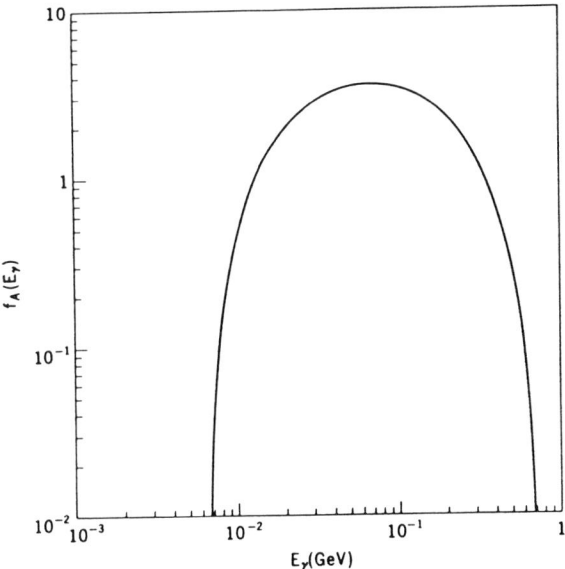

**Figure 6.** *Gamma-ray spectrum from proton-antiproton annihilation at rest (Stecker 1971).*

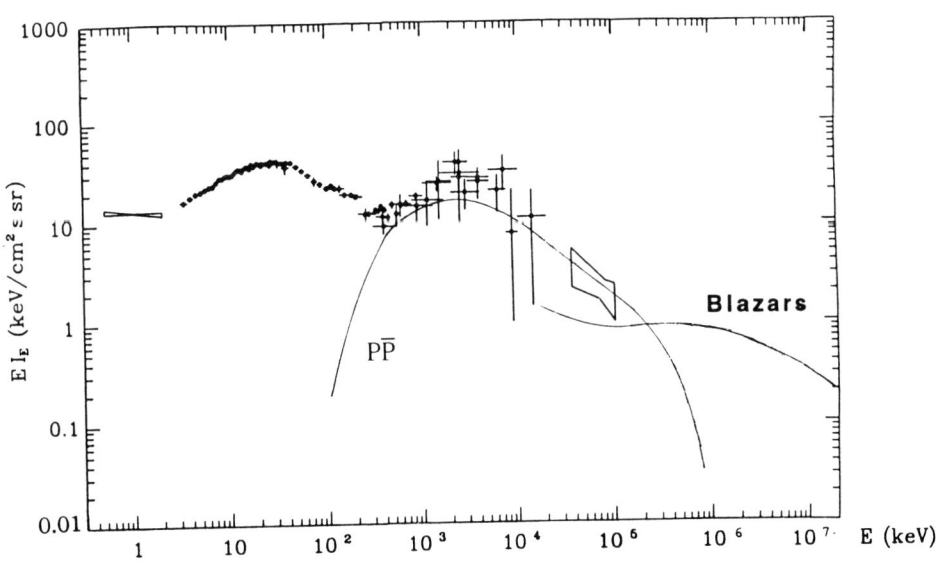

**Figure 7.** *Contribution to the extragalactic background by proton-antiproton annihilations in a baryon-symmetric universe (Stecker et al. 1971).*

## 6. *COMPTON* RESULTS AND IMPLICATIONS FOR GRB

### 6.1 *Compton* Observations of AGN

Before the launch of *Compton*, only four AGNs have been detected at gamma-ray energies: 3C 273, Cen A, NGC 4151 and MCG 8-11-11 (see, *e.g.,* Gehrels & Cheung 1991; Bassani *et al.* 1985). 3C 273 was the only quasar that has been detected above 10 MeV. It was during one of the early *Compton* pointings which included 3C 273 in the field of view that another quasar, 3C 279, was first detected by the EGRET instrument from 30 MeV to over 5 GeV (Hartman *et al.* 1992). The photon flux intensity of 3C 279 was very high, comparable to that of the Crab or Geminga above 100 MeV. The differential photon spectrum is well represented by a single power law of photon index $\sim 2$ over the entire observed energy range. The lack of detection by *COS-B* in 1976 implies that 3C 279 is highly variable. Later observations by EGRET confirmed that 3C 279 was initially detected in a flaring state and it is variable on the time scale of days with no significant change in spectral index (Kniffen *et al.* 1993). Subsequently, many more AGNs have been detected by EGRET. Table 2 lists the AGNs discovered as of November 1992 together with their characteristics (see *e.g.,* Fichtel 1993, Bertsch *et al.* 1993 and references therein). All the AGNs detected in high-energy gamma rays are core-dominated, flat spectrum radio quasars or radio-loud BL Lac objects (or "blazars"). Blazars generally exhibit strong variability, significant optical polarization and superluminal flow. The rapid variability indicates that the radiation is likely to be beamed in a relativistic jet, which also produces the radio emission. The BL Lac object Mrk 421 has also been detected in TeV gamma rays by the Whipple Observatory ground-based Cerenkov telescope (Punch *et al.* 1992). The data are consistent with a single power law of photon index $\sim 2$ from the 100 MeV to the TeV energy range, though it must be kept in mind that the Whipple observations were not simultaneous with the EGRET observations (Lin *et al.* 1992). Note that no Seyfert galaxies have been detected by EGRET in the 100 MeV range.

In the low-energy range of 50 keV to 10 MeV, the OSSE instrument has observed 21 AGNs and reported ten firm detections and four marginal ones (Cameron *et al.* 1993). Of these, nine are Seyfert 1 galaxies, with a detection rate close to 100%. Upper limits have been obtained for two Seyfert 2 galaxies (NGC 1068 and NGC 4593). The spectral indices for energies > 60 keV range from 1.6 to 3.2, with an average of 2.6. The spectrum of NGC 4151 obtained in July 1991 by OSSE (Fig. 8) is steeper than most previous observations (Maisak *et al.* 1993). Such a steep spectrum has been observed once before by the *GRANAT* instruments SIGMA and ART-P (Jourdain *et al.* 1992; Apal'kov *et al.* 1992). The data can be fitted by a broken power law or a thermal Comptonization spectrum (Sunyaev & Titarchuk 1980). There is evidence that the spectrum steepens at times of source brightening (Perola *et al.* 1986; Yaqoob *et al.* 1989). It is possible that Seyfert galaxies in general have variable spectral states correlated with luminosity (Paciesas *et al.* 1993). These results together with observations from *Ginga* (*e.g.,* Williams *et al.* 1993) and *ROSAT* (*e.g.,* Turner *et al.* 1993) at lower energies indicate that AGNs have more complex spectral characteristics than can be represented by a simple "canonical" power-law spectrum. For example, *Ginga* observations have revealed an additional flat component in the > 10 keV portion of Seyfert 1 spectrum (*e.g.,* Pounds *et al.* 1990; Matsuoka *et al.* 1990). This has been attributed to reflection or Comptonization of the incident power-law spectrum by an accretion disk.

A picture is emerging that the gamma-ray emitting AGNs can be divided into two classes: (1) the radio-quiet Seyferts, in which we observe gamma rays emitted from

Table 2.
AGN Discovered by EGRET (Fichtel 1993)
Status Nov. 1992

| Name | OVV | BL Lac | Super Lum | Radio Loud | Radio Flat Spect.[1] | Optical Pol. >3% | Position Diff.[2] | Position Uncert.[3] | Flux ($10^{-6}$ cm$^{-2}$ sr$^{-1}$) (E > 100 MeV) | Photon Spectral Index | Z | Relative Lum.[4] |
|---|---|---|---|---|---|---|---|---|---|---|---|---|
| 0202+149 (4C+15.05) | | | | • | • | | 0.3° | 0.4° | 0.3 ± 0.1 | −1.7 ± 0.1 | 1.00 | 2 |
| 0208−512 PKS | | | | • | • | • | 0.13° | 0.13° | 0.4 to 0.9 | −2.0 ± 0.2 | 0.94 | 2.0 |
| 0235+164 (OD+160) | | • | • | • | • | • | 0.10° | 0.3° | 0.8 ± 0.1 | | 0.92 | 0.4 |
| 0420−014 (OA 129) | • | | | • | • | • | 0.5° | 0.4° | 0.4 ± 0.1 | | 0.86 | 0.3 |
| 0454−463 PKS | | | | • | • | | 0.27° | 0.38° | 0.25 ± 0.1 | | 2.06 | 4 to 13 |
| 0528+134 PKS | | | | • | • | | 0.13° | 0.15° | 0.4 to 1.6 | −2.4 ± 0.1 | | |
| 0537−441 PKS | | • | • | • | • | • | 0.4° | 0.6° | 0.3 ± 0.1 | −2.0 ± 0.2 | 0.894 | 0.2 |
| 0716+714 | | • | ? | • | • | | 0.47° | 0.4° | 0.20 ± 0.06 | −1.8 ± 0.2 | | |
| 0836+714 (4C+71.07) | • | | • | • | • | | 0.58° | 0.50° | 0.15 ± 0.04 | −1.9 ± 0.1 | 2.17 | 1.1 |
| 1101+384 (Mrk 421) | | • | | • | | | 0.3° | 0.4° | 0.14 ± 0.03 | −2.4 ± 0.1 | 0.031 | 0.0002 |
| 1226+023 (3C273) | | | • | • | • | | 0.2° | 0.5° | 0.30 ± 0.05 | −2.0 ± 0.1 | 0.158 | 0.008 |
| 1253−055 (3C279) | • | | • | • | • | • | 0.083° | 0.08° | 0.6 to 4.9 | | 0.54 | 0.3 to 2 |
| 1406−076 | | | | | | | 0.2° | 0.4° | 1.0 ± 0.2 | | 1.49 | 2 |
| 1606+106 (4C+10.45) | | | | • | • | | 0.42° | 0.50° | 0.5 ± 0.2 | −2.0 ± 0.1 | 1.23 | 1.6 |
| 1633+382 (4C+38.41) | • | | | • | • | | 0.08° | 0.15° | 0.4 to 1.4 | | 1.81 | 3 to 11 |
| 2052−474 | | | | | | | 0.35° | 0.5° | 0.3 ± 0.1 | | 1.489 | 0.6 |
| 2230+114 (CTA 102) | | | • | • | ? | • | 0.3° | 0.4° | 0.24 ± 0.07 | −2.4 ± 0.07 | 1.037 | 0.5 |
| 2251+158 (3C454.3) | • | | • | • | • | • | 0.25° | 0.22° | 0.8 ± 0.1 | −2.0 ± 0.1 | 0.859 | 0.5 |

1. Flat spectrum radio sources: $\alpha_r > -0.5$ (2–5 GHz band)
2. Difference between gamma-ray determined position and known position of identified source. Most are preliminary.
3. There is a 68% probability that the source is within a circle of this radius. Most are preliminary.
4. The source luminosity (> 100 MeV) in $f \times 10^{48}$ erg/s, with $f$ = beaming factor.

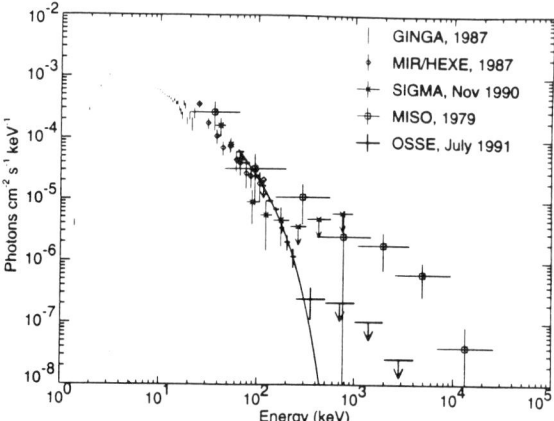

**Figure 8.** *The OSSE spectrum of NGC 4151 compared to selected previous observations. The solid curve is the best fit to OSSE data of a Sunyaev-Titarchuk thermal Comptonization model (Maisack et al. 1993).*

accretion disks; and (2) the radio-loud blazars, in which we observe gamma rays emitted from the relativistic jets and which are highly variable.

### 6.2 Implications for GRB

The *Compton* observations firmly identify AGNs as gamma-ray emitters that can contribute to the GRB. Several groups have proposed schemes to explain the low-energy GRB spectrum based on the "reprocessed" AGN model (Zycki et al. 1993; Zdziarski et al. 1993; Madau et al. 1993; Rogers & Field 1991; Terasawa 1991; Mereghetti 1990). The intrinsic AGN spectrum can be generated either through thermal Comptonization of low-energy photons by hot plasma or by non-thermal pair cascade process in a compact plasma (Svensson 1987). Both mechanisms produce a spectral break at $\sim 80$ keV. Zdziarski et al. (1993) used a Seyfert 1 thermal Comptonization model and the X-ray luminosity function derived by Boyle et al. (1993). They integrated the contributions over redshift and obtained a good fit to the GRB spectrum from 2 to 100 keV, including the 30 keV bump (Fig. 9). The dominant contribution comes from the AGNs at the highest redshift. However, the model fails to explain the MeV bump as the Seyfert spectrum cuts off above 80 keV. Zycki et al. (1993) examined the contribution by non-thermal pair models with different compactness. They found that a good fit to the MeV bump can be obtained by a population of unobserved low-compactness AGNs (Fig. 10), though they cannot explain the entire 30 to 100 MeV GRB radiation with the model.

Padovani et al. (1992), Dermer and Schlickeiser (1992), and Stecker et al. (1993) estimated the contribution of blazars to the high-energy GRB. Since the gamma-ray luminosity function of blazars has not yet been determined, they used the radio luminosity evolutionary function by Dunlop and Peacock (1990) instead. This is based on a significant (possibly linear) correlation between the 100 MeV gamma-ray luminosity and the 5-GHz radio luminosity seen in the gamma-ray blazars (Fig. 11). The computed background falls an order of magnitude below that observed by SAS-2 in the 100 MeV range (Stecker et al. 1993) and so fails to be a major contributor to the GRB

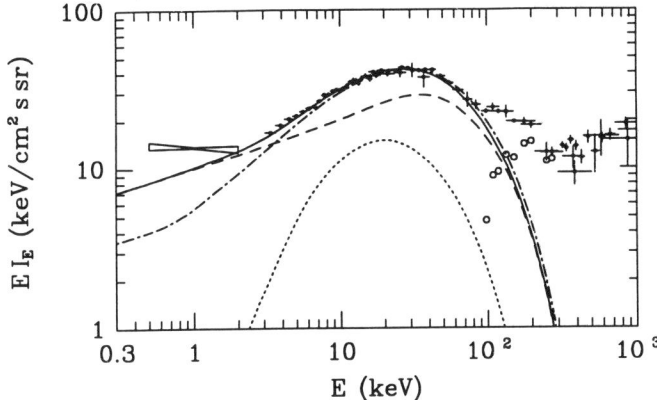

**Figure 9.** *Contribution to the GRB by thermal AGNs. The solid and dot-dashed curves are Comptonization models without and with absorption. The dashed curve gives the thermal component, and the dotted curve the reflection component (Zdziarski et al. 1993).*

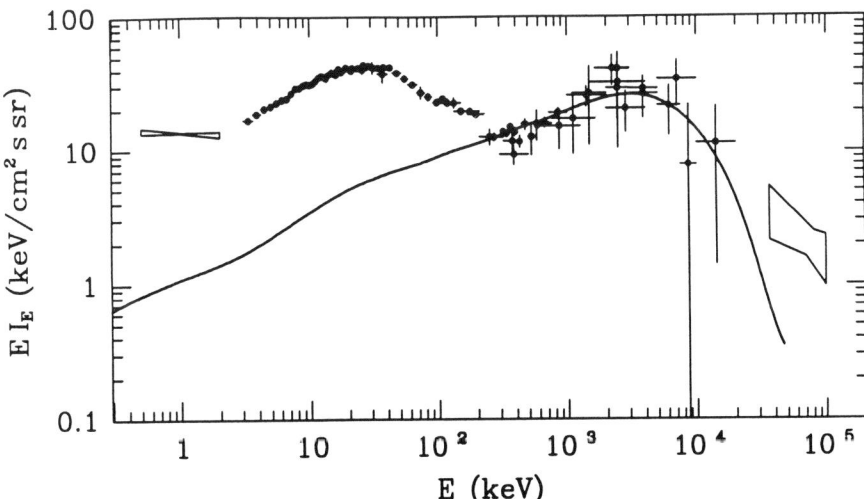

**Figure 10.** *Contribution to the GRB spectrum by a population of low-compactness non-thermal AGNs (Zycki et al. 1993).*

in this energy range. Moreover, the average blazar spectrum is a power law of photon index $\sim 2$, which is much harder than the observed GRB (Fig. 7). However, if the hard spectrum continues unattenuated to higher energies, blazars could be important contributors to the GRB in the multi-GeV energy range.

It may well be that the GRB is the sum of a number of different components with different origins. Each component's contribution varies as a function of energy. We need to keep in mind that when we study the gamma-ray background, we are looking at an energy span of over six decades. A particular physical mechanism probably dominates

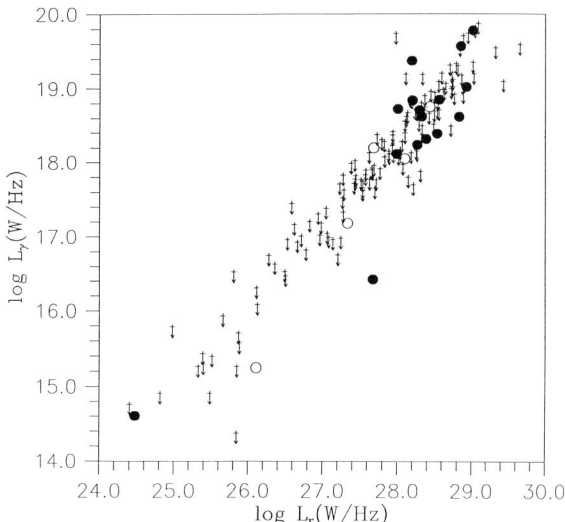

**Figure 11.** *Gamma-ray luminosity vs. radio luminosity plot for blazars in the EGRET survey. The filled circles represent solid detections, the hollow circles represent marginal detections, and the crossed arrows represent the upper limits for those radio sources which EGRET did not detect (Stecker et al. 1993).*

in a specific energy range, for example: the radio-quiet AGNs at $< 1$ MeV; the diffuse matter-antimatter annihilations at $\sim 1$ to 100's MeV; the SN Ia's at $\sim 1$ MeV (see Section 7.2); the primordial black holes at 100's MeV; and blazars at 100's GeV.

## 7. OTHER TOPICS

### 7.1 TeV Observations and Constraints on IR Background

The recent detection of Mrk 421 in TeV gamma rays by the Whipple observatory (Punch et al. 1992) suggests that other blazars may also be detectable in that energy range. Stecker et al. (1992) predicted that the observed blazar spectrum (photon index $\sim 2$) will soften to an index of $\sim 3.5$ at $\sim 300$ GeV due to absorption by pair production off the intergalactic infrared starlight. Stecker (1993) has computed the IR optical depths of the TeV photons for the EGRET blazars. The IR photons with wavelengths around 2 $\mu$m will contribute most to the absorption of TeV gamma rays. Ground-based Cerenkov detectors have improved their sensitivities such that this absorption break can now be studied. In fact, the measurements could be used to constrain the infrared intergalactic background which has not been well determined (Stecker et al. 1992).

### 7.2 GRB from Type Ia SN

The cumulative gamma-ray spectrum of Type Ia supernovae (SN Ia's) during the history of the Universe are expected to contribute significantly to the GRB. This idea was first advanced by Clayton & Silk (1969), followed by a more detailed discussion by Clayton & Ward (1969) and a recent improved calculation by The et al. (1993). SN Ia's emit strong gamma-ray lines at 847, 1238, 2599 and 3250 keV from the radioactive decay of $^{56}$Ni → $^{56}$Co → $^{56}$Fe produced in the explosion. This has been confirmed by the observations of SN 1987a (e.g., Leising & Share 1990). Integrating over the cosmological history of nucleosynthesis smears the lines and they become edges near the rest energies in the GRB spectrum. In particular, the SN Ia's may contribute a significant fraction of the GRB at ∼ 1 MeV (Fig. 12). If high resolution spectral measurements can be made of the GRB between 0.1 and 1.0 MeV, the model of The et al. (1993) may be used to derive the history of SN Ia nucleosynthesis in the Universe. Such high resolution spectral measurements of the GRB have never been made. The current generation of germanium gamma-ray spectrometers has sufficient spectral resolution to measure such edges in the GRB spectrum. For example the GRIS balloon instrument has successfully resolved the $^{56}$Co lines of SN 1987a (Tueller et al. 1989). Observations of the GRB by GRIS are planned for October 1994. Better spatial resolution in future instruments would allow determination of the GRB fine-scale anisotropy. This is essential to distinguish between the baryon-symmetric cosmological origin and the AGN origin of the GRB (Gao et al. 1990; Cline & GAO 1990), as these models predict different intensity fluctuation patterns in the GRB.

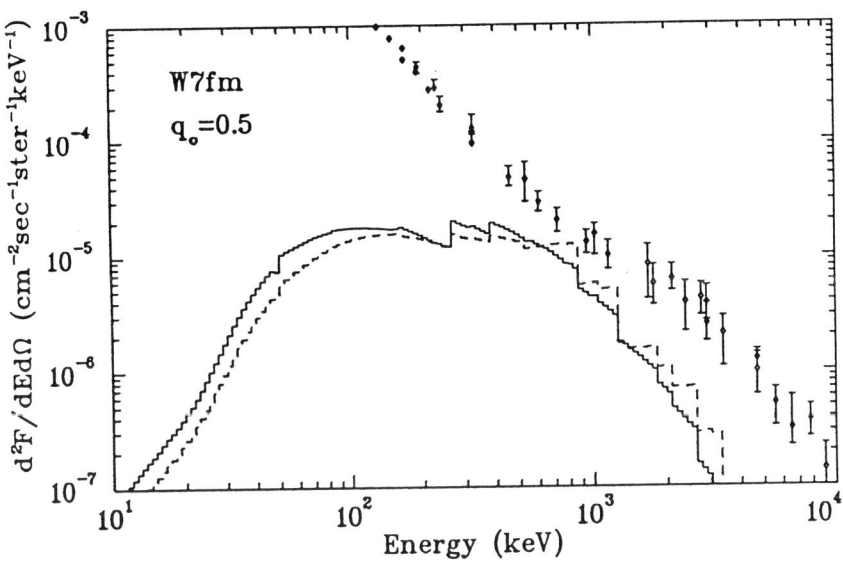

Figure 12. *The differential flux of cosmic Type Ia supernovae in the Einstein-de Sitter ($q_o = 0.5$) universe. The data points are measurements of the GRB (The, Leising & Clayton 1993).*

# REFERENCES

Apal'kov, Y., et al. 1992, in *Frontiers of X-Ray Astronomy*, ed. Y. Tanaka & K. Koyama (Tokyo: Universal Academy Press), 251.

Arnold, J. R., Mertzger, A. E., Anderson, E. C., & Van Dilla, M. A. 1962, *J. Geophys. Res.*, **67**, No. 12, 4878.

Avni, Y. 1978, *Astr. Ap.*, **63**, L13.

Bassani, L., Dean, A. J., Di Cocco, G., & Perotti, F. 1985, in *Active Galactic Nuclei*, ed. J. E. Dyson, (Manchester: Manchester University Press), p. 252.

Bertsch, D. L., et al. 1993, *Ap.J.*, **405**, L21.

Bignami, G. F., Fichtel, C. E., Hartman, R. C., & Thompson, D. J. 1979, *Ap.J.*, **232**, 649.

Boldt, E. 1987, *Phys. Reports*, **146**, No. 4, 215.

Boyle, B. J., Griffiths, R. E., Shanks, T., Stewart, G. C., & Georganopoulos, I. 1993, *MNRAS*, **260**, 49.

Brown, R. W., & Stecker, F. W. 1979, *Phys. Rev. Letters*, **43**, 315.

Cameron, R. A., et al. 1993, in *Proc. Compton Gamma Ray Observatory Symposium*, ed. M. Friedlander, N. Gehrels, & D. Macomb (AIP: New York), p. 478.

Clayton, D. D. & Silk, J. 1969, *Ap.J.*, **158**. L43.

Clayton, D. D. & Ward, R. A. 1975, *Ap.J.*, **198**, 241.

Cline, D. B. & Gao, Y.-T. 1990, *Ap.J.*, **348**, 33.

Daly, R. A. 1988, *MNRAS*, **232**, 853.

Daniel, R. R., Joseph, G., & Lavakare, O. 1972, *Astrophys. Space Sci.*, **18**, 462.

Dermer, C. D. & Schlickeiser, R. 1992, *Science*, **257**, 1642.

Dunlop, J. S. & Peacock, J. A. 1990, *MNRAS*, **247**, 19.

Fichtel, C. E. 1993, *Ann. NY Acad. Sci.* (Texas/PASCOS, ed. C. W. Akerlof & M. A. Srednicki), **688**, 136.

Fichtel, C. E. et al. 1993, *Astr. Ap. Suppl.*, **97**, 13.

Fichtel, C. E., Hartman, R. C., Kniffen, D. A., Thompson, D. J., Bignami, G. F., Ögelman, H., Özel, M. F., & Tümer, T. 1975, *Ap.J.*, **198**, 163.

Fichtel, C. E., Simpson, G. A., & Thompson, D. J. 1978, *Ap.J.*, **222**, 833.

Fukuda, Y., Hayakawa, S., Kasahara, I., Makino, F., & Tanaka, Y., 1975, *Nature*, **254**, 398.

Gao, Y. T., Cline, D. B., & Stecker, F. W. 1990, *Ap.J.*, **357**, L1.

Gao, Y. T., Stecker, F. W., Gleiser, M., & Cline, D. B. 1990, *Ap.J.*, **361**, L37.

Gehrels, N. & Cheung, C. 1991, in *Testing the AGN Paradigm*, ed. S. S. Holt, S. G. Neff & C. M. Urry (New York: AIP), 348.

Gehrels, N. & Tueller, J. 1993, *Ap.J.*, **407**, 593.

Griffiths, R. E. et al. 1993, *in preparation*.

Gruber, D. E. 1992, in *X-Ray Background*, ed. X. Barcons & A. C. Fabian (Cambridge: Cambridge Univ. press), 44 (G92).

Gruber, D. E., Matteson, J. L., & Jung, G. V. 1985, *Proc. 19th Intl. Cosmic Ray Conf.*, ed. F. C. Jones, **OG-1**, 349.

Hartman, R. C., et al. 1992, *Ap.J.*, **385**, L1.

Johnson, W. N., et al. 1993, *Astr. Ap. Suppl.*, **97**, 21.

Jourdain, E., et al. 1992, *Astr. Ap.*, **256**, L38.

Kogut et al. 1993, submitted to *Ap.J.*

Kniffen, D. A., et al. 1993, *Ap.J.*, **411**, 133.

Kraushaar, W. L., Clark, G. W., Garmire, G. P., Borker, R., Higbie, P., Leong, V., & Thorsos, T. 1972, *Ap.J.*, **177**, 341.

Lavigne, J. M., et al. 1982, *Ap.J.*, **261**, 720.

Leising, M. D., & Share, G. H. 1990, *Ap.J.*, **357**, 638.
Leiter, D. & Boldt, E. 1992, in *Proc. Compton Observatory Science Workshop* (NASA Conf. Pub. 3137), 359.
Levine, A. M., et al. 1984, *Ap.J. Suppl.*, **54**, 581.
Lin, Y. C., et al. 1992, *Ap.J.*, **401**, L61.
Madau, P., Ghisellini, G. & Fabian, A. C. 1993, *Ap.J.*, **410**, L7.
Maisack, M., et al. 1993, *Ap.J.*, **407**, L61.
Mandrou, P., Vedrenne, G., & Niel, M. 1979, *Ap.J.*, **230**, 97.
Marshall, E. F., et al. 1980, *Ap.J.*, **235**, 4.
Matsuoka, M. et al. 1990, *Ap.J.*, **361**, 440.
Mazets, E. P., Golenetskii, S. V., Il'inskii, V. N., Gur'yan, Y. A., & Kharitonova, T. V. 1975, *Astrophys. Space Sci.*, **33**, 347.
Mereghetti, S. 1990, *Ap.J.*, **354**, 58.
Mushotzky, R. F. 1982, *Ap.J.*, **256**, 92.
Mushotzky, R. F. 1984, *Astrophys. Space. Sci.*, **3**, Nos. 10–12, 157.
Olive, K. A., & Silk, J. 1985, *Phys. Rev. Letters*, **55**, 2362.
Paciesas, W. S., et al. 1993, in *Proc. Compton Gamma Ray Observatory Symposium*, ed. M. Friedlander, N. Gehrels, & D. Macomb (AIP: New York), p. 473.
Padovani, P., Ghisellini, G., Fabian, A. C. & Celotti, A. 1993, *MNRAS*, in press.
Page, D. N., & Hawking, S. W. 1976, *Ap.J.*, **206**, 1.
Perola, G. C. et al. 1986, *Ap.J.*, **306**, 508.
Piccinotti, G., Mushotzky, R. F., Boldt, E. A., Holt, S. S., Marshall, F. E., Serlemitsos, P. J., & Shafer, R. A. 1982, *Ap.J.*, **253**, 485.
Pinkau, K. 1973, in *Gamma Ray Astrophysics*, ed. F. W. Stecker & J. I. Trombka, NASA SP-339 (US Govt. Printing Office, Washington, DC), p. 133.
Pounds, K. A. et al. 1990, *Nature*, **344**, 132.
Puget, J. L. 1973, in *Gamma Ray Astrophysics*, ed. F. W. Stecker & J. I. Trombka, NASA SP-339 (US Govt. Printing Office, Washington, DC), p. 367.
Punch, M., et al. 1992, *Nature*, **358**, 477.
Rogers, R. D. & Field, G. B. 1991, *Ap.J.*, **370**, L57.
―――――.1991, *Ap.J.*, 378, L11.
Rothschild, R. E., Mushotzky, R. F., Baity, W. A., Gruber, D. E., Matteson, J. L. & Peterson, L. E. 1983, *Ap.J.*, **269**, 423.
Schönfelder, V., Graml, F., & Penningsfeld, F.-P. 1980, *Ap.J.*, **240**, 350.
Shanks, T. et al. 1991, *Nature*, **353**, 315.
Smoot, G. F., et al. 1992, *Ap.J.*, **396**, L1.
Stecker, F. W. 1971, *Cosmic Gamma Rays*, NASA SP-249 (US Govt. Printing Office, Washington, DC).
Stecker, F. W. 1973, in *Gamma Ray Astrophysics*, ed. F. W. Stecker & J. I. Trombka, NASA SP-339 (US Govt. Printing Office, Washington, DC), p. 211.
Stecker, F. W. 1985, *Nuclear Physics*, **B252**, 25.
Stecker, F. W. 1989, in *Proc. of Gamma Ray Observatory Science Workshop*, ed. W. N. Johnson, p. 4–73.
Stecker, F. W. 1993, preprint.
Stecker, F. W., Morgan, D. L., & Bredekamp, J. 1971, *Phys. Rev. Letters*, **27**, 1467.
Stecker, F. W. & Puget, J. L. 1972, *Ap.J.*, **178**, 57.
Stecker, F. W., De Jager, O. C. & Salamon, M. H. 1992, *Ap.J.*, **390**, L49.

Stecker, F. W., Salamon, M. H., & Malkan, M. A. 1993, *Ap.J.*, **410**, L71.
Strong, A. W. 1984, *Adv. Space Res.*, **3**, No. 10–12, 87.
Svensson, R. 1987, *MNRAS*, **227**, 403.
Sunyaev, R. A. & Titarchuk, L. G. 1980, *Astr. Ap.*, **86**, 121.
Terasawa, N. 1991, *Ap.J.*, **378**, L11.
The, L.-S., Leising, M. & Clayton, D. D. 1993, *Ap.J.*, **403**, 32.
Trombka, J. I., Dyer, C. S., Evans, L. G., Bielefeld, M. J., Seltzer, S. M., & Metzger, A. M. 1977, *Ap.J.*, **212**, 925.
Tueller, J., Barthelmy, S. D., Gehrels, N., Teegarden, B. J., Leventhal, M. & MacCallum, C. J. 1989, *Ap.J.*, **351**, L41.
Turner, T. J., George, I. M., & Mushotzky, R. F. 1993, *Ap.J.*, **412**, 72.
Vedrenne, G., Albernhe, F., Martin, I., & Talon, R. 1971, *Astr. Ap.*, **15**, 50.
Vette, J. I., Gruber, D. E., Matteson, J. L., & Peterson, L. E. 1969, in *IAU Symp. No. 37*, ed. L. Gratton (Dordrecht: Reidel), p. 335.
Weinberg, S. 1972, *Gravitation and Cosmology*, (New York: Wiley).
White, R. S., Dayton, B., Moon, S. H., Ryan, J. M., Wilson, R. B., & Zych, A. D. 1977, *Ap.J.*, **218**, 920.
Williams, O. R., et al. 1992, *Ap.J.*, **389**, 157.
Yaqoob, T., et al. 1989, *MNRAS*, **236**, 153.
Zdziarski, A. A., Zycki, P. T., Svensson, R. & Boldt, E. 1993, *Ap.J.*, **405**, 125.
Zdziarski, A. A., Zycki, P. T., & Krolik, J. H. 1993, *Ap.J.*, **414**, L81.
Zycki, P. T., Zdziarski, A. A., Ghisellini, G. & Svensson, R. 1993, preprint.

## DISCUSSION

**J. Krolik**: Is it possible to get limits on the number density of sources on the sky, in that map actually, from the fluctuations in photon counts?

**Gehrels**: Yes. There are papers that have talked about this for the X-ray background and I guess the same kind of thing could be done for the high energy background. In fact Rick Shafer's thesis is all about the fluctuations and trying to estimate number densities from that. I guess that the counting statistics from the EGRET map will allow us to get into that kind of game also. Carl Fichtel is the Principal Investigator for the EGRET instrument.

**C. Fichtel**: One of the things one has to remember is that one does have the galactic diffuse radiation which is extremely strong and dominates at mid-latitudes. This galactic background has to be subtracted first, and this complicates any fluctuation analysis. One of the things which was not mentioned here, is that if you really believe for example the AGN-blazar theory, and you realize that we are seeing all the way back to $z = 2$, then there are going to be fluctuations. There are not going to be that many blazars that contribute, so we have to look for fluctuations on the sky and that not only means spatial, but probably also energy spectral fluctuations, because we have a range of slopes from 1.6 to 2.5. If you like the shock acceleration theory and accept it should be 2 plus or minus a few tenths, then you are quite happy with it. But you could very well see something which is quite non-uniform, and this does mean that we have to get rid of the galactic diffuse component first, which is the problem we are working on right now.

**Gehrels**: The high-latitude, low-density region that we will observe with the EGRET instrument will be the best way of getting rid of that galactic component.

**J. Peebles**: Do I take it that the contours I see at the bottom and top here are, in fact, the residual gradient in the intensity of the diffuse galactic background?

**Gehrels**: Well, I think you cannot really do a quantitative study off this particular map. For example, the dark ridges at the top and the bottom are just artifacts of the way we put these data together. So I do not think you should believe this particular map, but contours like those will be, in fact, coming out of the real map.

**J. Peebles**: Do I take it that you expect the extragalactic component to be time variable?

**Gehrels**: No.

**J. Peebles**: Well, I mean 3C 279 is variable.

**C. Fitchel**: You really hit on something I think is extremely important. You could have variability in three ways. It could obviously be spatial just because you have a limited number; it could be an energy spectrum variability; and, as every AGN we have observed is time variable, you could also get a time variation, which would be really exciting. As Neil mentioned, I think we should really get this very deep exposure of the lowest-density region because that would eliminate the galactic diffuse component.

**Gehrels**: I think the time fluctuations will be largely averaged out, but the numbers may be small enough that there could be some residuals.

**F. Stecker**: Two comments in regard to this. The first one is that, if we are right, then blazars are basically the only thing producing gamma-rays in significant quantities above a few GeV and the galactic gamma-rays would be falling off with a spectrum like $\nu^{-2.7}$ which is basically the cosmic ray primary spectrum. Whereas the blazars have a hard spectrum, roughly $\nu^{-2}$ plus or minus a few tenths. As a matter of fact, the higher up in energy you go, the harder the sources that dominate are. You can always get fluctuations because the galactic high latitude matter is very choppy, as shown quite dramatically by the IRAS observations. So, one thing to do if you can get enough photons, and Carl might want to comment on that, is to concentrate on the multi-GeV part of the background spectrum. The other thing I would like to say is related to the number of sources contributing to the background. If you look a little carefully at our integration with the flat spectrum radio source luminosity function, which we have taken from Dunlop and Peacock, there are hundreds of faint blazars that would be contributing to the gamma-ray background, but are below the threshold of detectability for EGRET. And, of course, this is always the case with backgrounds, that is how you get a background instead of just looking at sources; so the angular fluctuations may not stand out quite as clearly as perhaps Carl indicated. But he might want to comment on that also.

**C. Fitchel**: I'll comment on both of your statements. First of all, with regard to looking at higher energies, that is certainly correct. The higher energies are going to tell you a lot because you also know the directions better and, if you have enough photons, that is certainly the place to look. With regard to the fluctuations, there are a lot of blazars that are just as strong as the ones we have seen that we do not observe, so I do not think you can assume that all of these weak ones are going to be contributing equally. I think there is still the possibility that at any one time it could not be that large. I still would say that your fluctuations are a thing you might really find.

**Gehrels**: You got to an interesting point here of asking what is really the cosmic gamma-ray background. If you can see individual blazars or AGNs of any type, and take them out, then they are sources. If you cannot, then maybe that is what you really call cosmic diffuse background. And you can see that this will be a very important issue from the gamma-ray data at high energies.

**C. Norman**: I wanted to ask about the constraint from pair production on the infrared background. I wonder if it is actually a constraint on the infrared background or a constraint on the fairly strong infrared intensity that you find in these AGNs, that is, does the effect occur in the central regions rather than in the propagation from just outside those regions to the observer?

**Gehrels**: This extragalactic attenuation that Floyd Stecker and others have been working on is on top of any internal absorption that there might be. I think you are right, the internal absorption could explain why WHIPPLE does not see some of these more distant blazars.

**C. Norman**: I guess the more extreme version of my question is that let me propose that it is all from the active nuclei.

**M. Urry**: From what I have seen of the 24 blazars with EGRET spectra, the gamma-ray luminosity is between 1 and 1000 times the luminosity in any other waveband including the infrared. And these are not particularly bright in the infrared, in any case, so I do not think that is a likely explanation and then of course how did the photons get out?

**Gehrels**: But it is a question of number of photons.

**C. Norman**: It is a question of photon number density, the photon column density.

**M. Urry**: And also you know they get out of Markarian 421.

**Gehrels**: I think that was an important point.

**F. Stecker**: What I wanted to say on this was, as we pointed out in our paper, you can always have intrinsic absorption in the source, but what makes this workable is the fact that Markarian 421 is seen at TeV energies. And that shows that there is at least some evidence that these blazars have basically $\nu^{-2}$ power law spectra at TeV energies and, of course, that upper graph there assumes that your source spectrum is a power law, that is the dotted line and that is what makes the whole thing work. If these source spectra are indeed power laws out to TeV energies, it means that the gamma-rays basically are not having trouble getting out of these sources and so you really can probe the intergalactic infrared radiation field that way.

**C. Canizares**: I wanted to ask a question myself about these spectra and wonder whether INTEGRAL, which I gather is now been selected by ESA, is going to make a contribution to those measurements.

**Gehrels**: Yes, INTEGRAL should make a nice contribution. It does not have as wide a field of view as you would ideally want for diffuse background. It has a 10° field of view. But it is a big enough field given the factor of more than 10 better sensitivity, in particular, if we are looking for spectral lines in the spectrum. We do have a blocking crystal on INTEGRAL and it is an important part of the science objectives to try to find the absolute cosmic diffuse spectrum. And we are in a high orbit. The spacecraft is fairly light and it will be launched on a Proton most likely; it will go away from the magnetosphere and from the earth. So it could make a really nice measurement.

**C. Canizares**: And for the model that you showed, what are the redshifts that you are integrating out to? Because this presumably touches Jim's point about where the heavy elements are being formed.

**Gehrels**: The mean redshift for these different calculations varies from about 1 or a little less than 1 to about 2, depending on how much luminosity evolution was assumed. I did not mention this, but most of the more recent calculations have luminosity evolution of $(1+z)^{2.5}$.

**M. Livio**: I just wanted to add a comment, since you mentioned type Ia supernovae as one of the possible sources. Some variety of novae, in particular of the ONeMg variety, can produce $^{26}$Al and $^{22}$Na at rather large quantities and that can also be a contributor.

**Gehrels**: That is a good point. Of course, we have never seen that line, but it is predicted.

# THE X-RAY BACKGROUND (OBSERVATIONS)

Giovanni Zamorani
Osservatorio Astronomico
Via Zamboni 33
40100 Bologna
Italy

## 1. INTRODUCTION

It is a heavy responsibility for me, as it would be for everybody else, to give a paper on the X-ray background (XRB) in this particular meeting, held in honour of Riccardo Giacconi. Everybody knows the enormous contribution that Riccardo has given to the development and better understanding of the subject from the very beginning 30 years ago up to now. A significant fraction of the results which I will describe in this paper either are his own results or are based on experiments which he conceived and led to success.

In Section 2 I will give a brief historical overview of the XRB problem, from its discovery up to the results obtained in the eighties with the HEAO-1 and EINSTEIN missions. During these years the origin of the XRB has been discussed mainly in terms of two alternative interpretations: the truly diffuse hypothesis (e.g. hot intergalactic gas) and the discrete source hypothesis. The existence of these radically alternative hypotheses has not been "neutral" with respect to devising experiments which wanted to study the XRB. In fact, if the XRB is mainly due to discrete sources, experiments aimed at studying the single sources responsible for it obviously need high angular resolution in order to study and resolve the large number of expected faint sources. Vice versa, if the XRB is mainly diffuse, source confusion is not a problem anymore and one could safely abandon the high angular resolution option. In this case the crucial experiment would be a measurement as accurate as possible of the spectrum in order to reveal the physical production processes. The two working hypotheses led various groups of scientists to design very different sets of experiments (Giacconi and Burg 1992). In Section 3 I will show some recent results from deep surveys with ROSAT. These surveys have already resolved into discrete sources $\sim 60\%$ of the measured XRB in the 1–2 keV band. The available optical identifications, still in progress, suggest that AGNs are the dominant population at these faint X-ray fluxes. Finally, in Section 4 I will discuss some recent results on the X-ray spectra of AGNs at higher energy and a few models which, making use of these data, are able to produce acceptable fits to the spectrum of the XRB up to about 100 keV.

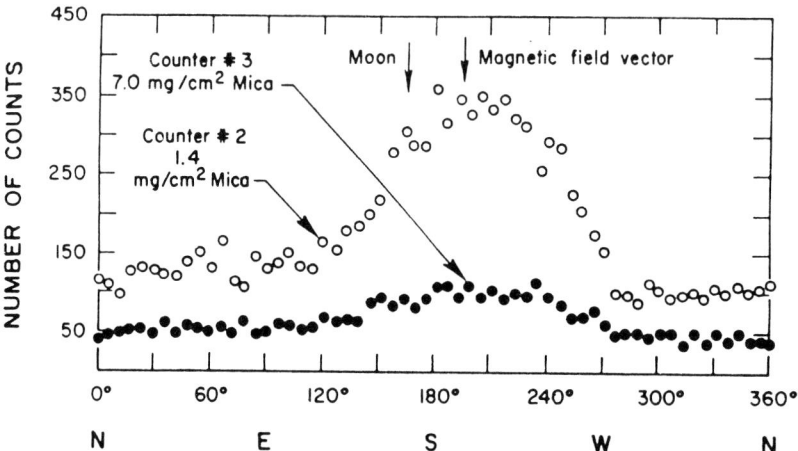

**Figure 1.** *Data from the rocket flight in which the XRB was discovered.*

## 2. EARLY HISTORY

The existence of a diffuse XRB was discovered more than thirty years ago (Giacconi et al. 1962). Figure 1 shows data from the discovery flight. It is interesting to note that in these data both the diffuse emission and a strong source (i.e. the two elements which became the basis for the two main hypotheses for the production of the XRB) are already present. The first important step with respect to our knowledge of the XRB has been made with the first all–sky surveys (UHURU and ARIEL V) at the beginning of the seventies. The high degree of isotropy revealed by these surveys led immediately to realize that the origin of the XRB has to be mainly extragalactic. Moreover, under the discrete source hypothesis, the number of sources contributing to the XRB has to be very large ($N > 10^6 \; sr^{-1}$; Schwartz 1980).

In the same years a number of experiments were set up to measure the spectrum of the XRB over a large range of energy. It was found that over the energy range 3–1000 keV the XRB spectrum is reasonably well fitted with two power laws with slopes $\alpha_1 \sim 0.4$ for $E \leq 25$ keV and $\alpha_2 \sim 1.4$ for $E > 25$ keV (see Figure 1 in Tanaka 1992).

At the beginning of the eighties two different sets of measurements led additional fire to the debate between supporters of the discrete source and diffuse hypotheses. On the one hand, the excellent HEAO–1 data showed that in the energy range 3–50 keV the shape of the XRB is very well fitted by an isothermal bremsstrahlung model corresponding to an optically thin, hot plasma with kT of the order of 40 keV (Marshall et al. 1980). Moreover, it was shown by Mushotzky (1984) that essentially all the Seyfert 1 galaxies with reliable 2–20 keV spectra ($\sim 30$ objects, mostly from HEAO–1 data) were well fitted by a single power law with an average spectral index of the order of 0.65, significantly different from the slope of the XRB in the same energy range. These two observational facts were taken as clear "evidences" in favour of the diffuse thermal hypothesis. On the other hand, the results of the EINSTEIN deep surveys showed that about 20% of the soft XRB (1–3 keV) are resolved into discrete sources at fluxes of the order of a few $\times 10^{-14} erg \, cm^{-2} s^{-1}$ (Giacconi et al. 1979, Griffiths et al. 1983, Primini et al. 1991, Hamilton et al. 1991). A large fraction of these faint X-ray sources have been identified with Active Galactic Nuclei (AGNs). Because of the difference between the spectra of the XRB and those of the few bright AGNs with good spectral data, the supporters of the diffuse, hot plasma hypothesis had to play down as much as possible the contribution of AGNs to the XRB to a limit which was close to be in conflict with an even mild extrapolation of the observed log N – log S. Actually, a number of papers were published in which it was "demonstrated" that even in the soft X-ray band AGNs could not contribute much more than what had already been detected at the EINSTEIN limit.

At that time I personally think that there were already evidences (for those who wanted to see them...) that the diffuse thermal emission as main contributor to the background was not tenable (see, for example, Setti 1985). Very simple arguments in this direction were given by Giacconi and Zamorani (1987). On the basis of reasonable extrapolations of the X-ray properties and the optical counts of known extragalactic X-ray sources (mainly AGNs and galaxies), they concluded that it is unlikely that their contribution to the soft X-ray background is smaller than 50%. Given this constraint, they then discussed two possibilities:

i) either faint AGNs have the so-called (at that time) "canonical" spectrum observed for brighter AGNs. In this case the residual XRB (i.e. the spectrum resulting after subtraction of the contribution from known sources) would not be fitted anymore by optically thin bremsstrahlung;

ii) or spectral evolution for AGNs is allowed. In this case, in order not to destroy the excellent thermal fit in the 3–50 keV data, diffuse emission could still be accommodated only if discrete sources have essentially the same spectrum as the XRB. On this basis, they concluded that "since in this scenario we would already require that the average spectrum of faint sources yielding 50% of the soft XRB is essentially the same as the observed XRB, there is nothing that prevents us from concluding that the entire background may well be due to the same class of discrete sources, at even fainter fluxes."

In other words, reversing the usual line of thought, the excellent thermal fit of the 3–50 keV XRB spectrum was shown by these arguments to be a point in favour of the discrete source hypothesis, rather than of the hot gas hypothesis! These conclusions, however, were not well received in a large fraction of the X-ray community; probably, they had the defect of being too simple and direct...

Thus, the debate between the supporters of the two hypotheses continued, until the final resolution of the controversy came from the incredibly neat results obtained with the FIRAS instrument on board COBE: the absence of any detectable deviation from a pure black body of the cosmic microwave background set an upper limit to the comptonization parameter $y < 10^{-3}$ (Mather et al. 1990), more than ten times smaller than the value required by the hot intergalactic gas model. The most recent upper limit for the comptonization parameter is now $y < 2.5 \times 10^{-5}$ (Mather et al. 1993). Discussing these data, Wright et al. (1993) conclude that a uniform, hot intergalactic gas produces at most $10^{-4}$ of the observed XRB!

## 3. ROSAT DEEP SURVEYS DATA

### 3.1 The log N – log S relation

Having the COBE data definitely eliminated the possibility of an important contribution of diffuse gas emission to the XRB, the important question to be addressed is

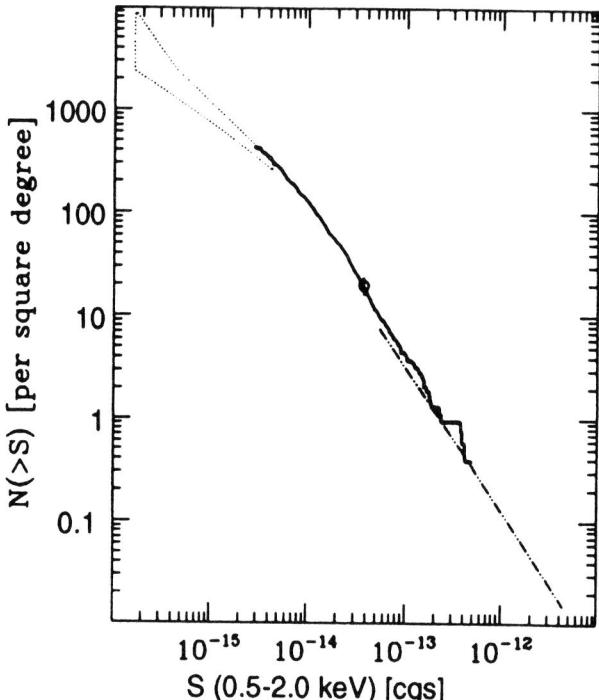

**Figure 2.** *Integral source counts for ROSAT data. The dash–dotted line represents the best fit to the EINSTEIN Medium Sensitivity Survey total sample (i.e. galactic and extragalactic). The open circle represents the EINSTEIN Extended Deep Survey point. The dotted area at faint fluxes shows the 90% confidence regions from the fluctuation analysis of the deepest ROSAT field in the Lockman Hole (Hasinger et al. 1993).*

now: what are the sources that are responsible for the observed XRB? In this Section I will discuss some recent results, relevant to this question, obtained with ROSAT.

The good angular resolution and sensitivity of the Position Sensitive Proportional Counter aboard ROSAT have allowed to extend to significantly lower fluxes the deep imaging studies first performed with EINSTEIN. The deepest ROSAT image has been obtained by Hasinger *et al.* (1993) in the direction of the *Lockman Hole*, characterized by an extremely low neutral hydrogen column density. A total of 152 ksec of PSPC observations have been accumulated in this pointing. Seventy–five sources have been detected in the hard (0.4–2.4 keV) ROSAT band in the inner 15.5 arcminutes, corresponding to a surface density of about 360 sources/sq.deg. These data have been used by Hasinger *et al.*, together with additional data from 26 other shallower ROSAT exposures, to obtain the log N – Log S relation shown in Figure 2. The total number of sources used in the construction and analysis of the log N – log S relation is 661 and they cover a range of more than two decades in flux.

The observed flux distribution of these sources has been fitted with a model in which the differential counts (N(S)) are represented by two power laws:

$$N(S) = N_1 \times S^{-\beta_1} \text{ for } S > S_b$$
$$N(S) = N_2 \times S^{-\beta_2} \text{ for } S < S_b.$$

After detailed Monte Carlo simulations aimed at understanding and correcting all possible systematic effects present in the source detection procedure, the best fit parameters obtained for the above parameterization are: $\beta_1 = 2.72 \pm 0.27, \beta_2 = 1.94 \pm 0.19, S_b = (2.66 \pm 0.66) \times 10^{-14} erg\, cm^{-2} s^{-1}$.

In order to obtain constraints on the shape of the log N – log S relation below the discrete source detection threshold, a fluctuation analysis of the intensity distribution in the inner region of the Lockman field has been performed. On the basis of extensive simulations, which took into account all known systematic instrumental effects, it has been obtained the 90% confidence region shown by the dotted area at faint fluxes in Figure 2.

In summary, the main results of this analysis of the ROSAT log N – log S relation are:

a) There is a reasonably good agreement between the ROSAT log N – log S and the EINSTEIN EMSS (Extended Medium Sensitivity Survey) source counts in the flux range where both surveys have good statistics.

b) There is a highly significant flattening of the log N – log S relation at a flux of $\sim 2.5 \times 10^{-14} erg\, cm^{-2} s^{-1}$. The need for such a flattening had already been inferred by fluctuation analyses of the EINSTEIN deep survey fields (Hamilton and Helfand 1987; Barcons and Fabian 1990).

c) The integral surface density of X-ray sources above a flux of $2.5 \times 10^{-15} erg\, cm^{-2} s^{-1}$, resulting from the integration of the log N – log S shown in Figure 2, is $\sim 410\, deg^{-2}$ and the corresponding integrated flux amounts to $\sim 60\%$ of the measured XRB in the 1–2 keV band.

d) The flattest power law extrapolation allowed by the fluctuation analysis resolves 85% of the background, while the steepest allowed slope resolves all of the background already at a flux of $\sim 10^{-16} erg\, cm^{-2} s^{-1}$, i.e. only a factor $\sim 20$ below the flux limit of the resolved sample.

## 3.2 The optical identifications

The X-ray log N – log S shown in Figure 2 includes all the X-ray sources, without any selection on the basis of the optical counterparts. The obvious questions now are: which fraction of these sources are extragalactic? what are the optical identifications of these sources? Systematic work aimed at identifying the optical counterparts of faint ROSAT sources is in progress. Such a work requires a large amount of telescope time because of the faintness of some of these counterparts. Typical magnitudes for various classes of sources with a ROSAT flux $\sim 10^{-14} erg\, cm^{-2} s^{-1}$ are shown in Table 1; these magnitudes have been estimated on the basis of the typical X-ray to optical ratios of the about 800 X-ray selected sources of the EMSS (Maccacaro et al. 1988).

At a flux limit of $S_x \sim 10^{-14} erg\, cm^{-2} s^{-1}$ there are at least four ROSAT fields with a high percentage of optical identifications already available. These fields are the Lockman and the Marano fields, studied by Hasinger and collaborators, and the QSF1 and QSF3 fields studied by Boyle, Shanks and collaborators. While the spectroscopic observations for the optical identifications of the Lockman field have been obtained after acquiring the ROSAT data (Schmidt et al., in preparation), the other three fields had already been studied spectroscopically before the ROSAT data in order to obtain complete optically selected samples of AGNs with $m_B \leq 22.0$ (Marano, Zamorani and Zitelli 1988; Zitelli et al. 1992; Boyle et al. 1990). X-ray data and a discussion of the optical identifications of the QSF1 and QSF3 fields have been presented by

## TABLE 1
Expected $m_v$ Range of Sources with $S_x \sim 10^{-14} erg\ cm^{-2}\ s^{-1}$

| Objects | $m_v$ |
|---:|:---|
| B - F stars | 10.0 – 14.5 |
| M stars | 13.5 – 19.5 |
| Normal Galaxies | 16.0 – 19.0 |
| AGNs | 18.5 – 23.5 |
| BL Lacs | 21.5 – 25.0 |

## TABLE 2
Optical Identifications of X–ray Selected Sources

| Sample | AGNs | BL Lacs | Galaxies | Clusters | Stars | No Id. |
|---|---|---|---|---|---|---|
| ROSAT Deep Surveys | 61% | 1% | 5% | — | 8% | 24% |
| EMSS | 51% | 4% | 2% | 12% | 26% | 4% |

Shanks et al. (1991) and Boyle et al. (1993). The total number of X–ray sources with $S_x \geq 10^{-14} erg\ cm^{-2} s^{-1}$ in the inner regions of these four ROSAT fields is 119; 90 of these sources ($\sim 76\%$) have already been classified spectroscopically. The results of this identification process are shown in Table 2, along with a comparison with the almost complete identifications of the EMSS survey (Stocke et al. 1991).

Most of the objects still without optical identifications are optically faint and therefore are likely to be extragalactic (see Table 1). In addition to AGNs, BL Lacs and galaxies, some of these sources will turn out to be clusters. Although a few possible cluster candidates have already been identified, no percentage for clusters in the Rosat deep surveys has been given in Table 2, because more spectroscopic data on faint galaxies are needed in order to establish the reliability of the proposed identifications. Since almost all the stars in the sample have probably already been identified, we can conclude that the final percentage of stars in the ROSAT deep surveys should be $\leq 10\%$, significantly smaller than the percentage of stars found in the brighter EMSS survey. Vice versa, already at this preliminary stage the fraction of AGNs in the ROSAT deep surveys (61%) is higher than in the EMSS survey, and could be as high as 86% in the extreme hypothesis that all the sources still to be identified are AGNs. This shows without any doubt that AGNs are the dominant population among the X–ray sources at this flux.

What are expected to be the X–ray sources at fluxes even fainter than the current ROSAT flux limit? The most "economic" hypothesis is that they are still AGNs, fainter than those detected so far. If so, which region of the redshift–luminosity plane are they expected to fill? Figure 3, which shows redshift versus X-ray luminosity for the EMSS (small dots) and ROSAT deep surveys (large dots) AGNs, can help us in defining the region of interest. The figure clearly shows the increase in the median redshift from $\sim 0.3$ for the EMSS AGNs to $\sim 1.4$ for the ROSAT AGNs. Shanks et al. (1991) have shown that the redshift distribution of the ROSAT deep survey AGNs is similar to that of faint optically selected AGNs. Because of the presence of the redshift cutoff at z $\sim$

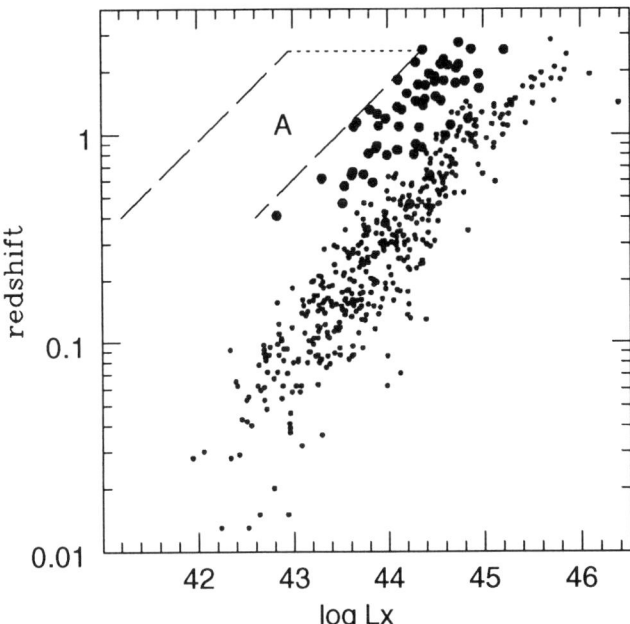

**Figure 3.** *Redshift versus X-ray luminosity for the EMSS (small dots) and ROSAT deep surveys (large dots) AGNs. The dotted line corresponds to $z = 3$, while the two dashed lines correspond to ROSAT fluxes $2 \times 10^{-16}$ and $5 \times 10^{-15}$ erg cm$^{-2}$ s$^{-1}$, respectively.*

2.5, however, we expect that AGNs at even fainter X-ray fluxes will mainly populate the region corresponding to the faint part of the X-ray luminosity function in the redshift interval 0.4–2.5 (see area "A" in Figure 3), rather than the higher redshift region. The best fit of the differential slope of the log N − log S in the fluctuation analysis is ∼ 1.8 (Hasinger et al. 1993); on the other hand, the slope of the faint part of the X-ray luminosity function, as derived by Boyle et al. (1993), is $1.7 \pm 0.2$. The agreement between these two slopes suggests that it is quite possible, or at least consistent with the presently available data, that AGNs would provide the bulk of the X-ray sources at least down to fluxes $2 \times 10^{-16}$ erg cm$^{-2}$ s$^{-1}$. It is also clear from the figure that in this case most of the AGN contribution to the XRB would come from objects with X-ray luminosities smaller than $10^{44}$ erg s$^{-1}$, similar to the X-ray luminosity of Seyfert galaxies.

As seen in Table 2, the percentage of galaxies in the ROSAT deep surveys (∼ 5%) is higher than the corresponding percentage in the EMSS (∼ 2%). At the limit of $S_x \sim 10^{-14}$ erg cm$^{-2}$ s$^{-1}$ the number of galaxies is only 10% of the number of AGNs. However, while the differential slope of the X-ray log N − log S at these fluxes is slightly flatter than two, the corresponding slope for the optical counts of galaxies in the B band is ∼ 2.1 (Tyson 1988) in the range of magnitudes 18–27. The steeper slope in the optical counts of faint galaxies implies that, even without evolution in the ratio of X-ray to optical blue fluxes, the ratio between galaxies and AGNs might increase toward fainter X-ray fluxes. Actually, Griffiths and Padovani (1990) have suggested that star-forming galaxies, with some evolution, may be a major component of the XRB. The expected X–ray log N – log S for these galaxies is strongly model dependent. From Figure 5 in Griffiths and Padovani (1990) it is seen that, under some assumptions, these objects might become a very substantial fraction of the faint X–ray sources at fluxes $S_x \leq 10^{-15} erg\, cm^{-2} s^{-1}$. Griffiths *et al.* (1993) are presenting some evidence from their preliminary optical identifications that the galaxy population is already becoming important and of the same order as the AGN population in the flux range $S_x = (5 - -10) \times 10^{-15} erg\, cm^{-2} s^{-1}$. Obviously, this result has to be confirmed with more extensive identifications of X–ray sources at faint ROSAT fluxes.

## 4. AGN SPECTRA AND FITS TO THE XRB SPECTRUM

In the last few years detailed spectral data of AGNs have been obtained by GINGA in the energy range 2–30 keV. These high quality data have changed substantially our views on the spectral characteristics of AGNs. As shown convincingly by Pounds *et al.* (1990) and Nandra (1991), the typical spectrum of Seyfert 1 galaxies shows a flattening at $\sim 10$ keV, with respect to the observed power law slope in the range 2–10 keV. Such a flattening has been interpreted either as a partial coverage of an underlying X–ray power law continuum or as reprocessed emission (reflection) from thick relatively cold matter, possibly in an accretion disk. These observations showed that the average spectrum for these objects is very similar to the shape of the spectrum hypothesized by Schwartz and Tucker (1988). In their illuminating paper they had shown that such a spectrum, integrated through redshift with reasonable assumptions on the cosmological evolution, could provide an adequate fit to the shape of the observed XRB above 3 keV.

The Ginga data have immediately led a number of groups to construct models for fitting the XRB spectrum with various combinations of AGN spectra (see, for example, Morisawa *et al.* (1990), Fabian *et al.* (1990), Terasawa (1991), Rogers and Field (1991)). Although qualitatively in agreement with the overall shape of the XRB in the energy range 3–100 keV, these first models have been shown not to be able to fit satisfactorily the position and the width of the peak of the XRB spectrum (Zdziarski *et al.* 1993a). In the same paper Zdziarski *et al.* discuss two models which produce improved fits to the XRB. In the first model the major contribution to the XRB is due to an as yet unobserved AGN population at high redshift, while in the second model most of the XRB emission comes from foreground AGNs. Neither model is, however, fully compatible with the observed XRB spectrum and/or with the available AGN spectral data; in particular, the average spectra of the required foreground AGNs are different from the observed ones.

Figure 4 shows the results of a fit to the XRB spectrum obtained by Comastri *et al.* (1993). This model takes into account the observed spectral properties of different classes of AGNs over a broad energy range and is based on the X–ray properties of AGN unified schemes (Setti and Woltjer 1989). The main ingredients of the model are the following:

a) The X–ray spectrum of Seyfert 1 galaxies is described by the reflection model, with about half of the flux of the primary spectrum reprocessed (Pounds *et al.* 1990).

b) As required by the adopted unified scheme, the Seyfert 2 galaxies are assumed to have the same intrinsic spectrum as the Seyfert 1 galaxies, but modified by absorption effects (Awaki *et al.* 1991). A break to a steeper power law ($\alpha_E \sim 2.0$) has been introduced in the spectrum of Seyfert galaxies, as indicated by recent OSSE observations (Cameron *et al.* 1993).

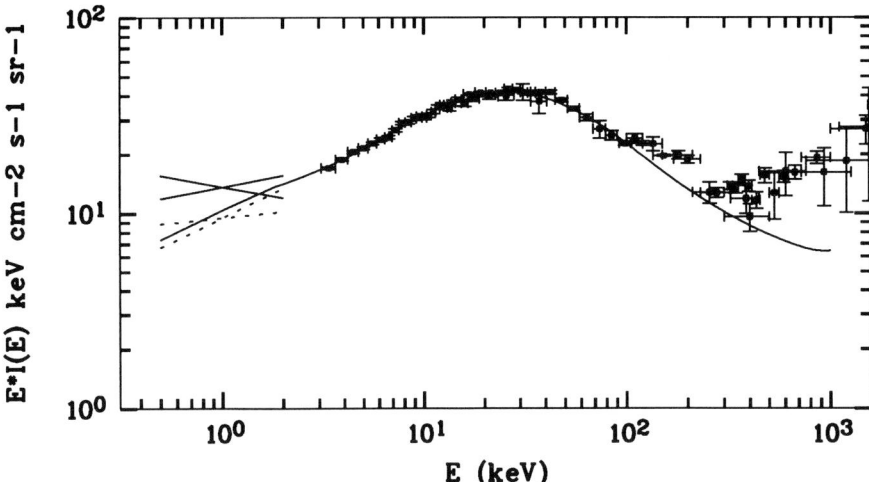

**Figure 4.** *The XRB spectrum: comparison between model (continuous line) and data. The soft (0.5–2.0 keV) XRB spectrum is from ROSAT (solid lines from Hasinger et al. 1993; dashed lines from Wang and McCray 1993), while the data above 3 keV are taken from a compilation of the best experimental results by Gruber (1992).*

c) For the high luminosity AGNs (i.e. quasars with $L_x > 5 \times 10^{44} erg\, s^{-1}$) a single power law spectrum ($\alpha_E = 0.9$) has been assumed (Williams et al. 1992).

Given these assumptions, all of them consistent with the available observational data, the fit shown in Figure 4 has been obtained assuming an evolving volume emissivity $(nL)_z = (nL)_0 \times (1+z)^\beta$, with $\beta = 2.75$ (Boyle et al. 1993) for $z \leq z_{max} = 3.0$. The number ratio between absorbed and unabsorbed Seyfert galaxies which is more consistent with the data is $\sim 2.5$, in good agreement with results from optical surveys (Huchra and Burg 1992). As shown in the Figure, the fit is really good over the energy range 3–100 keV; above 100 keV the computed model starts to departure significantly from the XRB data. It may be interesting to note, however, that while the data points in the energy range 20–100 keV derive essentially from the low energy experiment on HEAO–1 A4, most of the data between 100 and a few hundred keV are from the medium and high energy experiments on HEAO–1 A4: a difference in relative calibration of about (20–25)% between the low and high energy data would be enough to allow an acceptable fit at least up to $\sim 300$ keV. At even higher energies additional ingredients to the model are required in order to reproduce the observed data.

Given the good fit to the XRB spectrum shown in Figure 4, can we conclude that the problem of the production of the XRB is definitely solved? Unfortunately, the answer is still "no." In fact, equally good fits to the XRB spectrum in the energy range 3–100 keV have recently been obtained with significantly different assumptions on the dominating AGN population by Zdziarski et al. (1993b) and Madau et al. (1993). While one of Zdziarski et al. models does not include any contribution from self–absorbed AGNs and identifies the primary sources of the XRB with AGNs detectable by soft X–ray imaging, Madau et al. model is instead dominated by type 2 objects at all energies $> 3$ keV.

The somewhat paradoxical conclusion from these results is that using the most recent AGN spectral data it has become too easy to obtain good fits to the XRB spectrum: very different models give equally good fits! As a consequence, a model that produces an acceptable fit to the XRB spectrum may not be the correct model. Before accepting it, one has to compare its predictions with other observational constraints, such as the soft (ROSAT) and hard (GINGA) log N – log S, the redshift distributions and the average spectra of soft and hard X–ray selected AGNs as a function of flux (see, for example, Franceschini *et al.* 1993). Finally, also the optical classification of X–ray selected AGNs (i.e. type 1 versus type 2) as a function of the X–ray band and flux would provide additional constraints and would help in reducing the wide parameter space still acceptable.

## 5. CONCLUSIONS

The past history of the XRB has been characterized by the "hot" controversy between supporters of two alternative models: diffuse emission versus discrete sources. The recent COBE data, which have conclusively set strong upper limits to a substantial contribution of diffuse gas emission to the XRB, have finally solved this long–standing controversy.

The present controversy has now shifted focus. ROSAT deep surveys on the one hand, and high energy spectra on the other hand suggest that, indeed, AGNs can produce a very substantial fraction of the observed X–ray background over a large range of energy. However, substantially different models for the AGN population seem to provide very similar and almost equally acceptable fits to the XRB spectrum. The questions which have to be answered in the next years are therefore more related to the X–ray properties of these AGN populations rather than to the XRB itself. What are the X–ray luminosity functions and evolution of the different classes of AGNs? Are different classes of AGNs dominating different energy ranges? Are unified schemes required by or at least consistent with the X–ray data? Optical identifications of very faint ROSAT sources, coupled with the study of sources selected at higher energy, should provide some of the answers to these questions.

## REFERENCES

Awaki, H., Koyama, K., Inoue, H. and Halpern, J. P. 1991 *Publ. Astron. Soc. of Japan*, **43**, 195.
Barcons, X. and Fabian, A. C. 1990, *M.N.R.A.S.*, **243**, 366.
Boyle, B. J., Fong, R., Shanks, T. and Peterson, B. A. 1990, *M.N.R.A.S.*, **243**, 1.
Boyle, B. J., Griffiths, R. E., Shanks, T., Stewart, G. C. and Georgantopoulos, I. 1993, *M.N.R.A.S.*, **260**, 49.
Cameron, R. A. *et al.* 1993, in the *Proceedings of The Compton Symposium*, N. Gehrels ed., in press.
Comastri, A., Hasinger, G., Setti, G. and Zamorani, G. 1993, in the *Proceedings of The Physics of Active Galaxies*, G. Bicknell, M. Dopita and P. Quinn eds., July 1993, Canberra, Australia, in press.
Fabian, A. C., George, I. M., Miyoshi, S. and Rees, M. J. 1990, *M.N.R.A.S.*, **242**, 14P.
Franceschini, A., Martin–Mirones, J. M., Danese, L. and De Zotti, G. 1993, *M.N.R.A.S.*, in press.
Giacconi, R., Gursky, H., Paolini, F. R. and Rossi, B. B. 1962, *Phys. Rev. Letters*, **9**, 439.
Giacconi, R. *et al.* 1979, *Ap. J. Letters*, **234**, L1.

Giacconi, R. and Zamorani, G. 1987, *Ap. J.*, **313**, 20.
Giacconi, R. and Burg, R. 1992, in *The X-Ray Background*, X. Barcons and A. C. Fabian eds., (Cambridge: Cambridge Univ. Press), 3.
Griffiths, R. E. *et al.* 1983, *Ap. J.*, **269**, 375.
Griffiths, R. E. and Padovani, P. 1990, *Ap. J.*, **360**, 483.
Griffiths, R. E. *et al.* 1993, *ST ScI preprint*, in press.
Gruber, D. E. 1992, in *The X-Ray Background*, X. Barcons and A. C. Fabian eds., (Cambridge: Cambridge Univ. Press), 44.
Hamilton, T. T. and Helfand, D. J. 1987, *Ap. J.*, **318**, 93.
Hamilton, T. T., Helfand, D. J. and Wu, X. 1991, *Ap. J*, **379**, 576.
Hasinger, G., Burg, R., Giacconi, R., Hartner, G., Schmidt, M., Trumper, J. and Zamorani, G. 1993, *Astr. Ap.*, **275**, 1.
Huchra, J. and Burg, R. 1992, *Ap. J.*, **393**, 90.
Maccacaro, T., Gioia, I. M., Wolter, A., Zamorani, G. and Stocke, J. T., *Ap. J.*, **326**, 680.
Madau, P., Ghisellini, G. and Fabian A. C., 1993, *Ap. J. Letters*, **410**, L7.
Marano, B., Zamorani, G. and Zitelli, V. 1988, *M.N.R.A.S.*, **232**, 111.
Marshall, F. E. *et al.* 1980, *Ap. J.*, **235**, 4.
Mather, J. C. *et al.* 1990, *Ap. J. Letters*, **354**, L37.
Mather, J. C. *et al.* 1993, *Ap. J.*, in press.
Morisawa, K., Matsuoka, M., Takahara, F. and Piro, L. 1990, *Astr. Ap.*, **263**, 299.
Mushotzky, R. F. 1984, *Advances in Space Research*, Vol. **3**, no. 10–12, p. 157.
Nandra, K. 1991, *Ph.D. Thesis*, Leicester University.
Pounds, K. A., Nandra, K., Stewart, G. C., George, I. M. and Fabian, A. C. 1990, *Nature*, **344**, 132.
Primini, F. A. *et al.* 1991, *Ap. J.*, **374**, 440.
Rogers, R. D. and Field, G. B. 1991, *Ap. J. Letters*, **370**, L57.
Schwartz, D. A. 1980, *Phys. Scripta*, **21**, 644.
Schwartz, D. A. and Tucker, W. H. 1988, *Ap. J.*, **332**, 157.
Setti, G. 1985, in *Non-Thermal and Very High Temperature Phenomena in X-Ray Astronomy*, G. C. Perola and M. Salvati eds., p. 159.
Setti, G. and Woltjer, L. 1989, *Astr. Ap*, **224**, L21.
Shanks, T., Geogantopoulos, I., Stewart, G. C., Pounds, K. A., Boyle, B. J. and Griffiths, R. E. 1991, *Nature*, **353**, 315.
Stocke, J. T. *et al.* 1991, *Ap. J. Suppl. Ser.*, **76**, 813.
Tanaka, Y. 1992, in *X-Ray Emission from Active Galactic Nuclei and the Cosmic X-Ray Background*, W. Brinkmann and J. Trumper eds., MPE Report **235**, 303.
Terasawa, N. 1991, *Ap. J. Letters*, **378**, L11.
Tyson, J. A. 1988 *A. J.*, **96**, 1.
Wang, D. Q. and McCray, R. 1993, *Ap. J. Letters*, **409**, L37.
Williams, O. R. *et al.* 1992, *Ap. J.*, **389**, 157.
Wright *et al.* 1993, *Ap. J.*, in press.
Zdziarski, A. A., Zycki, P. T., Svensson, R. and Boldt, E. 1993a, *Ap. J.*, **405**, 125.
Zdziarski, A. A., Zycki, P. T. and Krolik, J. H. 1993b, *Ap. J. Letters*, in press.
Zitelli, V., Mignoli, M., Zamorani, G., Marano, B. and Boyle, B. J. 1992, *M.N.R.A.S.*, **256**, 349.

## DISCUSSION

**R. Giacconi**: I agree with most of what has been said, and I think that what is happening now is that we will use these deep surveys to study formation and evolution of structure, AGNs, and clusters; and this is the new field. These deep surveys are only mildly interesting to get the last few percent of the soft X-ray background, while they are truly interesting from the point of view of looking at the formation of structure and evolution.

**Q.**: So what is your position on the second bump in $\nu f(\nu)$?

**R. Giacconi**: I would like to make a couple of statements, one is that at night all cows look gray and if you keep studying cows at night you keep seeing them gray and they will give you no information about what type of cows you are looking at. The other one is that the predictive power of these theoretical, if you pardon the expression, models is very modest in that they are all done a posteriori and, therefore, I do not know if this advances the field immensely. So I think there is going to be another 20 years of study of the second bump which will be resolved essentially like the first one.

**M. Longair**: Riccardo, could you tell us something about the detection of clusters in the ROSAT survey, because many of us would love to know whether they exist at redshifts bigger than about 0.2 or 0.3.

**R. Giacconi**: This reminds me a story about Zichichi. At the University of Aquila he was with 12 students who kept saying "Zichichi, Zichichi, tell us, what is the truth?", and he went on talking for a while. Later the Jesuits felt that, perhaps, a report of this meeting done verbatim could have been a problem and a scandal for the Catholic faith, in that it was not quite clear who was Jesus Christ and who was Zichichi. So, I certainly do not want to repeat that problem.

**R. Griffiths**: Can I just show a viewgraph? This is a summary of the optical identifications in the ROSAT survey field that we have been doing with Brian Boyle and Tom Shanks and others. That is, 60 kiloseconds and shallower ROSAT surveys with optical identifications from multi-object spectroscopy at the AAT. I think the interesting point is that the fraction of galaxies, although it is at about the 15% level relative to AGNs at the brighter end, and down to the level of the Einstein deep survey at $10^{-14}\,erg\,cm^{-2}\,s^{-1}$ (0.5-3.5 keV), seems to be increasing from $10^{-14}$ down towards the $10^{-15}\,erg\,cm^{-2}\,s^{-1}$ level of the deepest surveys. The point is that galaxies in fact may be as important as the AGNs at a flux of about $10^{-15}\,erg\,cm^{-2}\,s^{-1}$, so I do not think it is a completely closed case that the AGNs account for all of those objects ...

**Zamorani**: This is not what I meant. I meant that, on the basis of the observed $\log N$–$\log S$ itself, discrete sources can account for ...

**R. Griffiths**: They are all discrete, that is true. When I say galaxies, some of those are elliptical galaxies where ROSAT may be detecting the X-ray corona. But these points also include starburst galaxies. They may also include some contamination from Seyfert 2 galaxies.

**Zamorani**: Is your optical identification complete?

**R. Griffiths**: There is some incompleteness at the faint end actually.

**Zamorani**: And the QSOs have on average fainter magnitudes than galaxies ...

**R. Griffiths**: Well, the QSOs are easier to find because of their strong equivalent widths; the galaxies are harder to find. I think that is what we have been finding.

**Q.**: Richard, how faint are those galaxies?

**R. Griffiths**: Typically about 21st or 22nd magnitude.

**Q.**: Do you have redshifts for them?

**R. Griffiths**: Yes, the redshifts are quite different from those of the AGNs. QSOs are at $z \sim 1.5$, while galaxies are at $z \sim 0.3$.

**V. Khersonsky**: My question is related to previous ones. A recent discussion of X-ray data showed that spiral galaxies may contribute substantially to the X-ray background, which is then definitely connected to star formation processes and galaxy merging. Do you think this process can be important in your model, if we keep in mind that at intermediate redshifts merging and interactions might be much more effective?

**Zamorani**: Yes, connecting your question to Richard Griffiths's viewgraph, I agree that it is quite possible that at some stage starburst galaxies become important in contributing to the faint X-ray counts; in fact, even without assuming any evolution in the X-ray to optical ratio, we already know that at least 10% of the background has to come from normal, so to speak, galaxies with some enhanced star formation processes. If these processes are stronger in the past, if there is some evolution, this 10% can well be a lower limit and can increase to some 20% or so easily.

**S. White**: To make a connection with what Jim Peebles was saying this morning, you have now 60% of the sources optically identified. If you wanted to estimate what the median redshift was for the generation of the background, what estimate would you come up with?

**Zamorani**: On the basis of the AGN distribution the average redshift is 1.3-1.5.

**S. White**: So half the observed flux comes from objects further away than 1.5.

**Zamorani**: Yes.

**Q.**: In light of the findings by Wilkes and Elvis a couple of years ago, that radio loud quasars have soft X-ray spectra that are flatter than the radio quiet quasars, and in the light of what we heard this morning about how blazars, which are really a rare species in the optical sky, may be important in the gamma-ray sky, do you have any general thoughts about the role of non-thermal processes in the x-ray sky?

**Zamorani**: Well, in our Lockman field we have some deep VLA observations at the level of 120 $\mu$Jy ($5\sigma$) which cover essentially the same area as the ROSAT field. The surface density of the radio sources is almost exactly the same as the surface density of the X-ray sources. This numerical agreement is a pure coincidence, of course. The cross-correlation between the two lists of sources shows that only about 15% of the radio sources coincide with the X-ray sources. We still do not have full optical identifications for these objects. Some of them are definitely radio loud quasars, as expected. Some of them appear not to be related to point–like, stellar images on our CCD images, so those might be some radio galaxies, but until we have spectroscopic data, I won't say more than that.

**G. Ghisellini**: I would like to make a comment. I think that one way to attack the problem, to explain the background is to find some class of sources that have the same shape as the background. But it may be that the more productive way to attack the problem is to ask what is the physical process which is able to give us a typical energy, which is the observed $30\,(1+z)$ keV bump. And I think the Cambridge group has some merit in this respect: for some time we kind of believed that the Compton reflection component could give us this characteristic energy, more or less independently of the input spectrum. Another possibility which is now confirmed by the new OSSE results is the thermal pair models, which give us a characteristic energy of 50–100 keV. So this way of attack can have a more predictive power than just to look for what we already know.

**Zamorani**: Ah, but also in that case, it would be nice to make predictions on other sets of data which can be tested.

**G. Ghisellini**: Yes, for example, in thermal plasma models, no annihilation emission line is predicted.

# EXTRAGALACTIC ULTRAVIOLET BACKGROUND RADIATION

Richard C. Henry
Center for Astrophysical Sciences
Henry A. Rowland Department of Physics and Astronomy
The Johns Hopkins University
Baltimore, MD 21218
and
Space Astronomy and Astrophysics Group
Los Alamos National Laboratory

Jayant Murthy
Center for Astrophysical Sciences
Henry A. Rowland Department of Physics and Astronomy
The Johns Hopkins University
Baltimore, MD 21218

**Abstract.** We describe observations of the cosmic background in the far ultraviolet. If the Voyager upper limit on the cosmic diffuse ultraviolet background at 1100 Å at some locations is accepted as correct, the spectrum of the high-galactic latitude background is most remarkable, featuring an abrupt rise at about 1216 Å. Such a rise suggests an origin in redshifted Lyman $\alpha$ recombination radiation, but that explanation requires the existence of an ionization source such as the radiative decay of massive neutrinos to maintain the ionization. We therefore explore a more conservative origin in the scattered light of galactic plane OB stars. This explanation is fraught with difficulties: a dust population specially invented for the purpose seems to be required. This *ad hoc* explanation may be preferred by some; a perhaps somewhat exotic extragalactic origin by others. Quite simple additional observations should clarify matters considerably.

## 1. INTRODUCTION

The subject of our paper is the *extragalactic* far ultraviolet background radiation. We start, therefore, by describing the observations at high galactic latitudes; and we also start by simply *assuming* that the observed signals are extragalactic in origin. Having explored various possible origins, in relation to the observations, we then consider the observations at lower galactic latitudes; at this point we examine carefully whether the high-latitude observations themselves might be galactic in origin, rather than extragalactic. Finally, we summarize our conclusions.

## 2. THE ULTRAVIOLET BACKGROUND AT HIGH LATITUDE

The observations of the diffuse ultraviolet background at high galactic latitudes are collected in Figure 1. The figure caption contains references to the sources of the various observations. We will, for now, assume that this radiation is extragalactic.

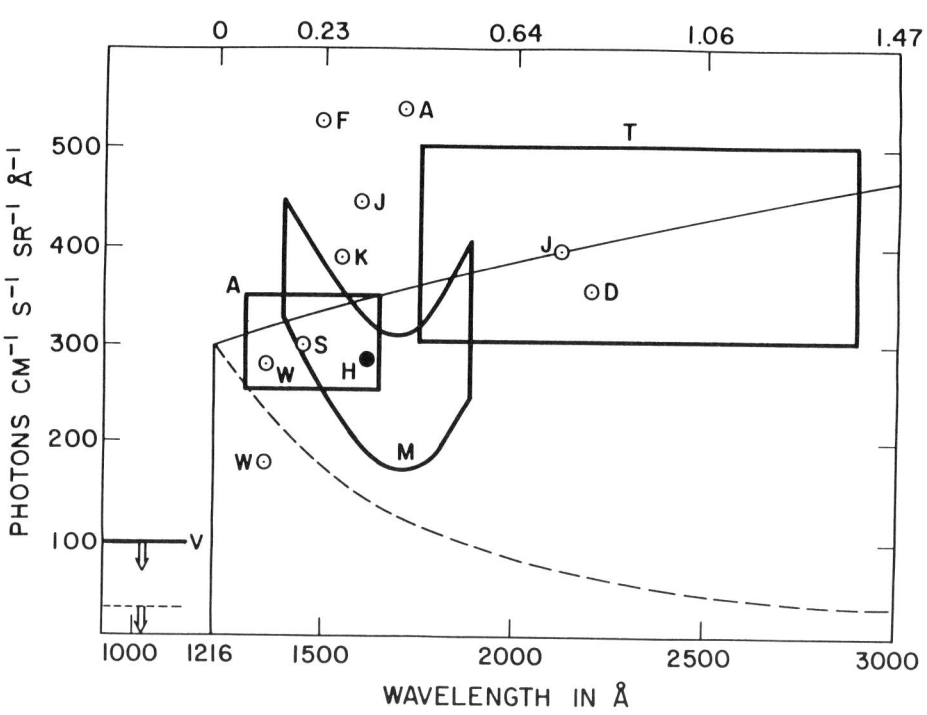

**Figure 1.** The plotted observations of (assumed) extragalactic diffuse cosmic ultraviolet background radiation (letters in Figure) have been reviewed by Henry (1991). The intensity in units is plotted against the wavelength of observation in Å. Superposed on the observations are predicted spectra of the recombination radiation from ionized intergalactic clouds (solid line, clouds that are expanding as the universe expands; dashed line, gravitationally bound clouds). Neutrino decay radiation would also produce a spectrum shaped like the dashed line. The observation V, shortward of Lyman $\alpha$ (1216 Å), is the Voyager upper limit of Holberg (1986; solid line: calibration of Holberg et al. 1982, dashed line: calibrations of Brune et al. 1979 and Cook et al. 1989). The observations longward of L$\alpha$ are all positive detections, not upper limits. They include spectroscopy (boxes, and filled point): A (Anderson et al. 1979), M (Martin and Bowyer 1990), T (Tennyson et al. 1988), H (Hurwitz et al. 1990), and photometric observations (open circles): W (Weller 1983), S (Paresce et al. 1979), J (Jakobsen et al. 1984), K (Onaka 1990), A (Anderson et al. 1979), D (Joubert et al. 1983), F (Fix et al. 1989). We have omitted the Apollo 17 observation of Henry et al. (1978), which agrees with the others at 1500 Å, but which shows a decline at longer wavelengths that is surely spurious.

First, we note that the extragalactic background shortward of L$\alpha$ (1216 Å) is less than 100 photons cm$^{-2}$ s$^{-1}$ sr$^{-1}$ Å$^{-1}$ (units that will be called "units" hereafter) if the calibration of Holberg et al. (1982) is used (solid line in figure), or still less (dashed line) if the calibration of Brune et al. (1979) or Cook et al. (1989) is used. In contrast, all of the observations in Figure 1 longward of 1216 Å are positive detections of $\geq$ 200 units, not upper limits.

We stress that the scatter that appears in the observations of Figure 1 demonstrably reflects a *spatial* variation in overall intensity of the ultraviolet background on the sky, rather than either uncertainties in the observations or gross (or detailed) *spectral* variations. This is demonstrated most clearly by examining individual spectra, for example Figure 10 of Anderson et al. (1979). Such examination makes clear that the ultraviolet background longward of 1216 Å *(at a given location)* shows *no trace* of a decline toward shorter wavelengths. The decline is *abrupt*, and occurs near 1216 Å.

What is the interpretation of the remarkable spectrum of Figure 1? The striking feature of course is the ledge: no detectable radiation short of about 1216 Å; strong, well-observed radiation longward of that wavelength. Now, a ledge of just this character has been searched for ever since de Rújula and Glashow (1980) suggested that the dark matter might be massive neutrinos that might decay with the emission of an ultraviolet photon. A ledge of exactly this character was predicted, and has been vigorously searched for. For example Stecker (1980) identified a ledge in the fragmentary observations of the time, and suggested the possibility that neutrino decay radiation had actually been observed. While that particular ledge is no more, interest in the general subject remains keen (Kimble et al. 1981; Henry and Feldman 1981; Murthy and Henry 1987).

However, while we have now located an excellent ledge, the shape of the spectrum toward longer wavelengths does not fit an origin in neutrino decay radiation. The expected spectrum is given in Equation 1, as a function of redshift $z$ (for the same equation in terms of wavelength, see Kimble et al. 1981).

$$I_{\lambda > 1216 \text{ Å}} = \frac{n_V c}{\lambda_{e_\text{Å}} 4\pi \tau H_o (1+z)^2 (1+\Omega z)^{1/2}} \text{ units} \qquad (1)$$

where $n_V$ is the local density of neutrinos of the type that decay with the emission of a photon, $\lambda_{e_\text{Å}}$ is the wavelength of the emitted radiation, $\tau$ is the lifetime of the neutrinos against radiative decay, $c = 3 \times 10^{10}$ cm s$^{-1}$, the Hubble parameter $H_o = 50\,h_{50}$ km s$^{-1}$ Mpc$^{-1}$, $1+z = \lambda/1216$ Å is the redshift due to the expansion of the universe, and $\Omega = 1$ (from inflation, but see also Dekel et al. 1993).

The spectrum of Equation 1 is shown in Figure 1 as a dashed line. The decline to longer wavelengths is $\sim \lambda^{-5/2}$ and clearly such a decline is not supported at all by the observations. The only hope of preserving this origin would be to hypothesize that some other (galactic?) source was filling in at longer wavelengths, but if this were the case, the observations (see the figure) would provide no evidence at all for the neutrino decay radiation itself. There is therefore no support whatever for the idea that neutrino decay radiation has been detected.

Had we been able to conclude that the observed radiation actually was neutrino decay radiation, we would have had to attribute the fact that the ledge occurred at a wavelength very close to that of Lyman $\alpha$ to chance. Let us next look at another potential origin where again we would ascribe the wavelength of the ledge to coincidence: an origin in the redshifted light of galaxies. In particular, could the ledge be the Lyman limit to the light from a tremendous burst of star formation in starburst galaxies?

To have the ledge occur near 1216 Å requires that the starbursts occurred at a redshift of 0.33, and the sharpness of the ledge requires that the activity have begun and ended rather abruptly. None of this is very palatable. Even more important, Martin, Hurwitz, and Bowyer (1991) have looked very carefully and critically at the idea that the diffuse background spectrum observed longward of 1216 Å could originate in starburst activity, and do not succeed in reconciling the observations with a dominant contribution from unclustered starburst galaxies at low redshift.

Finally, we consider the idea that we are seeing redshifted extragalactic hydrogen recombination radiation and that the fact that the ledge occurs near 1216 Å is not a coincidence. The expected spectral shape for the case of gravitationally-bound clouds of ionized hydrogen is identical to that of neutrino decay radiation, and so that case may be rejected immediately on observational grounds.

The spectrum for intergalactic clouds that are expanding as the universe expands is given in Equation 2.

$$I_{\lambda > 1216 \text{ Å}} = \frac{\alpha \, x^2 \, n_o^2 \, c \, C \, (1+z)}{\lambda_{e_{\text{Å}}} 4\pi H_o (1 + \Omega z)^{1/2}} \text{ units} \qquad (2)$$

where $\alpha = 2.8 \times 10^{-13}$ cm$^3$ s$^{-1}$ is the effective recombination coefficient (recombination to n = 1 generates no L$\alpha$ radiation), $x = 0.746$ is the fraction of baryons that are hydrogen nuclei, the local density of ionized hydrogen nuclei $n_o = 2.83 \times 10^{-6}$ h$_{50}^2$ $\Omega_g$, $\lambda_{e_{\text{Å}}} = 1216$, C = the clumping factor (which is independent of z in this model). We take the gas temperature (which affects $\alpha$) to be 8000 K, for reasons that are given below; our results are very insensitive to the temperature.

The spectrum of Equation 2 appears in Figure 1 as the solid line, which fits the data acceptably well, considering that the data are obtained at many different locations at high galactic latitudes. The parameter $\Omega_g$ is the contribution to $\Omega$ that is due to ionized intergalactic gas. If we attribute the jump at $z = 0$ of 300 units that is shown by the observations (Figure 1) to intergalactic recombination radiation, we find that

$$\Omega_g^2 h_{50}^3 C = 180 \qquad (3)$$

describes the observations.

At this point, we recapitulate facts that bear directly on the possibility of detecting redshifted Lyman $\alpha$ recombination radiation from ionized intergalactic gas.

The very small spatial fluctuations observed by COBE indicate that the dark matter was an essential ingredient in the formation of structure among the baryons following recombination. The structured dark matter was already there, and following recombination the neutral hydrogen and helium fell into the potential wells, creating the structure we observe today.

Intergalactic space was left free of neutral hydrogen. Indeed, intergalactic space is astonishingly free of neutral hydrogen, the density being $< 4.5 \times 10^{-14}$ h$_{50}$ cm$^{-3}$ (Steidel and Sargent 1987). This means that in a volume of 50 cubic megaparsecs, where there is on average one galaxy, of mass $8 \times 10^{10}$ solar masses (Allen 1973), there are $< 50,000$ solar masses of (smoothly distributed) neutral intergalactic gas. Galaxy formation gathered up all of the baryons except a fraction $< 6 \times 10^{-7}$, an efficient process indeed—unless intergalactic hydrogen is highly ionized. Observations of the cosmic microwave background (Mather et al. 1990) show that such intergalactic hydrogen cannot be at very high temperatures (Rogers and Field 1991). At lower temperatures, recombination becomes more efficient, especially if the intergalactic gas is clumped. The clumping $C = \langle n_e^2\rangle/\langle n_e\rangle^2 = 1.5 \times 10^7$ for galaxies, and is unknown for ionized intergalactic gas.

The final fact that bears directly on the search for recombination radiation is that only a small part of the expected baryonic matter is accounted for by matter that has already been detected. For example Persic and Salucci (1992) estimate the baryon mass density of the universe due to the stars in galaxies and hot gas in clusters and groups of galaxies. They find $\Omega_b = 0.003$, which is less than 10 percent of the lower limit predicted by standard primordial nucleosynthesis which implies that the great majority of the baryons in the universe are as yet unseen.

We are now prepared to explore the consequences of equation 3. Persic and Salucci quote Kolb and Turner (1990) and Peebles et al. (1991) in giving $\Omega_b h_{50}^2 = 0.06$ as the most probable value for the baryon density from nucleosynthesis. We take $\Omega_g = \Omega_b$, that is, essentially all of the baryons are intergalactic ionized hydrogen. Insertion into equation 3 then gives $C/h_{50} = 50{,}000$. Notice that this clumping is vastly less than for the visible matter (galaxies) in the universe. The clumps of ionized baryons in the universe are much larger in relation to their separation than are galaxies.

We have seen that the intergalactic medium is unquestionably highly ionized. What causes this ionization? There is some controversy over this. Meiksin and Madau (1993) provide several models in which the observed QSOs can provide the required ionizing photons at early epochs, and we accept their conclusion.

However, our highly clumped ionized intergalactic medium has much more severe problems. For $h_{50} = 1$, the recombination time of our clouds is only $1.3 \times 10^7$ years, and for $h_{50} = 2$ the recombination time is $6.3 \times 10^6$ years. Thus, if this interpretation of the diffuse high galactic latitude diffuse background is correct, a strong additional source of ionizing photons is required. Just such a source, radiative decay of neutrinos, has been proposed by Sciama (1993, e.g., in which Sciama references defenses of his neutrinos against the conclusions of Davidsen et al. 1991, who failed to observe a neutrino decay line from the cluster of galaxies A665). Sciama's neutrinos decay with the emission of photons that are just capable of ionizing hydrogen (hence our assumed temperature, above, of $\sim 8000$ K). Sciama's neutrinos, in our present picture, would be the dark matter into which the baryons all fell following recombination. In every potential well, most (or rather, in most cases, all) of the hydrogen was re-ionized by Sciama's neutrinos. In exceptional cases dissipation occurred and quasars and galaxies formed. In most cases, in contrast, the hydrogen simply re ionized and expanded out of the well, forming our present clouds that are expanding with the universe. Such clouds would be extremely hard to detect, either in emission or absorption, because of the very large velocity dispersion that is expected.

In the present picture there is no way of avoiding Sciama's neutrinos (or their equivalent). If we assume that all of closure density ($\Omega = 1$) is ionized gas (that is, we ignore the nucleosynthesis argument), we obtain present-day recombination times that are of the order of the Hubble time, but the recombination time earlier (when the clouds were denser) would still be too short.

We conclude that if the high-galactic latitude diffuse ultraviolet background is extragalactic in origin, then its observed spectrum implies all of: a) detection of the baryonic dark matter; b) detection of the effects of the non-baryonic dark matter (Sciama neutrinos); and c) evidence for new physics beyond the standard model of elementary particle physics (Sciama neutrinos). As that would be an important set of discoveries indeed, we now turn to the observations at lower galactic latitudes in an effort to account for the high galactic latitude signal as being galactic, rather than extragalactic, in character.

## 3. OBSERVATIONS OF DIFFUSE BACKGROUND GENERALLY

The overall observational situation concerning ultraviolet background radiation has been reviewed recently by Henry (1991), and also by Bowyer (1991). Here we will survey progress since 1991, and try to resolve differences in interpretation of the existing data which have occurred.

Galactic sources that have been posited for diffuse background include fluorescence of interstellar molecular hydrogen (Martin, Hurwitz, and Bowyer 1990), atomic emission lines from hot gas in the interstellar medium and/or galactic halo (Feldman *et al.* 1981; Martin and Bowyer 1990), two-photon emission from the ionized component of the interstellar medium (Deharveng *et al.* 1982), and the light of hot stars scattering from interstellar dust grains (many authors).

There has been no recent change in the situation regarding molecular hydrogen fluorescence. The spectrum (Jakobsen 1986; Sternberg 1989) is sufficiently structured that there is no possibility that the signal at high galactic latitudes could be dominated by this source, which in any case continues strongly shortward of Lyman $\alpha$.

There has also been no change in the situation regarding two-photon emission. The expected intensity from galactic sources can be accurately predicted by using the H$\alpha$ measurements of Reynolds (1986). No sharp break is expected at 1216 Å of course, and the average intensity is predicted to be well below the observed level longward of 1216 Å.

For the other two sources there is more to say on developments since 1991, and we devote separate sections, below, to each.

## 4. LINE EMISSION FROM THE INTERSTELLAR MEDIUM

Until recently, the situation concerning detection of line emission from hot gas has been murky. The tentative detection by Feldman *et al.* (1981) was not confirmed by Murthy *et al.* (1989). Martin and Bowyer (1990) present a spectrum for one of their UVX targets that contains a quite impressive and convincing CIV 1549 Å line, and a much less impressive O III] 1663 Å line. For their other targets, for some of which similar line emission was claimed, Martin and Bowyer present only tiny portions of their spectra, at the location of the claimed lines. The full spectra of all targets should be published.

The recent development on this topic is detection of very strong line emission in the Eridanus region by Murthy, Im, Henry, and Holberg (1993) using the Voyager spacecraft. Figure 2 shows the spectrum of their Target B, which shows strong emission lines of O VI 1032/1038 Å and C III 977 Å radiation. The lines are extremely strong, and predicted associated lines (see figure) should be detectable even with IUE. The region involved is one where there is a very strong soft X-ray enhancement (Burrows *et al.* 1993) and based on our other Voyager spectra is not typical of the general interstellar medium. What this suggests is that when a high-sensitivity sky survey (*e.g.*, Kimble *et al.* 1990) is finally carried out, what will be revealed is a highly patchy structured hot interstellar medium.

Of course the highly structured character of the spectrum of Figure 2 shows that there is no possibility that the background at high latitudes is line emission. (To verify this, the reader should consult the individual spectra from the relevant references in Figure 1, rather than rely on Figure 1 itself.)

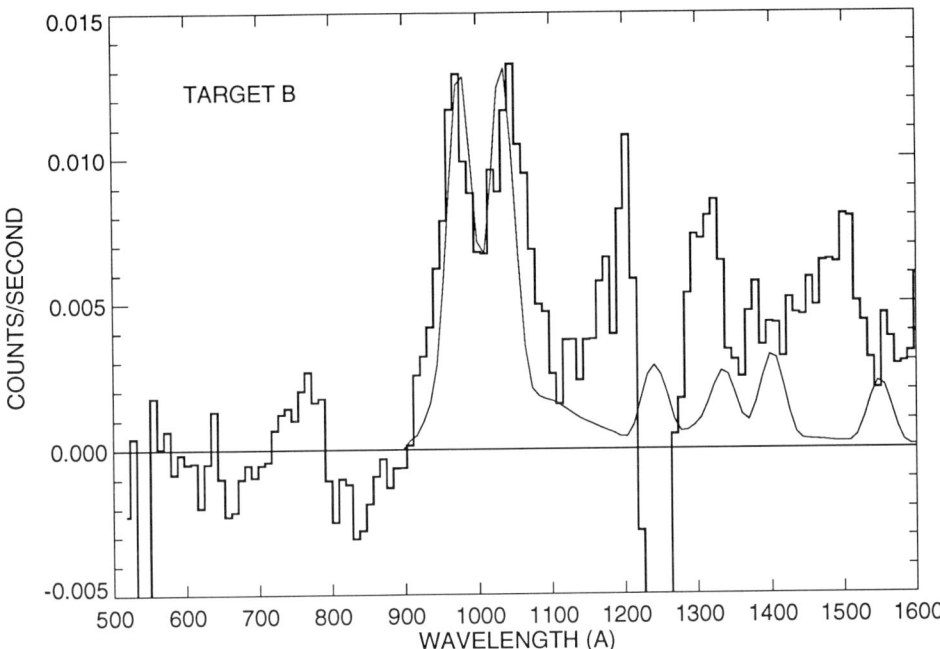

**Figure 2.** Spectrum of a region in Eridanus observed by Murthy, Im, Henry, and Holberg (1993). Strong solar system Lyman α (1216 Å) has been subtracted. Emission lines of C III (977 Å) and O VI (1032/1038 Å) are seen. The solid line shows the emission that is expected (Hartigan et al. 1987) from a shock with a velocity of 180 km s$^{-1}$, including two photon emission, plus appropriate dust-scattered light. The sensitivity of Voyager above 1200 Å is too low to allow detection of additional predicted lines of N V, C II, Si IV, O IV, and C IV, but those lines should be accessible to IUE and to the Hopkins Ultraviolet telescope.

## 5. STARLIGHT SCATTERING FROM INTERSTELLAR DUST

This is a critical topic in regard to our effort to identify the source of the diffuse cosmic background at high galactic latitudes. The interstellar radiation field in the far ultraviolet has been directly measured by Henry, Anderson, and Fastie (1980), and it is found to be flat when expressed in units. Our spectrum, Figure 1, is also consistent with being flat longward of 1216 Å, and if the Voyager measurement is dismissed (we will briefly discuss its likely validity below), then on spectral grounds there obviously is strong reason to hope that the light that is seen at high galactic latitudes is simply starlight scattering from interstellar dust.

The two recent reviews of the diffuse ultraviolet background (Bowyer 1991; Henry 1991) reached very different conclusions concerning the subject of diffuse galactic light (starlight scattering from dust). New developments, and careful reconsideration of earlier discussions, allow us to resolve most of the controversy.

The most important new development is the measurement of the Henyey-Greenstein

(1941) scattering parameter g in the ultraviolet by Witt et al. (1992). This parameter g characterizes the scattering pattern of the interstellar grains: $g = 1$ means complete forward scattering, $g = 0$ is isotropic scattering, and $g = -1$ represents complete back-scattering. The Henyey-Greenstein function has no physical basis, it is simply an heuristic tool for model-building. The reason that the value of g is so critical to our present concern is that if the albedo of the grains is high and if g is zero or at least has not too large a positive value, then the dust which is known to exist at high galactic latitudes (e.g., from IRAS cirrus observations) would backscatter sufficient light to account for the high latitude observations longward of 1216 Å.

The state of our knowledge of the value of g in 1991 can be gleaned from the excellent summary by Bowyer (1991, his Table 2). Values are widely divergent. The new observation by Witt et al. is of the nebula NGC 7023. The authors show an ultraviolet photograph, taken using UIT on the *Astro* mission, that shows the scattered light. Their analysis produces a fairly model-independent measurement of $g = 0.75$, which corresponds to very strong forward scattering. Their value for the albedo at 1400 Å is 0.65. In the light of all the controversy there has been over the value of g, it is important to note that Witt et al. indicate that their conclusion that $g_{uv} > g_{vis}$ is based on general radiative transfer principles and on the observational data alone. The value therefore should be quite secure. Let us now develop a simple model to use these values to predict what we should see at the highest galactic latitudes.

There is considerable dust at high galactic latitudes; for example Hauser et al. (1984) report, from their study of IRAS cirrus observations, that $A_V = 0.1$ mag at high latitudes. Stark et al. 1992 show $2 \times 10^{20}$ cm$^{-2}$ as a typical column density of neutral hydrogen at the highest galactic latitudes. Use of $E_{B-V} = N_{HI}/5 \times 10^{21}$ (Knapp and Kerr 1974) then gives $A_V = 0.12$. If $E_{1500-V}/E_{B-V} = 5.3$ (Bless and Savage 1972, for $\zeta$ Oph), we then get $\tau_{1500\,\text{Å}} = 0.921\, A_\lambda = 0.3$. We adopt $A_V = 0.1$ and $\tau_{1500\,\text{Å}} = 0.255$. Also, there are many bright OB stars in or near the galactic plane. For our simple model for the scattered light of these stars, we integrate the Henyey-Greenstein (1941) scattering function

$$H(\theta) = (1 - g^2)/4\pi(1 + g^2 - 2g\cos\theta)^{-3/2} \quad (4)$$

over the back-scattering directions, $\pi/2$ to $\pi$, obtaining

$$B = 1/2 - 1/(2g) + (1 - g^2)/\left(2g\sqrt{(1+g^2)}\right) \quad (5)$$

for B, the fraction of the scattered light that is backscattered. Our model is, then, that at high latitudes we expect a scattered intensity $S = B\, G\, a\, \tau$, where G is the local far-ultraviolet interstellar radiation field ($\sim$ 10,000 units: Henry, Anderson, and Fastie 1980), a is the grain albedo ($\sim$ 0.65 at 1500 Å: Witt et al. 1992), and $\tau = 0.255$ is the far-ultraviolet optical thickness of the high galactic latitude scattering layer.

For detailed study of scattered light at any particular region of the sky, one unquestionably wants to use a detailed model. However, such models are often complex and not generally available. The only competing simple model is that of Jura (1979), which predicts the scattered light as a function of five variables: the source function in the disk (roughly our G), $\tau_0$ ($\sim$ 0.85, Joubert et al. 1983) the optical thickness of the galaxy in the ultraviolet, a, g, and the galactic latitude b. Use of Jura's model is illustrated nicely in Joubert et al. 1983. We prefer our model: because of its simplicity (no evaluation of an exponential integral is required); because it is valid for all values of g (Jura's model fails for large values of g); and because it does *not* give a galactic latitude dependence:

the source function shows asymmetry in galactic longitude that is very strong (Henry 1977), equaling that in latitude (just more than 78% of the source function originates at $|b|, < 21°$ while 78% of the source function originates at $180° < l < 360°$).

A crude estimate using our model is, however, very revealing. In Table 1 we present (as a function of the Henyey-Greenstein scattering parameter $g$) the predicted high galactic latitude flux, from our model and from that of Jura (for $b = 90°$), using the values that were specified above for the necessary parameters.

Table 1.
Backscattered Light $S$ (Units) at $b = 90°$ as a Function of $g$

| g | B | S (present) | S (Jura) | S (Onaka & Kodaira) |
|---|---|---|---|---|
| 0.98 | 0.004 | 7 | (−53) | |
| 0.90 | 0.023 | 38 | (7) | 41 |
| 0.80 | 0.051 | 84 | 81 | 90 |
| 0.75 | 0.067 | 110 | 118 | |
| 0.70 | 0.084 | 139 | 155 | 151 |
| 0.60 | 0.124 | 205 | 229 | 222 |
| 0.40 | 0.225 | 372 | 378 | 394 |
| 0.30 | 0.286 | 473 | 452 | 494 |
| 0.00 | 0.500 | 828 | 675 | 741 |
| −0.90 | 0.977 | 1619 | 1342 | |

The observed level of cosmic background reported at moderate and high latitudes by large numbers of observers is about 300 units (Henry 1991, and also Figure 1). A glance at Table 1 shows that if the value of $g$ in the ultraviolet is, say, 0.7 or greater, then the cosmic high-latitude background is not scattered starlight and is presumably extragalactic.

The final column in Table 1 is the prediction at high galactic latitudes of the sophisticated model of Onaka and Kodaira (1991), which takes into account the variation with galactic *longitude* of the galactic source function. We have used $a = 0.65$ and $\tau = 0.255$ in this application of the Onaka and Kodaira model. Note the excellent agreement among all these closely-related models.

That there is a very significant variation with galactic longitude of the source function for scattered light is of the greatest importance for interpretation of the diffuse galactic light. If one simply considers the distribution of the TD-1 stars (Figure 3), it is easy to mistakenly conclude that the source function is dependent on galactic latitude, and is independent of galactic longitude. That both of these conclusions are wrong, is demonstrated in Figure 4, which shows the integrated 1565 Å emission from the same stars that appear in Figure 3. There are profound effects due to absorption by the interstellar medium: also, the presence of Gould's Belt, which is tipped 19° with respect to the galactic plane, is very apparent. The model of Onaka and Kodaira, which takes some of these effects explicitly into account, will be very useful to us below.

If the background at high latitudes is not the back-scattered light of galactic plane stars, our rather exotic extragalactic model must perhaps be taken seriously. Before we do so, however, we must ask why Bowyer (1991) came to an opposite conclusion, concluding that most of the light, even at the highest galactic latitudes, is galactic in origin. Bowyer relied mostly on the result of Hurwitz, Bowyer, and Martin (1991) in

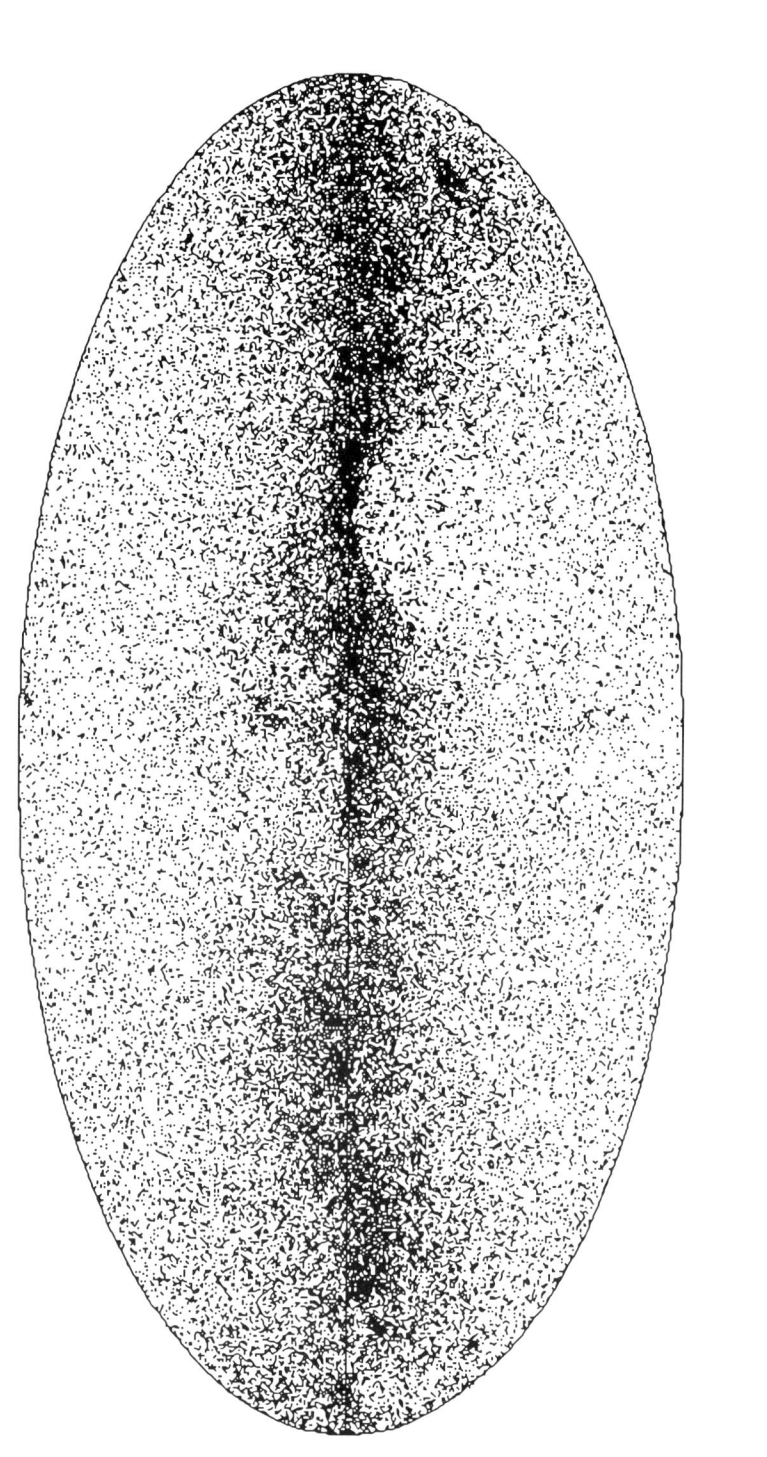

**Figure 3.** The ultraviolet (1656 Å) stars from the TD-1 catalog. The north galactic pole is at the top, and the galactic center at the center. Galactic longitude increases to the left. The TD-1 stars are concentrated to the galactic plane, and they are reasonably evenly distributed in galactic longitude, when simply number of stars (shown) is considered. When the flux from these same stars is considered, instead, the result is dramatically different, as is shown in Figure 4 and 5.

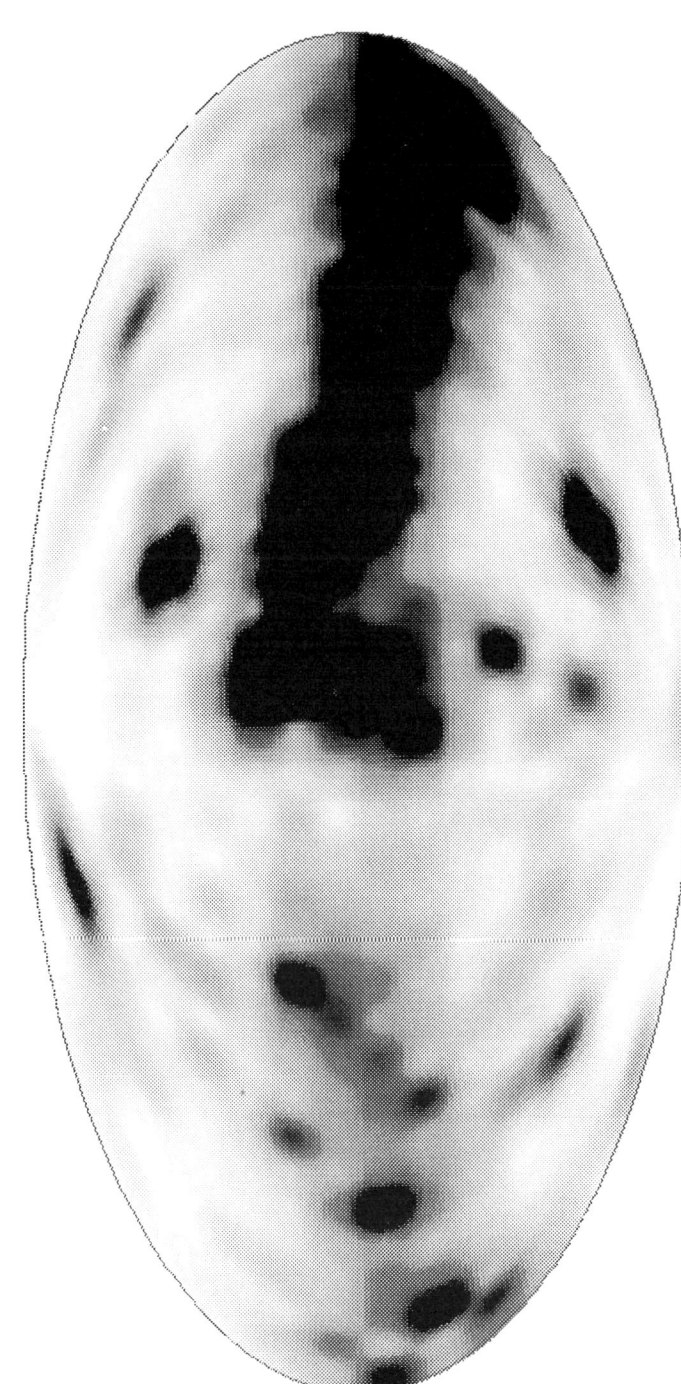

**Figure 4.** A linear, but saturated by a factor of ten, "photograph" of the sky at 1565 Å constructed from the TD-1 observations. This image contains only the light of the stars; that is, of the source function for scattering. The north galactic pole is at the top, and the galactic center is at the center. Just more than 78% of the source function originates between galactic longitudes 180° and 360°.

reaching this conclusion. The result of Hurwitz, Bowyer, and Martin was that in the far ultraviolet, the interstellar grains have albedo $0.1 < a < 0.3$ and Henyey-Greenstein scattering parameter is $0 < g < 0.4$, which numbers differ drastically from the values quoted above, and which if correct would give a scattered starlight signal at high galactic latitudes of as much as 340 units. As long as there is any possibility that Hurwitz et al. are correct, our extragalactic model must be rejected.

Unfortunately it is easy to show that the Hurwitz et al. analysis is suspect. It is based on the Berkeley UVX measurements. The locations of the nine UVX pointings are shown in Figure 5, superimposed on an unsaturated map of the source function for scattered light, the TD-1 stars (that is, Figure 5 is simply an unsaturated version of Figure 4, shifted in longitude). The Hurwitz et al. determination of the albedo relies mostly on their analysis of the signal seen during scan number 6 (see Figure 5), which was a scan from moderate galactic latitude to low galactic latitude in which the signal was interpreted as being saturated. We have organized Figure 5 so that this critical scan (at $l = 135°$) is centered in the figure. The potential flaw in their analysis is their assumed source function: they assumed that the interstellar radiation field arises from a smooth galactic-plane-parallel distribution of emitting (and absorbing) media. That this is not the case is dramatically apparent from Figures 4 and 5. The Hurwitz et al. model was scaled to match the TD-1 results for the sky-averaged interstellar radiation field at 1550 Å in the galactic plane. Now in fact 78% of the source function occurs in the hemisphere 180°–360°, far removed from scan 6. Thus, the light illuminating the dust of scan 6 was coming from quite different angles than assumed in their model. In particular, if $g$ were large and positive (as the Witt et al. result suggests) then the dust in the direction of scan 6 could not be expected to backscatter significant amounts of light *regardless* of the value of the albedo. The Hurwitz et al. determination of $g$ follows directly from their determination of the albedo $a$: they infer $g$ from the high galactic latitude UVX observations, after fixing $a$. If $a$ is low, then of course scattering must be isotropic if it is to provide the observed high latitude flux. If instead, the albedo is high as the analysis of Witt et al. 1992 suggests, then it follows from their data (and from our own UVX data, Murthy et al. 1990) that $g$ must be large, or too high a flux would be seen at high latitudes.

We emphasize that the asymmetry of the source function that is shown in Figure 5 is not in the least controversial (see, *e.g.*, Gondhalekar et al. 1980). It has been clear for quite some time that to extract accurate values of $a$ and $g$ from mapping of the scattered light at moderate and high galactic latitudes will require rather sophisticated models. A beginning for such a model was used by Murthy, Henry, and Holberg (1991) in interpreting Voyager observations of the diffuse background at 1100 Å. In Figure 6 we show a preliminary version of their model, to re-emphasize our point that a sophisticated model is required. In particular, for $g$ large (and the observation of Witt et al. suggests that $g$ is indeed large), models that take into account individual stars clearly will be necessary.

[Since the above was written, we have performed a reanalysis of the UVX data (Henry and Murthy 1993) in which we show that these data are in fact quite compatible with values of the albedo $\sim 0.65$, and of the scattering asymmetry parameter $g \sim 0.75$, if an extragalactic component of $300 \pm 100$ units exists.]

A number of works (Bowyer 1991; Henry 1991) report correlations between purported measurements of the diffuse ultraviolet background and either galactic latitude or hydrogen column density. On this front, there has been a certain amount of progress since 1991. Wright (1992) has criticized the approach, often used, of noting that such

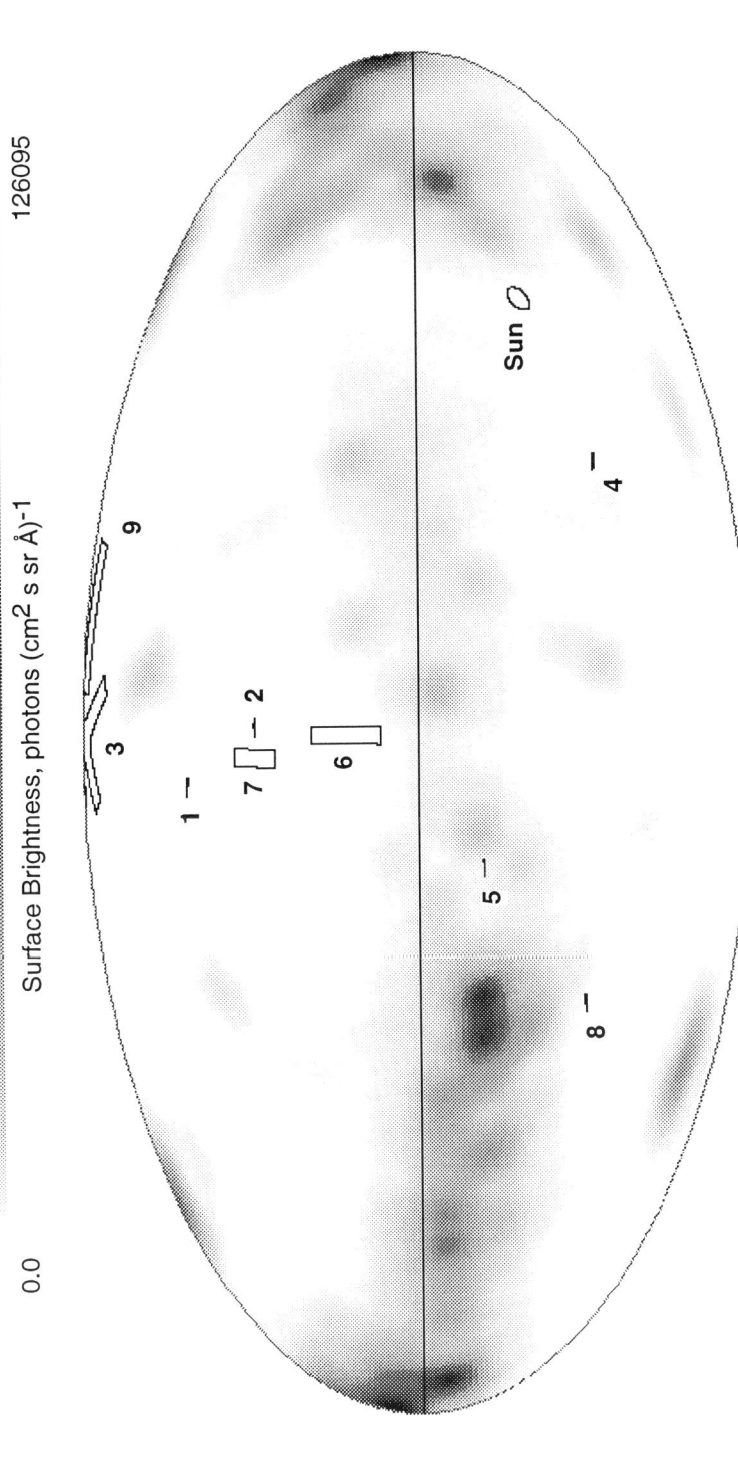

**Figure 5.** Integrated emission from the TD-1 stars. Dark regions in the figure are bright regions in the sky. The figure is linear and is just saturated at the darkest point. Also shown (numbered) are the regions observed by Johns Hopkins and Berkeley during the UVX mission. Targets are identified by number as specified in Murthy et al. (1989). Target 4 was a failed attempt to observe comet Halley, and produced no data useful for the present analysis. The north galactic pole is at the top, while the center of this Aitoff all-sky image is at galactic longituae $l = 135°$ so as to demonstrate clearly how poorly illuminated the dust is at the location of UVX scan number six, called GRADIENT by Murthy et al. (1989).

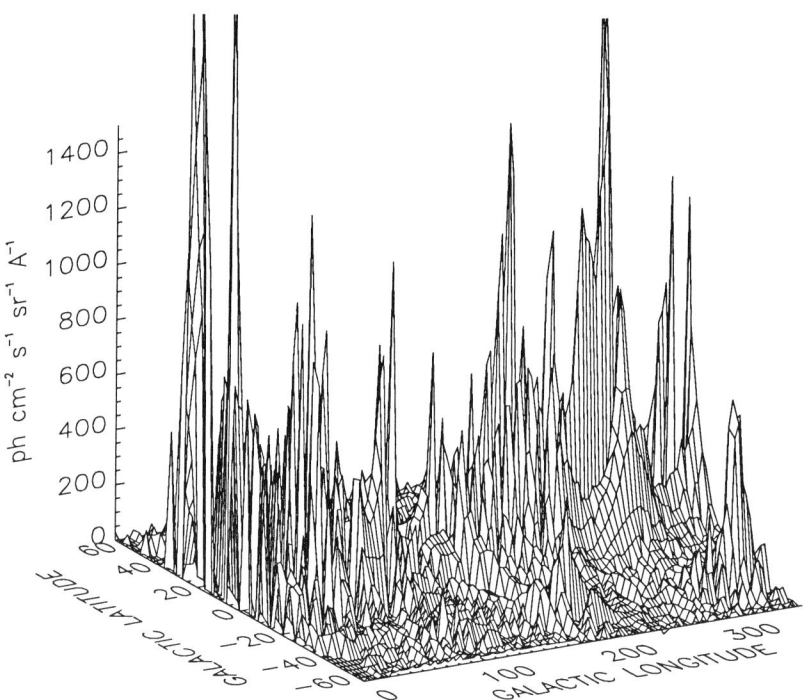

**Figure 6.** *The model of ultraviolet scattered light of Murthy, Henry, and Holberg (1991) for the case* a = 0.1, g = 0.9. *The model scales linearly with* a. *The very sharp spikes are diffuse halos around stars that are predicted by this model. The figure shows only the scattered light; the source (the stars) is not shown (the source appears in Figures 3 and 4).*

correlations when extrapolated to zero column density always leave an unexplained residual. He points out that ionized and molecular hydrogen are also present and should have associated dust. Wright's reanalysis of the data of Fix *et al.* (1989) yields $a = 0.42 \pm 0.06$, $g = 0.44 \pm 0.18$, and an extragalactic component of no larger than 500 units. However, Witt and Petersohn (1994) have reconsidered Wright's analysis, and they find that the Fix *et al.* data show, instead, that the albedo $a \sim 0.5$, $g \sim 0.9$, and the extragalactic component is $300 \pm 80$ units.

Onaka and Kodaira (1991) have now reported in detail their rocket study of diffuse far-ultraviolet radiation at high galactic latitudes. They find, using their model that contains a dependence on galactic longitude of the source function for the scattered starlight, that

$$a(1 - 1.04\, g) = 0.18 \pm 0.03 \tag{6}$$

which agrees at the $2\sigma$ level with the $a$ and $g$ values of *Witt et al.* (1991), and which also agrees within $2\sigma$ with the values of Wright (1991). They find that their regression line of intensity against hydrogen column density intercepts the ordinate at 200 to 300 units, a somewhat lower value than appears in Figure 1. A plot of their data against $\csc b$ (as recommended by Wright 1991) yields an "extragalactic" component of about 400 units.

Finally, Pérault *et al.* (1991) have re-examined the D2B-AURA measurements of the diffuse ultraviolet background of Joubert et al. (1983). The reexamination strongly justifies the skeptical attitude that was taken toward these data by Henry (1991). In the new analysis, the strong galactic latitude dependence of the "diffuse" flux is found to be dominated by direct starlight and starlight diffused in the instrument. They estimate that 1/2 to 2/3 of the ultraviolet flux is due to factors other than single scattering off dust. Pérault et al. believe that the major additional factor is light scattering in the instrument.

We now turn to the Voyager observations, and in particular, to the question of the reality of the jump in the high latitude background at 1216 Å.

## 6. VOYAGER OBSERVATIONS OF THE DIFFUSE BACKGROUND

In his review of the diffuse background Henry (1991) consistently took an extremely skeptical attitude toward claims of the detection of diffuse ultraviolet radiation, and particularly toward works that claimed to understand the physical source of the radiation absent a spectrum. In particular, it was only in the case of the Voyager observation of extended diffuse emission in Ophiuchus by Holberg (1990) that Henry felt that there was a very strong case for the assertion that ultraviolet starlight scattered from dust had been detected. There is now a second very strong case: Murthy, Henry, and Holberg (1993) have detected extremely strong scattered starlight in the direction of the Coalsack nebula (see Figure 7). Detailed modeling shows that this is not light backscattered from the Coalsack, but rather is the forward-scattered light of three very bright ultraviolet stars near the Coalsack. The spectral dependence of this diffuse emission is (as can be seen by our model fit) exactly that of the illuminating early B stars. Unless $g$ is varying with wavelength in such a way as to fortuitously exactly cancel changes in $a$, we can conclude that the albedo of the grains is as high at 1000 Å as it is at 1350 Å. In fact, exactly the same phenomenon can be seen in Figure 2 of Holberg (1990), where the geometry of the dust relative to the source star is not so clear, but is unlikely to be the same as for the Coalsack observation.

This is an important result, highly relevant to the question of what is happening at high galactic latitudes. Voyager, with its low sensitivity longward of Lyman $\alpha$, could not be expected to detect the background that appears in Figure 1. But it should definitely detect 300 units in the range 1000 Å to 1100 Å if it is there, and furthermore, the data of Figure 7 suggest that if what is being seen at high latitudes at longer wavelengths by many independent observers is starlight scattered from dust, the spectrum should continue strongly down to 912 Å. Voyager shows that it does not.

We have seen, in Figure 1, the Voyager upper limit of 100 units at 1100 Å. The extragalactic radiation field at slightly shorter wavelengths has been measured by Kulkarni and Fall (1993) by applying the proximity effect to Lyman $\alpha$ forest lines in the spectra of nearby quasars. They find a intensity of one unit, well below the Voyager upper limit at 1100 Å.

Holberg (1986, 1990) and Murthy, Henry, and Holberg (1991) present the evidence for the validity of the Voyager upper limit. There is a great deal more that can be done with the Voyager archive, and Murthy, Henry, Hall, and Holberg have been funded in an archival research program to carry out this project, which is under way. In Figure 12 of his review Henry (1991) indicated that every Voyager diffuse background observation above 20° latitude was only an upper limit. We have subsequently learned (Holberg, private communication) that the Voyager targets selected for analysis included only those that showed no evidence of a signal. That does not change any conclusions:

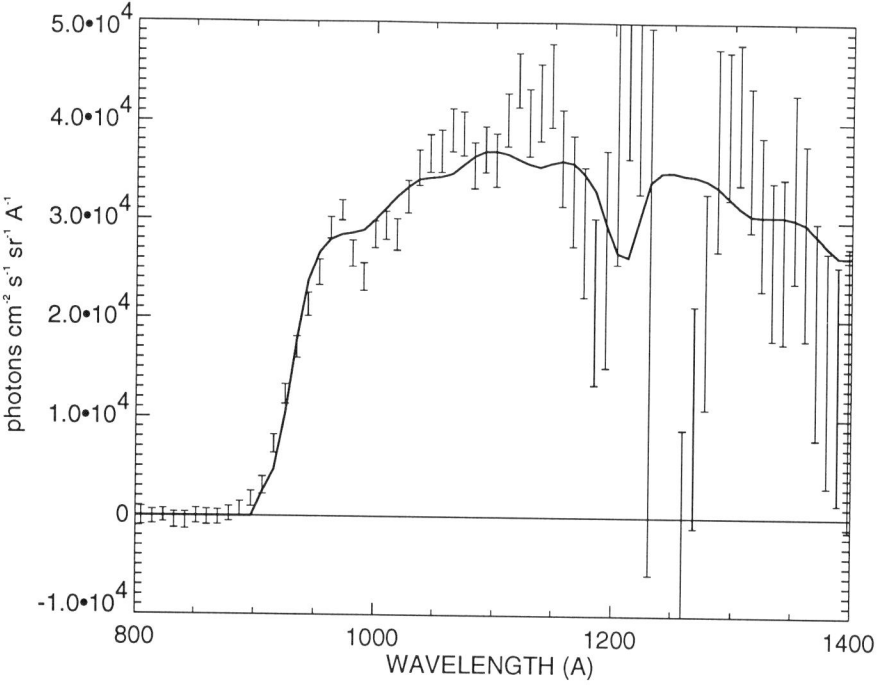

**Figure 7.** *Spectrum of the Coalsack nebula as observed by Murthy, Henry, and Holberg 1993. This is the brightest cosmic diffuse ultraviolet radiation ever reported in the night sky. The radiation is the forward-scattered light of three extremely bright ultraviolet-emitting stars, α Cru, β Cru, and β Cen. A large subtraction has occurred at Lyman α (1216 Å). The dark solid line represents our best fit model, after subtraction of interplanetary lines. Notice that no break occurs in this spectrum of dust-scattered starlight between wavelengths longward of Lyman α and wavelengths shortward of Lyman α.*

there is still the same considerable number of locations at moderate and high galactic latitudes that show only an upper limit (which does *not* occur at wavelengths longward of Lyman α, Figure 1), and those locations where a signal *is* present may all contain point sources (the overwhelming majority of Voyager pointings were toward known point sources). The new situation does leave open the possibility, however, that diffuse emission might still be detected at high latitudes near 1100 Å using Voyager. Indeed, if $g$ and $a$ are both large, we would predict a significant signal at high latitudes near the location of bright ultraviolet stars (see Figure 6).

Figure 8 shows the Voyager upper limits of Holberg (1990) and of Murthy, Henry, and Holberg (1991) superposed on a map of the expected scattered light above $b = 40°$ predicted using the model of Onaka and Kodaira (1991). We used $\tau = 0.4 \csc b$ in making this plot, with the albedo taken as 0.65 and $g = 0.9$ (northern hemisphere) and $g = 0.8$ (southern hemisphere), to illustrate the model. We have evaluated the reduced $\chi^2$ for these data (treated as detections, each with a standard deviation of 100 units) against this model as a function of $a$ and $g$. A plot of the reduced $\chi^2$ appears in Figure 9. An albedo of 0.65 is seen to require that $g > 0.8$. A sufficiently low albedo will also explain the data.

The new work already done on Voyager data that bears most directly on the present discussion is a new observation by Murthy, Henry, and Holberg of the extended dust patch at high galactic latitudes that was discovered by Sandage (1976). Sandage's Plate 1 shows very clear evidence for dust at $b = +38°$. Our Voyager observation shows nothing but an upper limit of 100 units, and our measurement is so clean that we have included it as part of our data-reduction template for "no astrophysical signal" (Murthy, Im, Henry, and Holberg 1993). The importance of this observation is that Sandage deduces that $A_V = 0.3$ mag, from the 21 cm observation of Heiles (1975). This translates into an optical depth $\tau$ at 1100 Å of 1.0 mag, using the $E_{1100-V}/E_{B-V}$ of York et al. (1973). Use of the model of Onaka and Kodaira for Sandage's location and value of t shows that for an albedo of 0.65 we require $g > 0.9$ to explain our Voyager result. (Another possible explanation would be a low albedo for the grains: this of course would also suggest that the high galactic latitude signal at longer wavelengths is extragalactic.) The Sandage region was also scanned at longer ultraviolet wavelengths by Murthy et al. (1989, 1990), and also by Martin, Hurwitz, and Bowyer (1990). No enhancement of background as the line of sight passed over the Sandage region was noted by any of the various UVX spectrometers.

## 7. DISCUSSION

The absence of a signal from the Sandage region discussed in the last section is perhaps the most powerful indication that the signal at longer wavelengths is not back scattered starlight, and hence is extragalactic. It is not proof, however. The new information that we presented in the previous section showing that the albedo of grains continues high into the farthest astronomical ultraviolet is not conclusive, as $g$ could be varying with wavelength to compensate. If one is desperate to avoid concluding that the radiation is extragalactic, one can postulate that for *backscattered* light the shape of the scattering function changes abruptly near 1216 Å. There is nothing sacred, after all, about the Henyey-Greenstein scattering form. One could also simply postulate an arbitrary additional previously-unknown dust population having the needed optical properties.

In this context it is interesting to note that Martin, Hurwitz, and Bowyer (1991) concluded (ignoring entirely the Voyager data) that what is observed at the highest galactic latitudes indicates "either the existence of a hitherto unidentified dust component, or ... a large enhancement in dust scattering efficiency in low-density gas." They reached their conclusion from the striking resemblance between their highest-latitude UVX spectrum and a lower-latitude spectrum that they believed was dust-scattered light. That is, their evidence for dust reflection at high latitudes is its exact resemblance to dust at low latitudes, but then they are forced to conclude that it is anomalous dust, and that the resemblance is therefore fortuitous.

What would it take to resolve the matters that we have discussed in this paper? A good deep ultraviolet image of the Sandage region at 1500 Å would be a great step forward. If the Sandage region is not clearly seen, that would be virtually conclusive evidence that the high-latitude background is not due to scattering by normal dust. Also, with the Hopkins Ultraviolet Telescope (Davidsen 1993) it ought to be possible to obtain a spectrum of the dust-scattered light in the direction of the Coalsack that has high enough signal to noise that a comparison is possible with the rather structured

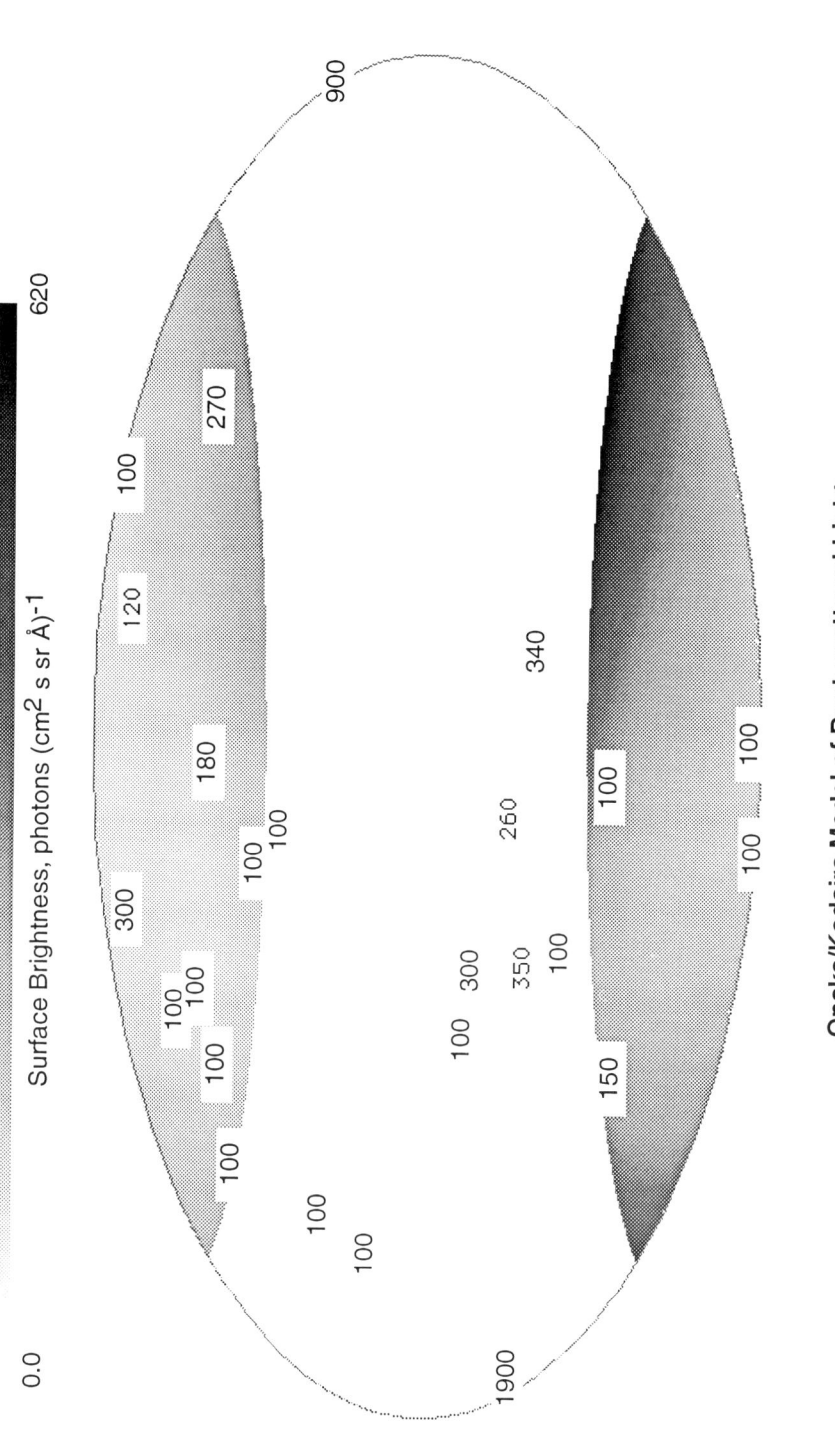

**Onaka/Kodaira Model of Dust-scattered Light**

**Figure 8.** The shading shows the predicted scattered light as a function of galactic longitude and latitude as predicted using the model of Onaka and Kodaira (1991). We have used albedo $a = 0.65$, and $\tau = 0.4\ csc(b)$, and we have used $g = 0.9$ for northern galactic latitudes, and $g = 0.8$ for southern galactic latitudes, simply to exhibit the results of the model.

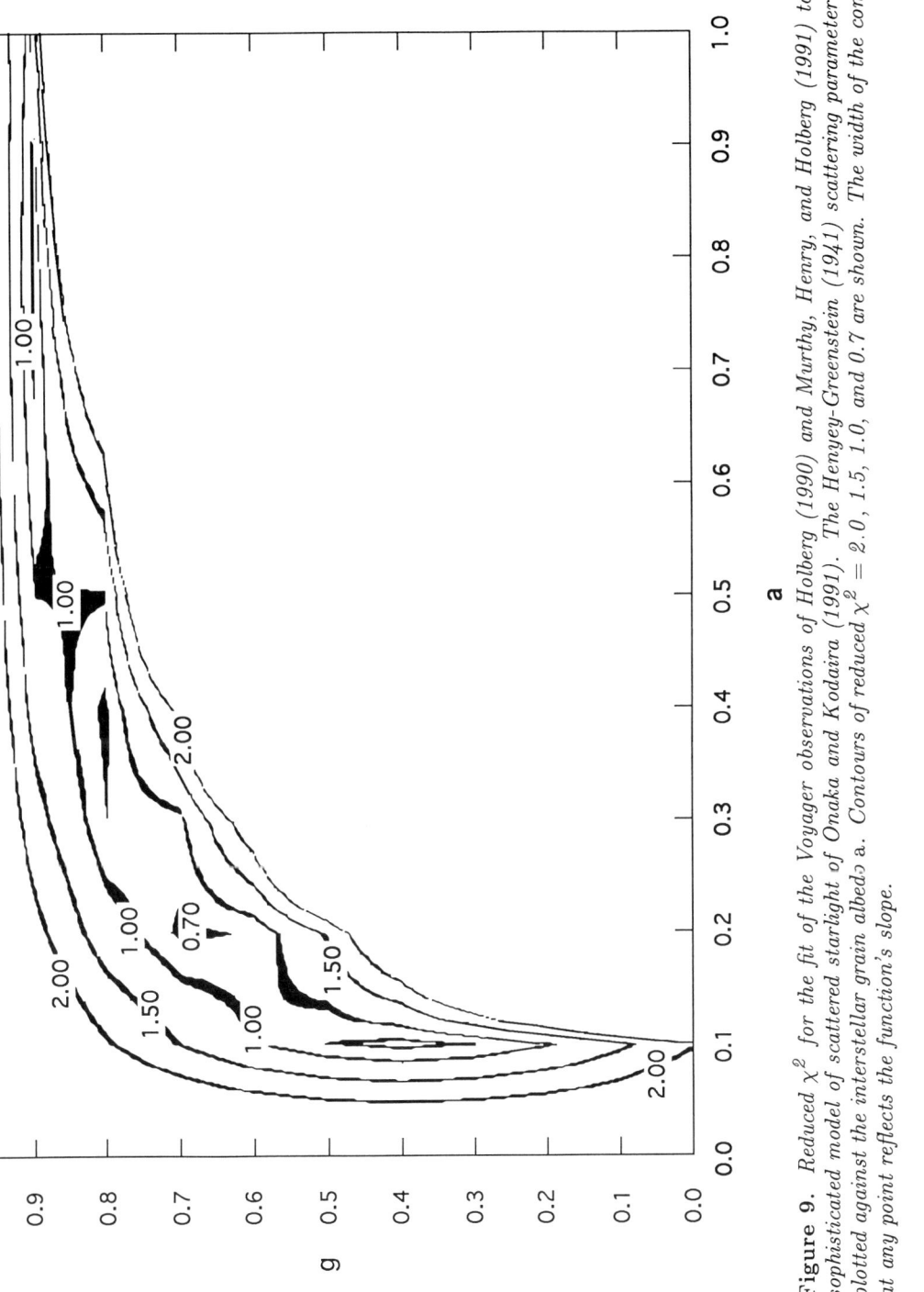

**Figure 9.** Reduced $\chi^2$ for the fit of the Voyager observations of Holberg (1990) and Murthy, Henry, and Holberg (1991) to the sophisticated model of scattered starlight of Onaka and Kodaira (1991). The Henyey-Greenstein (1941) scattering parameter g is plotted against the interstellar grain albedo a. Contours of reduced $\chi^2 = 2.0, 1.5, 1.0,$ and 0.7 are shown. The width of the contour at any point reflects the function's slope.

high galactic latitude spectrum of Martin and Bowyer (1990). Certainly the interstellar radiation field (Henry et al. 1980) that is incident on the putative high latitude dust does not have the observed cosmic background spectral character reported by Martin and Bowyer; the Coalsack observation, while not decisive (it would not be *back* scattered light), could be extremely suggestive one way or the other.

To disprove the existence of an anomalous dust component specially designed to account for the observations would be difficult. A very long exposure indeed might, if the light is scattered starlight, reveal the spectral structure (absorption lines) that is expected in an integrated B-star dust-reflection spectrum.

## 8. CONCLUSION

No definite conclusion is possible. An interesting possibility is raised, however, that deserves observational and theoretical exploration. The intergalactic clouds of ionized hydrogen that we postulate might be related to the objects that produce the Lyman $\alpha$ forest, and the extrapolated spectrum of the radiation approximates the integrated light of the faint blue "galaxies" discovered by Tyson (1988). Perhaps Tyson's objects are not galaxies at all, but our gaseous objects, radiating and dissipating.

Diffuse ultraviolet background study today is in the same state that study of the diffuse infrared background was before IRAS and COBE: fragmentary, often conflicting observations. What is needed is an all-sky survey in the ultraviolet, involving both imaging and spectroscopy.

This work was supported by United States Air Force Contract F19628-93-K-0004, and by National Aeronautics and Space Administration grant NASA NAG5-619. We are grateful for the encouragement of Dr. Stephan Price.

## REFERENCES

Allen, C. W., 1973, Astrophysical Quantities (Athlone Press: London).
Anderson, R. C., Henry, R. C., Brune, W. H., Feldman, P. D., and Fastie, W. G. 1979, *Ap. J.*, **234**, 415.
Bless, R. C., and Savage, B. D. 1972, *Ap. J.*, **171**, 293.
Bowyer, S. 1991, *Ann. Rev. Astron. and Ap.*, **29**, 59.
Brune, W. H., Mount, G. H., and Feldman, P. D. 1979, *Ap. J.*, **227**, 884.
Burrows, D. N., Singh, K. P., Nousek, J. A., Garmire, G. P., and Good, J. 1993, *Ap. J.*, **406**, 97.
Cook, T. A., Cash, W., and Snow, T. P. 1989, *Ap. J. Lett.*, **347**, L81.
Davidsen, A. F., Kriss, G. A., Ferguson, H. C., Blair, W. P., Bowers, C. W., Dixon, W. V., Durrance, S. T., Feldman, P. D., Henry, R. C., Kimble, R. A., Kruk, J. W., Long, K. S., Moos, H. W., and Vancura, O. 1991, *Nature*, **351**, 128.
Davidsen, A. F. 1993, *Science*, **259**, 327.
Deharveng, J. M., Joubert, M., and Barge, P. 1982, *Astron. Astrophys.*, **109**, 179.
Dekel, A., Bertschinger, E., Yahil, A., Strauss, M. A., Davis, M., and Huchra, J. P. 1993, *Ap. J.*, **412**, 1.
de Rújula, A. D., and Glashow, S. 1980, *Phys. Rev. Letters*, **46**, 80.
Feldman, P. D., Brune, W. H., and Henry, R. C. 1981, *Ap. J. Lett.*, **249**, L51.
Fix, J. D., Craven, J. D., and Frank, L. A. 1989, *Ap. J.*, **345**, 203.

Gondhalekar, P. M., Phillips, A. P., and Wilson, R. 1980, *Astron. Astrophys.*, **85**, 272.
Hartigan, P., Raymond, J., and Hartmann, L. 1987, *Ap. J.*, **316**, 323.
Hauser, M. G., Gillett, F. C., Low, F. J., Gautier, T. N., Beichman, C. A., Neugebauer, G., Aumann, H. H., Baud, B., Boggess, N., Emerson, J., P., Houck, J. R., Soifer, B. T., and Walker, R. G. 1984, *Ap. J.*, **278**, L15.
Heiles, C. 1975, *Astron. Astrophys. Suppl.*, **20**, 37.
Henry, R. C. 1977, *Ap. J. Supplement Series*, **33**, 451.
Henry, R. C., Anderson, R. C., and Fastie, W. G. 1980, *Ap. J.*, **239**, 859.
Henry, R. C., Anderson, R., Feldman, P. D., and Fastie, W. G. 1978, *Ap. J.*, **222**, 902.
Henry, R. C., and Feldman, P. D. 1981, *Phys. Rev. Lett.*, **47**, 618.
Henry, R. C., Feldman, P. D., Fastie, W. G., and Weinstein, A. 1978, *Ap. J.*, **223**, 437.
Henry, R. C. 1991, *Ann. Rev. Astron. and Ap.*, **29**, 89.
Henry, R. C., and Murthy, J. 1993, *Ap. J. (Letters)*, in press.
Henyey, L. G., and Greenstein, J. L. 1941, *Ap. J.*, **93**, 70.
Holberg, J. B. 1986, *Ap. J.*, **311**, 969.
Holberg, J. B., Forrester, W. T., Shemansky, D. E., and Barry, D. C. 1982, *Ap. J.*, **257**, 656.
Holberg, J. B. 1990, in *Proc. IAU 139, The Galactic and Extragalactic Background Radiation*, ed. S. Bowyer, Ch. Leinert (Dordrecht: Kluwer Academic), p. 220.
Hurwitz, M., Bowyer, S., and Martin, C. 1990, in *Proc. IAU 139, The Galactic and Extragalactic Background Radiation*, ed. S. Bowyer, Ch. Leinert (Dordrecht: Kluwer Academic), p. 229.
Hurwitz, M., Bowyer, S., and Martin, C. 1991, *Ap. J.*, **372**, 167.
Jakobsen, P. 1986, *Adv. Space Res.*, **6**, 59.
Jakobsen, P., Bowyer, S., Kimble, R., Jelinsky, P., and Grewing, M., Krmer, G., and Wulf-Mathies, C. 1984, *Astron. Astrophys.*, **139**, 481.
Joubert, M., Masnou, J. L., Lequeux, J., Deharveng, J. M., and Cruvellier, P. 1983, *Astron. Astrophys.*, **128**, 114.
Jura, M. 1979, *Ap. J.*, **227**, 798.
Kimble, R., Bowyer, S., and Jakobsen, P. 1981, *Phys. Rev. Lett.*, **46**, 80.
Kimble, R. A., Henry, R. C., and Paresce, F. 1990, in *Proc. IAU 139, The Galactic and Extragalactic Background Radiation*, ed. S. Bowyer, Ch. Leinert (Dordrecht: Kluwer Academic), p. 441.
Knapp, G. R., and Kerr, F. J. 1974, *Astron. Astrophys.*, **35**, 361.
Kolb, E. W., and Turner, M. S. 1990, in *The Early Universe*, (California: Addison-Wessley Publishing Company).
Kulkarni, V. P., and Fall, S. M. 1993, *Ap. J.*, **413**, L63.
Martin, C., and Bowyer, S. 1990, *Ap. J.*, **350**, 242.
Martin, C., Hurwitz, M., and Bowyer S. 1990, *Ap. J.*, **354**, 220.
Martin, C., Hurwitz, M., and Bowyer S. 1991, *Ap. J.*, **379**, 549.
Mather, J. C., Cheng, E. S., Eplee, R. E., Jr., Isaacman, R. B., Meyer, S. S., Shafer, R. A., Weiss, R., Wright, E. L., Bennett, C. L., Boggess, N. W., Dwek, E., Gulkis, S., Hauser, M. G., Janssen, M., Kelsall, T., Lubin, P. M., Moseley, S. H., Jr., Murdock, T. L., Silverberg, R. F., Smoot, G. F., and Wilkinson, D. T. 1990, *Ap. J.*, **354**, L37.
Meiksin, A., and Madau, P. 1993, *Ap. J.*, **412**, 34.
Murthy, J., and Henry, R. C. 1987, *Phys. Rev. Lett.*, **58**, 1581.
Murthy, J., Henry, R. C., Feldman, P. D., and Tennyson, P. D. 1989, *Ap. J.*, **336**, 954.

Murthy, J., Henry, R. C., and Holberg, J. B. 1991, *Ap. J.*, **383**, 198.
Murthy, J., Henry, R. C., Feldman, P. D., and Tennyson, P. D. 1990, *Astron. Astrophys.*, **231**, 187.
Murthy, J., Im, M., Henry, R. C., and Holberg, J. B. 1993, *Ap. J.*, in press.
Murthy, J., Henry, R. C., and Holberg, J. B. 1993, *Ap. J.*, submitted.
Onaka, T. 1990, in *Proc. IAU 139, The Galactic and Extragalactic Background Radiation*, ed. S. Bowyer, Ch. Leinert (Dordrecht: Kluwer Academic), p. 379.
Onaka, T., and Kodaira, K. 1991, *Ap. J.*, **379**, 532.
Paresce, F., Margon, B., Bowyer, S., and Lampton, M. 1979, *Ap. J.*, **230**, 304.
Peebles, P. J. E., Schramm, D. N., Turner, E. L., and Kron, R. G., 1991, *Nature*, **352**, 769.
Pérault, M., Lequeux, J., Hanus, M., and Joubert, M. 1991, *Astron. Astrophys.*, **246**, 243.
Persic, M., and Salucci, P. 1992, *Mon. Not. R. Astr. Soc.*, **258**, 14P.
Reynolds, R. J. 1986, *Ap. J.*, **309**, L9.
Rogers, R. D., and Field, G. B. 1991, *Ap. J.*, **370**, L57.
Sandage, A. 1976, *Astron. J.*, **81**, 954.
Sciama, D. W. 1993, *Ap. J.*, **409**, L25.
Stark, A. A., Gammie, C. F., Wilson, R. W., Bally, J., Linke, R. A., Heiles, C., and Hurwitz, M. 1992, *Ap. J. Supplement Series*, **79**, 77.
Stecker, F. W. 1980, *Phys. Rev. Lett.*, **45**, 1460.
Steidel, C., and Sargent, W. L. W., 1987, *Ap. J.*, **318**, L11.
Sternberg, A. 1989, *Ap. J.*, **347**, 863.
Tennyson, P. D., Henry, R. C., Feldman, P. D., and Hartig, G. F. 1988, *Ap. J.*, **330**, 435.
Tyson, J. A. 1988, *Astron. J.*, **96**, 1.
Weller, C. S. 1983, *Ap. J.*, **268**, 899.
Witt, A. N., Petersohn, J. K., Bohlin, R. C., O'Connell, R. W., Roberts, M. S., Smith, A. M., and Stecher, T. P. 1992, *Ap. J.*, **395**, L5.
Witt, A. N., and Petersohn, J. K. 1994, in "*The First Symposium on The Infrared Cirrus and Diffuse Interstellar Clouds,*" eds. R. Cutri & W. Latter, PASP Conference Series, in press.
Wright, E. L. 1992, *Ap. J.*, **391**, 34.
York, D. G., Drake, J. F., Jenkins, E. B., Morton, D. C., Rogerson, J. B., and Spitzer, L. 1973, *Ap. J. Lett.*, **182**, L1.

## DISCUSSION

**J. Peebles**: Very simply, what sky coverage would you have [using the Hopkins Ultraviolet Background Explorer]?

**Henry**: The Hopkins Ultraviolet Background Explorer (HUBE; see Kimble, Henry, and Paresce 1990), which was described by me briefly in my verbal presentation, is a candidate Small Explorer mission that was under consideration by NASA for possible selection. What we would have had in this proposed experiment was a complete survey of the entire sky, as far as imaging is concerned, in the wavelength range longward of Lyman $\alpha$. Spectroscopically, HUBE involved a partial survey of the sky: each time we took an image, we would get a spectrum of the central region, typically with 5 Å resolution. The spectrum would have covered from 800 Å all the way to 1800 Å. Unfortunately, HUBE was not selected by NASA. HUBE is currently one of seven "finalists" to fly on the SAC-D spacecraft of Argentina.

**V. Khersonsky**: If I understand correctly, your conclusion about light scattering is strongly dependent on the size of the dust particles. In particular, the properties of scattering are dependent on grain sizes, and we know that the sizes of grains cover approximately 3 orders, maybe even 4 orders of magnitude. How do you take this into account in your considerations? Maybe it is not important.

**Henry**: It is certainly important in terms of actually understanding the grains. I am afraid I do tend to regard the interstellar grains as simply a source of noise or interference, and so I tend to take a rather pragmatic view, which is to treat the grains, regardless of their actual nature, as being simply governed by two numbers: an albedo $a$ and a Henyey–Greenstein scattering phase parameter $g$. It is true that the Henyey–Greenstein scattering pattern is totally arbitrary. It was picked for its rather beautiful mathematical properties, and it might be that the actual pattern is very different. For example, Adolf Witt et al. (1992) get $g=0.7$ in NGC 7023, and they may be completely correct; but it may also be that there is three or four times more backscattering with that same forward pattern than there is in the Henyey–Greenstein function, and that is just the way the grains are, in which case some of what I said just does not mean anything. So, ultimately one is going to have to understand what is going on, on the basis of observations over the galactic cap itself. Peebles' question concerning how thoroughly I hoped to map the sky is therefore very apropos because what you want to do is really see the scattered light; you should see it! If it is backscattered light, we should see it in spades, because the cirrus is up there, right? So we know where the dust is. And if we see this coincidence with the cirrus, there will be absolutely no question: it is backscattered light from dust. Now in connection with that, Alan Sandage made an observation published in $A.J.$ in 1976. From the ground he photographed a nebulosity about 50° north and interpreted it as starlight in the visible, scattered from interstellar dust. So this is a nice prototype area you would like to look at. Now, one of those four Voyager observations we made at the highest galactic latitudes was of that region, and we see absolutely nothing.

**C. Norman**: Can I just come back to a simple question of units and numbers? That is, there are other estimates of the background UV flux from the integrated light of quasars and also estimates by Carrie et al. from high velocity clouds, etc. Are these 300 photon units roughly equivalent to what they estimate?

**P. Jakobsen**: The 300 photon units that people like to use in the far UV are, in comparison with the result deduced from the proximity effect for local ionizing flux, about an order of magnitude higher. I'll get into that in my talk. Quasars can provide nothing to the extragalactic far-UV background. The point is that 300 photon units is an enormous flux, even compared to galaxies.

# ULTRAVIOLET BACKGROUND (THEORY)

Peter Jakobsen
Astrophysics Division
Space Science Department of ESA
ESTEC
NL-2200 AG Noordwijk
The Netherlands

## 1. INTRODUCTION

Many astronomical sources are capable of emitting UV light with photon energies in the range $h\nu \simeq 10-20$ eV through various thermal and non-thermal emission processes. The diffuse cosmic background radiation at ultraviolet $\lambda 1000 - 2500$ Å wavelengths is therefore of interest for a wide variety of topics. These include the study of the properties of interplanetary and interstellar dust grains, thermal line emission from the interstellar and intergalactic gas, the integrated light of galaxies, and radiative decay of exotic cosmological particles.

Reviews of the topic of the diffuse UV background have been given by Davidsen, Bowyer & Lampton (1974), Paresce & Jakobsen (1980), and most recently by Bowyer (1992) and Henry (1992). Not least thanks to the experimental efforts of the Berkeley group, the rather confusing observational situation surrounding the far-ultraviolet background has lately become much clearer. There is today good evidence that the diffuse ultraviolet background is largely dominated by dust scattering and other interstellar emission processes occurring within the Milky Way (Martin & Bowyer 1990; Martin, Hurwitz & Bowyer 1990; Hurwitz, Bowyer & Martin 1991). Although there remains some disagreement as to the relative strengths of the various interstellar contributors (cf. Henry, this conference), there is general consensus that the intensity of any quasi-isotropic and therefore possibly *extragalactic* component to the background in the $\lambda 1300 - 2000$ Å range is approximately $I_\lambda \approx 100$ photons s$^{-1}$ cm$^{-2}$ sr$^{-1}$ Å$^{-1}$, give or take a factor of $\sim 2 - 3$.

There are at present three candidate extragalactic sources of UV radiation that could conceivably produce a diffuse background flux of this intensity: i) diffuse thermal emission from the intergalactic medium; ii) the integrated UV light of galaxies and quasars; and iii) radiative decay of massive neutrinos. In what follows, each of these emission sources is discussed in turn.

## 2. DIFFUSE UV EMISSION FROM THE INTERGALACTIC MEDIUM

Historically, one of the driving motivations behind attempts to measure the diffuse background radiation at far-ultraviolet wavelengths has been the realization that such observations could potentially reveal the existence of a cosmologically significant ($\Omega_b \approx 1$), "lukewarm" ($10^3$ K$< T < 10^6$ K) intergalactic medium (cf. Kurt & Sunyaev 1967; Davidsen, Bowyer & Lampton 1974; Paresce & Jakobsen 1980).

If a baryonic intergalactic medium (IGM) does exist, standard Big Bang nucleosynthesis calculations predict that it must consist primarily of a mixture of $\simeq 90\%$ hydrogen and $\simeq 10\%$ helium atoms. The dominant emission from such a mixture at temperatures between $T \simeq 10^3$ K and $T \simeq 10^6$ K is recombination and collisionally excited line radiation in the HI and HeII Ly$\alpha$ lines at $\lambda_l = 1216$Å and $\lambda_l = 304$Å, respectively. This line radiation, smeared by the redshift, will give rise to a diffuse background at wavelengths $\lambda \geq \lambda_l$ of intensity †

$$I_\lambda(\lambda_0) = \left[\frac{c}{H_0}\right] \frac{\epsilon_l(\tilde{z})}{4\pi} \frac{1}{\lambda_l} (1+\tilde{z})^{-5} (1+\Omega\tilde{z})^{-\frac{1}{2}} \qquad (1)$$

where $\lambda_0 \geq \lambda_l$ is the observed wavelength, $H_0$ is the Hubble constant and $\epsilon_l(z)$ is the line volume emissivity (in units of photons s$^{-1}$ cm$^{-3}$) evaluated at the appropriate redshift, $\tilde{z} = \lambda_0/\lambda_l - 1$ (see the Appendix for a derivation of this expression). In the case of a smooth IGM emitting through recombination or collisional excitation, where the emission goes as density squared, the line emissivity can be written

$$\epsilon_l(z) = n_H^{0\,2} (1+z)^6 \gamma_l(T) \qquad (2)$$

where $n_H^0 = 7.8 \times 10^{-6} \Omega_b h^2$ cm$^{-3}$ is the IGM hydrogen density at $z = 0$, $\Omega_b$ is the baryonic IGM contribution to the cosmological density parameter, $H_0 = 100 h$ km s$^{-1}$ Mpc$^{-1}$, and $\gamma_l(T)$ is a suitably normalized emission coefficient.

Figure 1 shows the run of $\gamma_l(T)$ as a function of temperature for a cosmological mixture of hydrogen and helium in collisional and thermal equilibrium. It is seen that collisionally excited emission in the HI and HeII Ly$\alpha$ lines is especially intense near the two "thermostat" temperatures $T \simeq 2 \times 10^4$ K and $T \simeq 8 \times 10^4$ K, i.e. at the temperatures at which the dominant ionization states change from HI to HII and HeII to HeIII, respectively. If, following recombination, the IGM was re-heated and re-ionized by dissipative processes, such as shock heating, the gas would have had to pass through these two temperatures. Depending on the redshift and duration of the re-heating process, the IGM emission could give rise to observable redshifted Ly$\alpha$ and HeII $\lambda$304Å spectral signatures in the ultraviolet background. Several detailed IGM models giving rise to such features can be found in the literature (Weymann 1967; Sherman 1979, 1982).

Of course, the intergalactic medium is not only constrained by limits on its possible *emission*, but also by its possible *absorption*. By far the most stringent limit on the IGM comes from the classical Gunn & Peterson (1965) test, which severely constrains the density of intergalactic neutral hydrogen at high redshift, $n_{HI}(z)$, from the observed lack of an intense redshift-smeared Ly$\alpha$ absorption trough seen just shortward of emitted Ly$\alpha$ in the spectra of high redshift quasars:

$$n_{HI}(z) = \frac{\tau(z)}{\sigma_l} \left[\frac{c}{H_0}\right]^{-1} (1+z)(1+\Omega z)^{\frac{1}{2}} \qquad (3)$$

---

† Throughout this paper $I_\lambda$ refers to the specific intensity of the background expressed in the units preferred by the observers; namely photons s$^{-1}$ cm$^{-2}$ sr$^{-1}$ Å$^{-1}$.

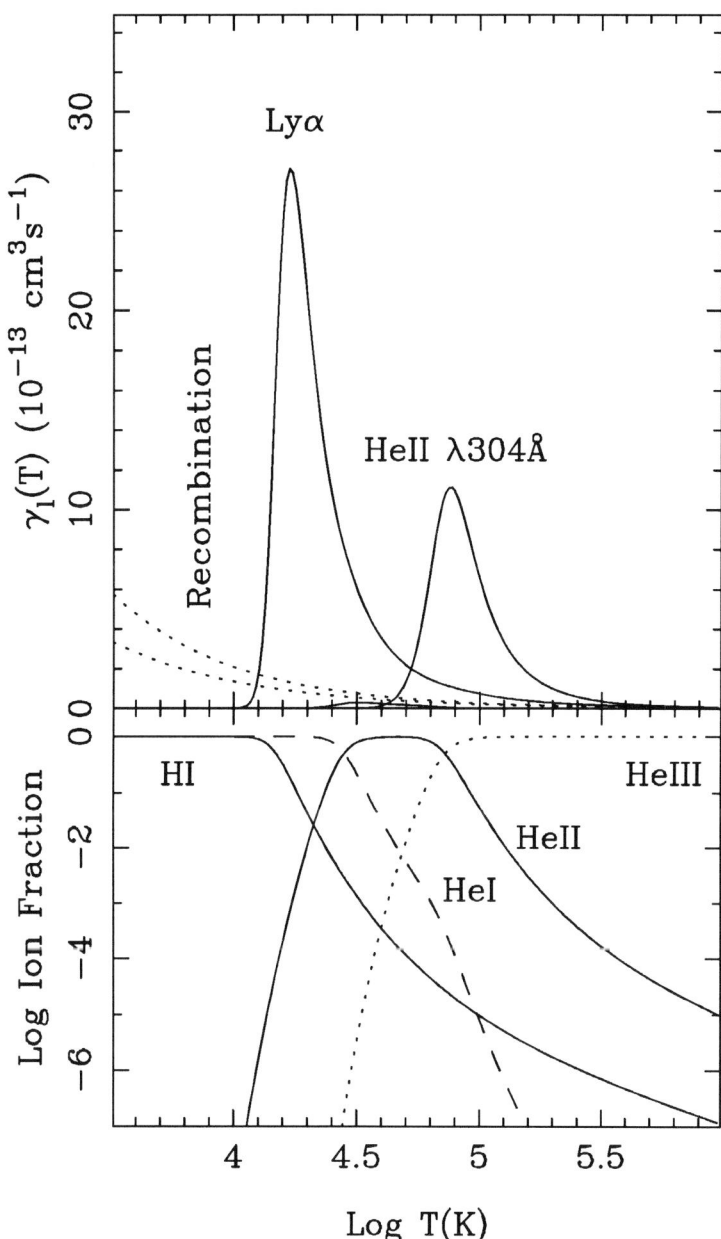

**Figure 1.** *Lower frame: Ionization structure of a cosmological mixture of hydrogen and helium in collisional equilibrium as a function of temperature. Upper frame: Corresponding emissivity per hydrogen atom of collisionally excited HI and HeII Lyα emission. The dotted lines show the equivalent line emission due to recombination in a fully photoionized gas. Emission from HeI is negligible compared to that of HI and HeII.*

In this expression, $\tau(z)$ is the optical depth of the absorption and $\sigma_l = \lambda_l \frac{\pi e^2}{m_e c^2} f_{ij} = 4.5 \times 10^{-18}$ cm$^2$ is the integrated Ly$\alpha$ absorption cross section. Both ground-based (Steidel & Sargent 1987) and *IUE* observations (Kinney et al. 1991) of the spectra of quasars show no signs of absorption troughs over the redshift range $0 < z < 4$ at the $\tau(z) \lesssim 0.1$ level. The corresponding upper limit on the intergalactic neutral hydrogen density is of order $n_{HI}(z) \lesssim 10^{-12}$ cm$^{-3}$. This stringent limit on the residual neutral hydrogen component of the IGM (and its HeI equivalent; Green et al. 1980; Reimers et al. 1989, 1992; Tripp, Green & Bechtold 1990; Beaver et al. 1991) implies that the IGM—if it exists at all—must be highly ionized and have a temperature greater than $T \approx 10^5$ K (compare lower frame of Figure 1).

Alternatively, if photoionization, due for example to the integrated ionizing flux of quasars, is the process responsible for re-ionizing the IGM, then the IGM will still emit primarily in the HI and HeII Ly$\alpha$ lines, albeit somewhat less efficiently (Figure 1). On the other hand, since the Gunn-Peterson test suggests that the IGM is transparent, the redshifted ionizing radiation source itself should be directly observable in the ultraviolet background. Absolute measurements of the ultraviolet background intensity can therefore in principle be used to derive the maximum IGM density that can be photoionized by the observed flux to the level required by the Gunn-Peterson test. However, the available observational limits on the extragalactic UV background intensity are generally far brighter than anticipated for realistic intergalactic fluxes and the corresponding constraints placed on the IGM density from the UV background observations are therefore not particularly confining, i.e. typically $\Omega_b^2 h^3 \lesssim 10^2$.

The dramatic developments that have taken place during the last decade in the new field of quasar absorption lines—which can be regarded as the "high resolution" refinement of the classical Gunn-Peterson test—call for the above considerations to be revisited. The discovery of the dense "Lyman forest" of weak Ly$\alpha$ absorption lines seen in the spectra of all quasars has revealed the existence of a highly clumped component of the IGM consisting of an abundant and evolving population of intergalactic clouds of possibly primordial gas. Surveys of the more massive metal-containing absorption systems associated with galaxies have also led to new understandings concerning the transparency of the universe in the ultraviolet out to large redshift.

## 2.1 Quasar Absorption Lines in Brief

In the following, a few key results of quasar absorption line studies of particular relevance for the topic of the diffuse ultraviolet background are briefly highlighted. A series of excellent and more comprehensive reviews that do the far-reaching topic far better justice can be found in the compilation of Blades, Turnshek & Norman (1988)

If classified according to their HI column density, there are two classes of intervening quasar absorption line systems: the numerous "Lyman forest" systems, whose column densities fall in the range $N_{HI} \simeq 10^{13} - 10^{17}$ cm$^{-2}$ and the scarcer, but denser, "Lyman limit" systems having column densities $N_{HI} \simeq 10^{17} - 10^{22}$ cm$^{-2}$. Lyman limit systems nearly always show matching absorption from heavy elements, and are therefore thought to be associated with the gaseous halos of galaxies. The Lyman forest systems, on the other hand, show little or no evidence for heavy elements and are therefore believed to be due to intergalactic clouds of possibly primordial material.

The Lyman forest systems are extremely numerous and evolve rapidly with redshift. Their line-of-sight density evolution is usually parameterized in the form (Sargent et al.

1980; Murdoch et al. 1986; Hunstead 1988)

$$E\left[\frac{dn}{dz}\right] = N(z) = A(1+z)^\gamma \qquad (4)$$

with $A \simeq 10$ and $\gamma \simeq 2-3$. Lyman limit absorbers are roughly ten times scarcer ($A \simeq 1$) and evolve less rapidly ($\gamma \simeq 1$) than Lyman forest systems (Tytler 1982; Bechtold et al. 1984; Lanzetta 1988; Sargent, Steidel & Boksenberg 1989; Bahcall et al. 1993). The column density spectrum of both classes of absorber is approximately a power law, $dP/dN_H \propto N_H^{-s}$, with index $s \simeq 1.2 - 1.6$ (Tytler 1988). The fact that the detailed statistics of the various types of HI containing quasar absorption systems are now reasonably well known permits a re-assessment of the question of the overall transparency of the UV universe out to high redshift (Section 2.3). These results have important implications for the interpretation of the UV background.

Of special relevance for the topic at hand is that the Lyman forest clouds are believed to be kept highly photoionized by a metagalactic flux of ionizing radiation. One key piece of evidence for this is found in the so-called "proximity effect" (Carswell et al. 1982; Murdoch et al. 1986; Bajtlik, Duncan & Ostriker 1988; Lu, Wolfe & Turnshek 1991). As the emission redshift is approached, the Lyman forest absorbers in a given quasar show a gradual under-density of absorbers with respect to the global density given by equation (4). This effect is interpreted as being caused by the radiation field of the background quasar enhancing the total ionizing flux above the metagalactic level. Since the quasar flux can be estimated from the magnitude and spectrum of the quasar, an estimate of the background metagalactic ionizing background intensity can be derived from the measured contrast of the proximity effect. Through this technique one infers that the intergalactic ionizing background in the redshift range $1.7 \lesssim z \lesssim 3.8$ is consistent with an $I_\nu \propto \nu^{-\alpha}$, $\alpha \simeq 0.5$ power law with intensity at the Lyman limit of order $I_{\nu_H} \approx 10^{-21}$ ergs s$^{-1}$ cm$^{-2}$ sr$^{-1}$ Hz$^{-1}$.

The origin of this ionizing flux is a topic of some debate (Bechtold et al. 1987; Miralda-Escudé & Ostriker 1990, 1992; Madau 1992). One obvious candidate is the integrated flux from quasars. However, only by pushing current uncertainties in our knowledge of the quasar luminosity function and evolution can the intensity required by the proximity effect barely be approached (Bajtlik et al. 1988; Lu et al. 1991; Meiksin & Madua 1993). Moreover, taking into account absorption from the Lyman forest and Lyman limit systems themselves (Section 2.3), worsens the discrepancy further by a factor of 4–5 (Figure 2). Several authors (Bechtold et al. 1987; Miralda-Escudé & Ostriker 1990, 1992; Songaila, Cowie & Lilly 1990) have suggested that an alternative candidate may be radiation from primordial galaxies. Regardless of its origin, this estimate of the ionizing flux at high redshift has important implications for the possible contributions to the extragalactic UV background from a photoionized IGM.

A final relevant constraint stemming indirectly from the quasar absorption line studies concerns the information on the properties any pervasive IGM gas surrounding the Lyman forest clouds that result from considerations of Lyman forest cloud confinement and survival (Ostriker & Ikeuchi 1983; Ikeuchi & Ostriker 1986). The classical Gunn-Peterson test limits, combined with constraints on cloud/IGM pressure and cloud lifetimes to evaporation, generally constrain the ambient IGM to have at density corresponding to $\Omega_b \approx 10^{-1}$ and a temperature $T \approx 10^5$ K. These constraints also translate into limits on possible IGM emission contributions to the UV background.

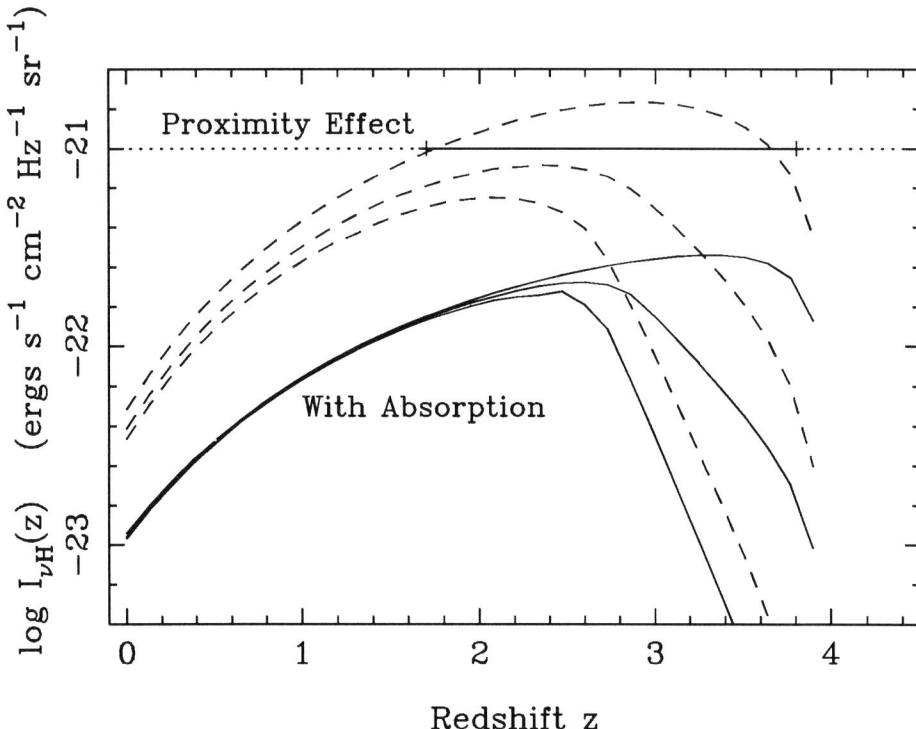

**Figure 2.** *Predicted integrated ionizing background due to quasars for several quasar evolution models. The upper dashed curves show the integrated background intensity at the Lyman limit as a function of redshift according to the model of Bajtlik, Duncan & Ostriker (1988). The lower full curves show the same fluxes attenuated by the accumulated Lyman continuum absorption from the Lyman forest and Lyman limit classes of quasar absorption line systems (see Section 2.4). The background level required to explain the proximity effect is also indicated.*

### 2.2 Redshifted Lyman Alpha Emission from the IGM Revisited

Diffuse emission in Ly$\alpha$ $\lambda 1216$Å from the IGM at redshifts $0 \lesssim z \lesssim 0.7$ will give rise to diffuse background radiation at far-UV wavelengths $\lambda\lambda 1200 - 2000$ Å. Two emission processes could produce such diffuse Ly$\alpha$ radiation: radiative recombination in case of a photoionized IGM, and collisional excitation in the case of a shock-heated IGM.

As mentioned in Section 2.1, the Lyman forest clouds and the smooth ambient IGM are believed to be held highly photoionized by a metagalactic ionizing radiation field, presumably due to the integrated light of quasars and primeval galaxies. What is the possible contribution to the UV background from redshifted Ly$\alpha$ recombination emission from this gas?

Consider first the case of recombination radiation from a smooth photoionized IGM. The line emissivity to be inserted into equation (1) is (Osterbrock 1989)

$$\epsilon_l(z) = n_{HII}(z) n_e(z) \alpha_l(T) \tag{5}$$

where $\alpha_l \simeq 2 \times 10^{-13}$cm$^3$ s$^{-1}$ is the effective Ly$\alpha$ recombination coefficient, and $n_{HII}$ and $n_e$ are the proton and electron density in the gas. With an $I_\nu \propto \nu^{-\alpha}$ ionizing background of average intensity $I_{\nu_H}(z)$ at the Lyman limit, the equation for hydrogen photoionization equilibrium is

$$n_{HII}(z)n_e(z)\alpha_H = \frac{2}{\hbar}I_{\nu_H}(z)\sigma_H^0 \frac{1}{(\alpha+3)}n_{HI}(z) \tag{6}$$

where $\alpha_{H_2} \simeq 3 \times 10^{-13}$cm$^3$ s$^{-1}$ is the total recombination coefficient and $\sigma_H^0 = 6.3 \times 10^{-18}$ cm$^2$ is the HI cross section at the Lyman limit. Combining equations (5) and (6) with the expression (3) for $n_{HI}$ from the Gunn-Peterson test and inserting into equation (1) leads to the following expression for the intensity of the resulting redshift-smeared Ly$\alpha$ background due to recombination emission

$$I_\lambda(\lambda_0) = \left[\frac{\lambda_H}{\lambda_l}\right]\left[\frac{\alpha_l}{\alpha_H}\right]\left[\frac{\sigma_H^0}{\sigma_l}\right]\frac{\tau(z)}{(\alpha+3)}\left[\frac{I_{\nu_H}(z)}{2\pi\hbar\lambda_H(1+z)^4}\right] \tag{7}$$

In this expression, the last factor in brackets is simply the intensity of the ionizing flux, $I_{\nu_H}(z)$ converted to suitable units and redshifted to zero redshift; that is, the background intensity that would be seen today if the ionizing flux propagated freely from redshift $z$. Since the product of the remaining factors in equation (7) is of order unity or less, this equation is merely reminding us that since $\alpha_l/\alpha_H \simeq 0.7$ Ly$\alpha$ photons are emitted per photoionization event, the intensity of the resulting redshift-smeared Ly$\alpha$ recombination background can never be greater than that of the ionizing input flux itself. This principle of photon conservation, of course, applies equally well in the clumped case. The equivalent expression for the redshift-smeared recombination emission from the Ly$\alpha$ forest clouds can be obtained through the substitution

$$\tau(z) \to E\left[\frac{dn}{dz}(z)\right]\sigma_l\langle N_{HI}\rangle(1+z) \tag{8}$$

Note that since equation (6) implicitly assumes optically thin conditions, equations (7) and (8) are as they stand only valid in that limit. However, the fact that the redshifted recombination line emission is bounded by the intensity of the ionizing input flux is inherent to the nature of the recombination process and true *regardless* of the amount and detailed spatial distribution of the matter being photoionized and the recombination line in question. In other words, since the photoionized intergalactic gas is simply acting as a simple photon down-converter, the task of estimating the possible contribution to the UV background from line emission from photoionized intergalactic matter boils down to the task of estimating the intensity of the ionizing input flux at high redshift, $I_{\nu_H}(z)$.

The proximity effect displayed by the Ly$\alpha$ forest implies that $I_{\nu_H} \approx 10^{-21}$ ergs s$^{-1}$ cm$^{-2}$ sr$^{-1}$ Hz$^{-1}$ within the $1.7 \lesssim z \lesssim 3.8$ redshift range that can be probed with ground-based telescopes. Judging from Figure 2, one anticipates an even lower ionizing background of intensity $I_{\nu_H}(z) \approx 10^{-22}$ ergs s$^{-1}$ cm$^{-2}$ sr$^{-1}$ Hz$^{-1}$ at the redshifts $z \lesssim 1$ of interest here. This expectation borne out observationally by recent observations of the proximity effect at low redshift carried out with *HST* (Kulkarni & Fall 1993) and by limits on the local extragalactic ionizing flux derived from H$\alpha$ observations of high-latitude and extragalactic HI clouds (Reynolds et al. 1986; Songaila, Bryant & Cowie 1989; Kutyrev & Reynolds 1989). This ionizing flux, if redshifted from $z \simeq 0.5$, corresponds to an equivalent far-UV background of $I_\lambda \simeq 0.4$ photons s$^{-1}$ cm$^{-2}$ sr$^{-1}$ Å$^{-1}$—an intensity an order of magnitude below the observational limits on a possible extragalactic contribution. It follows that redshifted Ly$\alpha$ recombination emission from the Lyman forest clouds or a smooth photoionized IGM is not likely to be a significant contributor to the far-UV background.

A similar conclusion can be reached in the alternative scenario of Ly$\alpha$ emission from a shock-heated IGM. In the case of collisionally excited emission from a smooth IGM component, the line emissivity can be written

$$\epsilon_l(z) = n_{HI}(z) n_e(z) \beta_l(T) \tag{9}$$

where $\beta_l(T)$ is the Ly$\alpha$ collisional excitation rate, and $n_{HI}$ and $n_e$ are the neutral hydrogen and electron densities in the IGM. The HI density is constrained observationally by the Gunn-Peterson limit, equation (3), while the electron density is constrained by the total baryonic density of the nearly fully ionized IGM. Combining, equations (1), (3) and (9), and introducing $n_e(z) = n_e^0 (1+z)^3$ where $n_e^0 \simeq 1.2 n_H^0 = 9.3 \times 10^{-6} \Omega_b h^2$ cm$^{-3}$ is the present epoch electron density, one obtains the following expression for the redshift-smeared Ly$\alpha$ background

$$I_\lambda(\lambda_0) = \frac{\tau(z)}{\sigma_l} \frac{n_e^0}{4\pi} \frac{\beta_l(T)}{\lambda_l} (1+z)^{-1} \tag{10}$$

For an ambient IGM temperature of $T \approx 10^5$ K as inferred from considerations of quasar absorption line cloud survival, the Ly$\alpha$ collisional excitation rate is $\beta_l \simeq 4 \times 10^{-9}$ cm$^3$ s$^{-1}$. This value, together with the Gunn-Peterson limit of $\tau(z) < 0.1$, yields a predicted smeared Ly$\alpha$ intensity of $I_\lambda \lesssim 4 \times 10^{-2} \Omega_b h^2$ photons s$^{-1}$ cm$^{-2}$ sr$^{-1}$ Å$^{-1}$ in the far-UV. It follows that collisionally excited Ly$\alpha$ emission from a smooth IGM component at $0 \lesssim z \lesssim 0.7$ is a negligible contributor to the far-UV background for any reasonable value of $\Omega_b h^2$. The reason for this firm conclusion is simply that it takes intergalactic HI atoms to produce collisionally excited Ly$\alpha$ emission, and $n_{HI}(z)$ is severely constrained by the Gunn-Peterson test.

Similar conclusions are reached concerning collisionally excited emission from the Lyman forest clouds. The equivalent expression to equation (10) for the redshift-smeared Ly$\alpha$ background in the clumped case can be obtained through the substitutions (8) and $n_e(z) \to n_e$. This leads to

$$I_\lambda(\lambda_0) = E\left[\frac{dn}{dz}(z)\right] \langle N_{HI} \rangle \frac{n_e}{4\pi} \frac{\beta_l(T)}{\lambda_l} (1+z)^{-3} \tag{11}$$

where $n_e$ is now the *in situ* electron density in the Lyman forest clouds. Based on studies of correlated Lyman forest absorption in quasar pairs combined with considerations of reasonable ionization levels (Carswell 1988; Sargent 1988), the gas density in Lyman forest absorbers is believed to be of order $n_e \approx 10^{-3}$ cm$^{-3}$. The observed line widths of the Lyman forest systems limit their temperatures to $T \lesssim 6 \times 10^4$ K, in which case $\beta_l \lesssim 3 \times 10^{-10}$ cm$^3$ s$^{-1}$. With these numbers and the values $E[dn/dz(z)] \simeq 30$ and $\langle N_{HI} \rangle \simeq 10^{15}$ cm$^{-2}$, appropriate to the Lyman forest at $z \simeq 0.3$, equation (11) yields $I_\lambda \lesssim 0.3$ photons s$^{-1}$ cm$^{-2}$ sr$^{-1}$ Å$^{-1}$. Again, we conclude that collisionally excited Ly$\alpha$ emission from the Lyman forest cannot a significant contributor to the far-UV background.

## 2.3 Redshifted HeII $\lambda 304$Å Emission from the IGM Revisited

As illustrated in Figure 1, the second most important emission line from a luke warm photoionized or shock-heated intergalactic primordial plasma is HeII Ly$\alpha$ emission at

$\lambda 304$Å. The discussion of the possible far-UV background contribution due to this source from very high redshifts ($3 \lesssim z \lesssim 5$) is slightly more complicated with respect to that of HI Ly$\alpha$ for several reasons. For one very little is currently known the intergalactic abundance of the HeII ion, since the HeII $\lambda 304$Å equivalent of the Gunn-Peterson test has yet to be carried out in the far-UV with *HST* or *Lyman/FUSE* (cf. Jakobsen et al. 1993). The expectation is that if the Lyman forest clouds and an ambient IGM are indeed photoionized by an $I_\nu \propto \nu^{-0.5}$ power law, then the HeII ion should be an order of magnitude more abundant than HI, in which case the HeII Gunn-Peterson effect should be extremely strong (and HeI absorption very weak). A second important difference with respect to HI Ly$\alpha$ is that HeII $\lambda 304$Å emission falls below the photoionization edge of neutral hydrogen and is therefore subject to absorption by intergalactic HI in the Lyman forest and especially the Lyman limit classes of quasar absorption systems. This last topic is addressed in detail in the following section. In spite of these complications, it is still possible to draw several reasonably firm conclusions concerning the possible contribution of HeII $\lambda 304$Å emission to the far-UV background.

In the previous section, the intensity of redshifted far-UV Ly$\alpha$ recombination radiation from photoionized intergalactic gas was constrained on the basis of estimates on the metagalactic ionizing background derived from the proximity effect and more local observations of diffuse H$\alpha$ emission from high-latitude and extragalactic HI clouds. The fundamental constraint expressed by equation (7), namely that the redshifted line background from a photoionized IGM can never be greater than that of the input ionizing flux, obviously applies equally well in the case of redshifted recombination HeII $\lambda 304$Å emission. In particular, with a flat $I_\nu \propto \nu^{-0.5}$ spectrum for the metagalactic ionizing background, the intensity of the ionizing flux at the HeII ionization edge at $\lambda 228$Å is of the same order of magnitude as the flux at the Lyman limit at $\lambda 912$Å of $I_{\nu_H}(z) \approx 10^{-21}$ ergs s$^{-1}$ cm$^{-2}$ sr$^{-1}$ Hz$^{-1}$ derived from the proximity effect at $1.7 \lesssim z \lesssim 3.8$. This intensity, if redshifted from $z \approx 4$ will give rise to a far-UV background of intensity $I_\lambda \approx 1$ photons s$^{-1}$ cm$^{-2}$ sr$^{-1}$ Å$^{-1}$. Since this background limit falls far below the observational limits—even without including the effects of intervening Lyman continuum absorption—redshifted recombination HeII $\lambda 304$Å emission from photoionized intergalactic gas can also be ruled out as a significant source of far-UV background radiation.

Collisionally excited HeII $\lambda 304$Å from a shock-heated IGM on the other hand is not quite as easily dismissed. Since the amount of HeII present in intergalactic space has not yet been measured by observations of the HeII version of the Gunn-Peterson test and the anticipated "helium forest" matching that seen in Ly$\alpha$, the HeII equivalents of equations (10) and (11) cannot be used to bracket the possible background contribution from this source. Instead, we are forced back to the more theoretical predictions described by equation (2). As shown in Figure 1, the net HeII $\lambda 304$Å emissivity per HI atom peaks at $\gamma_l \simeq 1.2 \times 10^{-12}$ cm$^3$ s$^{-1}$ at a temperature $T \simeq 8 \times 10^4$ K. Inserting this maximum emissivity into equations (1) and (2) yields a predicted far-UV background intensity at $\lambda 1600$Å of $I_\lambda \simeq 400 \Omega_b^2 h^3$ photons s$^{-1}$ cm$^{-2}$ sr$^{-1}$ Å$^{-1}$. Hence depending on the values of $\Omega_b$ and $H_0$, redshifted collisionally excited HeII $\lambda 304$Å radiation from $z \simeq 3-5$ could in principle yield a significant far-UV background flux. On the other hand, considerations of the survival of quasar absorption line clouds and primordial nucleosynthesis both point toward $\Omega_b \lesssim 0.1$, in which case the HeII line flux is insignificant. In any event, as discussed in the following section, the census of absorbing HI gas present in the Universe represented by the statistics of quasar absorption lines implies that the far-UV Universe is opaque in the Lyman continuum out to high redshift. This absorption will attenuate any diffuse HeII λ304Å radiation emitted at $z \gtrsim 3$ by about two orders of magnitude, thereby reducing even the most optimistic HeII background flux to an unobservable level.

## 2.4 Accumulated Lyman Continuum Opacity of the Universe

Any contribution to the far-UV background at observed wavelength $\lambda_0$ emitted originally at a wavelength below the Lyman limit, $\lambda_H = 912$Å, at a redshift $z_e \geq \lambda_0/\lambda_H - 1$ will be subject to photoelectric absorption by neutral hydrogen encountered along at least part of its path. Although it has been known for some time that the classical Gunn-Peterson test demonstrates that the Lyman continuum opacity of any *smoothly* distributed IGM is negligible, it has only recently been fully appreciated that the statistics of quasar absorption lines imply the that the accumulated absorption out to moderate and high redshift from the *clumped* component is quite substantial.

The character and magnitude of the accumulated Lyman continuum absorption from the Lyman forest and Lyman limit classes of quasar absorption lines has been discussed in detail by Møller & Jakobsen (1990). The general expression for the average transmission through a clumpy medium experienced by a photon emitted at wavelength, $z_e$, and received at wavelength, $\lambda_0$, is (Paresce, Bowyer & McKee 1980)

$$E[q(\lambda_0, z_e)] = \exp(-\int_0^{z_e} N(z)(1 - \langle q_c(\lambda_0, z)\rangle)dz) \tag{12}$$

where $\langle q_c(\lambda_0, z)\rangle = \langle \exp(-N_H \sigma_H(\lambda_0/(1+z)))\rangle$ is the average individual cloud transmission and $\sigma_H(\lambda)$ is the HI photoelectric cross section given by

$$\sigma_H(\lambda) \simeq \begin{cases} \sigma_H^0 \left(\frac{\lambda}{\lambda_H}\right)^3 & \text{if } \lambda \leq \lambda_H \\ 0 & \text{if } \lambda > \lambda_H \end{cases} \tag{13}$$

where $\sigma_H^0 = 6.3 \times 10^{-18}$ cm$^2$ is the photoionization cross section of neutral hydrogen at the Lyman limit.

As outlined in Section 2.1, the statistics of quasar absorption systems are today sufficiently well known to permit a reasonably accurate evaluation of equation (12). Figure 3 shows, as a function of wavelength, the resulting accumulated average residual transmission out to various redshifts from the combined total absorption due to the Lyman forest and Lyman limit systems. The characteristic "Lyman valley" shape of the accumulated absorption spectrum is caused by the interplay between the $\sigma \propto \lambda^3$ dependence of the HI photoelectric cross section and redshift evolution and pathlength effects.

The main point to be read from Figure 3 is that the anticipated net absorption out to high redshift in the ultraviolet is rather high. As an specific example, the lower panel of Figure 4 shows the average residual absorption at received wavelength $\lambda_0 \simeq 1600$Å as a function of emission redshift. From this figure it is seen that any HeII λ304Å radiation emitted at $z_e \simeq 4.3$ will be attenuated by a factor of order $\approx 10^{-2}$. This high opacity effectively implies that even if the IGM did go through a phase of intense HeII emission during re-heating, the resulting far-UV background radiation will in all likelihood remain forever hidden from our view.

It is important to stress that the accumulated intergalactic absorption given in Figures 3 and 4 refers to the *average* transmission. Since the dominant contributor to the opacity is the scarcer but optically thick Lyman limit systems, large fluctuations

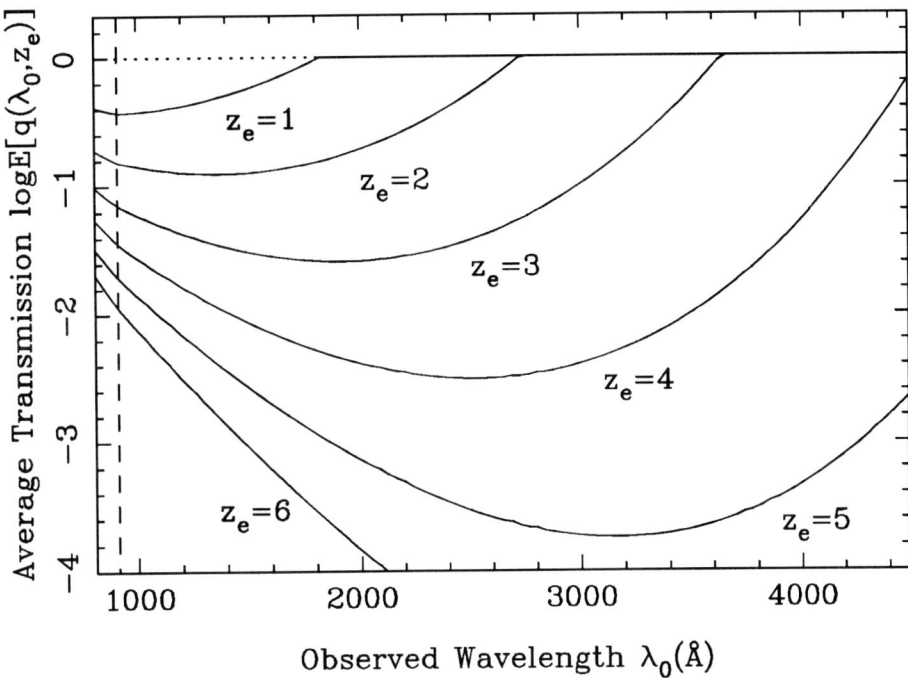

**Figure 3.** Average opacity of the UV universe out to high redshifts as a function of wavelength. The shown "Lyman valley" absorption spectra include the accumulated Lyman continuum opacity from both the Lyman forest and Lyman limit classes of quasar absorption systems. The same absorber parameters as used by Møller & Jakobsen (1990) are assumed.

around this average are expected along any given line of sight (i.e. in any given quasar spectrum). The magnitude of the fluctuations can be calculated from the expression for the second moment of the accumulated transmission

$$E[q^2(\lambda_0, z_e)] = \exp\left(-\int_0^{z_e} N(z)\left(1 - \langle q_c^2(\lambda_0, z)\rangle\right) dz\right) \quad (14)$$

The upper panel of Figure 4 shows the predicted relative transmission fluctuations, $(\delta q/q) = (E[q^2] - E[q]^2)^{\frac{1}{2}}/E[q]$, as a function of $z_e$, again for $\lambda_0 = 1600\text{Å}$. In the example of HeII Ly$\alpha$ emission from $z_e \simeq 4.3$ quoted above, the $1\sigma$ level fluctuations amount to a factor $\approx 5$. In other words, a characteristic signature of any contributor to the far-UV background originating at high redshift should be an extremely patchy background component.

This leads to the interesting prospect of detecting or constraining a high-$z$ component to the far-UV background—regardless of its origin—through measurements of background intensity fluctuations. In fact, from the observations of Martin and Bowyer (1989), it is known that the far-UV background at $\lambda_0 \simeq 1600\text{Å}$ is very smooth on angular scales of $\theta = 8'$ and larger: $(\delta I/I)_\theta \simeq 6\%$. If it is assumed that the background consists of the sum of a smooth local component and an attenuated distant high redshift component of average intensity $E[q(z)]I_z$, the fractional contribution to the total average emission from the distant component, $\phi_z$, can be estimated from the observed

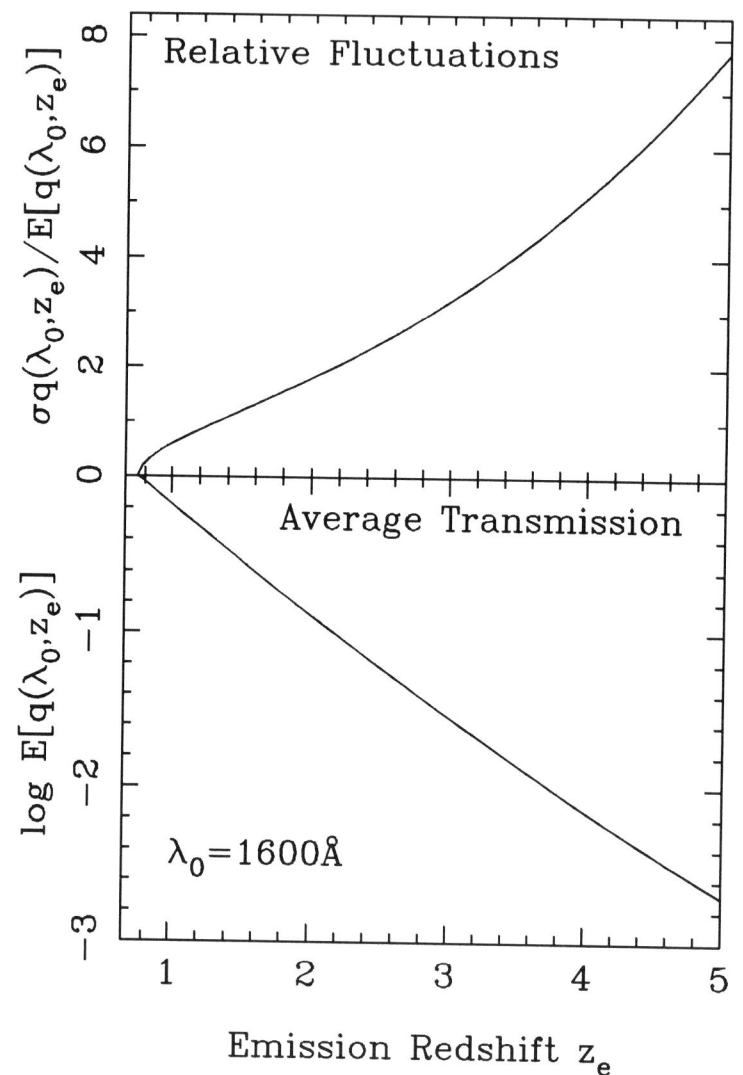

**Figure 4.** *Lower frame: Average transmission of the Universe at received wavelength $\lambda 1600\text{Å}$ as a function of emission redshift. Upper frame: Predicted relative fluctuations around the average transmission for an infinitely narrow pencil beam due to fluctuations in the composition and location of the absorbing HI clouds along the line of sight.*

dilution of the opacity fluctuations

$$\phi_z = \frac{E[q(z)]I_z}{I_{Total}} \simeq \left(\frac{\delta I}{I}\right)\left(\frac{\delta q}{q}\right)^{-1} \tag{15}$$

Taken at face value, the Martin and Bowyer (1989) fluctuation limit implies that less than $\phi_z \approx 6\%/5 \simeq 1\%$ of the far-UV background can originate from $z \simeq 4.3$. However, a slightly subtle point has been overlooked; namely that the $(\delta q/q)$ values in Figure 4 refer statistically to the absorption sampled along an infinitely narrow pencil beam (namely the line of sight to a quasar), whereas the UV background observations have been obtained with a finite $\theta = 8'$ beam size. Since the opacity is dominated by the Lyman limit absorbers, which are assumed to be associated with galaxy halos of, say, $D \approx 50$ kpc size, the absorption-generated background fluctuations on the sky are expected to be correlated only on very small scales of order $\theta_c \approx cD/H_0 \approx 10''$. Such small scale fluctuations would only appear in the finite beam measurements of Martin and Bowyer (1989) diluted by a factor of order $\theta/\theta_c \approx 50$. The upper limit on $\phi_z$ derived above therefore has to be relaxed by the same amount to $\phi_z \lesssim 50\%$. Given the severe practical problems of measuring UV background fluctuations on such small scales as $\theta \approx 10''$, it is unlikely that this constraint will be tightened much further in the foreseeable future. On the other hand, the above considerations do serve to demonstrate that, because of the very small size of the anticipated absorption coherence patch on the sky, use of the average attenuation given by equation (12) is well justified when dealing with the radiative transfer of diffuse background light.

## 3. THE INTEGRATED UV LIGHT OF GALAXIES AND QUASARS

As discussed in detail by others at this conference, the absolute brightness of the extragalactic sky at visible and infrared wavelengths carries important information on the history of star formation and galaxy evolution throughout the age of the Universe. As first emphasized by Tinsley (1973), the diffuse background at the shorter UV wavelengths is also an important element of this modern day version of "Olbers' paradox"—but with several key differences with respect to the situation in the adjacent spectral regions.

For one, the observed integrated far-UV spectre of the different classes of galaxies are not particularly well modelled or understood at present. This is especially true of the UV emission from ellipticals and the bulges of spirals, which is believed reflect complex stages of late stellar evolution (cf. Burnstein et al. 1988; King et al. 1993 and references therein). However, the global galaxy luminosity function at UV wavelengths is almost entirely dominated by the emission from massive O and B stars contained in the star-forming regions of spirals and irregulars. Consequently, in the UV late type galaxies most likely account for close to 90% of the local galaxy luminosity density and therefore dominate the integrated galaxy background (cf. Milliard et al. 1992 for a recent discussion).

Another important difference with respect to the case at visible and longer wavelengths is that the integrated UV light of galaxies is accumulated over a relatively modest cosmological pathlength. Late type galaxies are generally surrounded by large halos of neutral hydrogen which permit little or no radiation to escape below the Lyman limit at $\lambda \leq 912$ Å. At an observed wavelength of $\lambda 1500$ Å, the Lyman limit is reached at $z = 0.64$, corresponding to a look-back time of $\Delta t \simeq 3h^{-1}$ Gyr. This modest look-back time, combined with the short life times of main sequence OB stars, leads to the integrated UV light of galaxies being primarily a measure of the level of on-going star formation in the relatively local Universe rather than a measure of its total accumulated stellar content, as is the case at longer visible and infrared wavelengths.

Three rather different observational and theoretical approaches have so far been employed in assessing the possible contribution of galaxies to the extragalactic UV background. Although all three methods have considerable uncertainties associated with them, they nevertheless give answers that agree to better than a factor of 3-4.

The integrated light of galaxies is almost certainly that of the three sources of diffuse extragalactic UV background light considered in this review that is most likely to generate an observationally significant flux.

## 3.1 Integration of Theoretical Galaxy Evolution Models

Following the pioneering work of Tinsley (1972), several groups have in recent years developed very elaborate and physically self-consistent models for galaxy evolution based on stellar formation and evolution theory (e.g. Guiderdoni & Rocca-Volmerange 1991; Bruzual & Charlot 1993). Although the primary motivation for these models is to explain faint galaxy counts and colors observed in the visible, as emphasized by Tinsley (1973), the predicted integrated background spectrum provides an important observational constraint on the models.

The general expression for the background at received wavelength $\lambda_0$ due to the integrated light of galaxies is

$$I_\lambda(\lambda_0) = \frac{1}{4\pi} \left[\frac{c}{H_0}\right] \int_0^\infty \epsilon_\lambda^0(\lambda(z), z) \, (1+z)^{-3} (1+\Omega z)^{-\frac{1}{2}} dz \qquad (16)$$

where $\lambda(z) = \lambda_0/(1+z)$ and $\epsilon_\lambda^0(\lambda, z)$ is the total (co-moving) volume emissivity due to all classes of galaxies at all wavelengths $\lambda < \lambda_0$ and all redshifts $z \geq 0$. The function $\epsilon_\lambda^0(\lambda, z)$ encapsulates the luminosity and number evolution of all galaxies in the Universe at all wavelengths at all epochs. Needless to say, this function is not very well determined at present.

One complication is that the predicted UV fluxes for models of late type star-forming galaxies are less forgiving than at visible wavelengths in the sense that the emergent UV flux is very sensitive to not only the details of the assumed star formation history (see Fig. 4 of Bruzual & Charlot 1993 for a nice illustration), but also to the amount, detailed geometry and properties of any absorbing dust present in the galaxy (e.g. Bruzual, Magris, & Calvet 1988). Given these complications (the latter of which is usually ignored), and the fact that the faint optical counts still present a considerable challenge to the models (Koo & Kron 1992), too much faith cannot presently be placed in their extrapolation to far-UV wavelengths.

A recent discussion with emphasis on the ultraviolet has been given by Martin, Hurwitz & Bowyer (1991), who show that the current evolution models predict a rather wide range of spectral shapes for the background. Nonetheless, the predicted far-UV intensities generally span the $I_\lambda \simeq 40 - 240$ photons s$^{-1}$ cm$^{-2}$ sr$^{-1}$ Å$^{-1}$ range—suggesting that the integrated UV light of galaxies should be within reach observationally.

## 3.2 UV Background Fluctuation Measurements

An almost heroic observational attempt to measure the galaxy contribution to the UV background has been undertaken by Martin & Bowyer (1989), who by means of a sounding rocket experiment searched for the small-scale fluctuations in the far-UV background expected from the integrated light of galaxies. This technique, which was first pioneered by Schectman (1973; 1974) at visible wavelengths, consists of fitting the measured power spectrum of the UV background to that expected from galaxies calculated on the basis of observed visible light correlation functions and assumed models for galaxy spectral evolution. According to Martin & Bowyer (1989) the observed radial power spectrum of the UV background on angular scales of 6–12′ is consistent with the $P \propto \eta^{-1.2}$ power law signature anticipated from galaxies (where $\eta$ is the inverse angular scale). However, as mentioned in Section 2.3, since the observed background fluctuations in the $1.5 \times 3.0°$ area sampled by Martin & Bowyer (1989) were of very low amplitude ($\delta I/I \approx 5\%$) compared to the average background of $<I_\lambda> \simeq 220$ photons s$^{-1}$ cm$^{-2}$ sr$^{-1}$ Å$^{-1}$, the amplitude of the power spectrum is also small: $P(\eta = 200 \text{ rad}^{-1}) = 3 \pm 1\ 10^{-4}$(photons s$^{-1}$ cm$^{-2}$ sr$^{-1}$ Å$^{-1}$)$^2$ rad$^{-2}$. Consequently, Martin & Bowyer (1989) concluded that the integrated light of galaxies can at most contribute $\approx 20\%$ of the total (galactic and extragalactic) background observed, corresponding to an intensity of $I_\lambda \simeq 40 \pm 13$ photons s$^{-1}$ cm$^{-2}$ sr$^{-1}$ Å$^{-1}$.

Although subsequent analysis has questioned the validity of the fluctuation results (due to contamination from UV starlight scattered off the *IRAS* cirrus; cf. Sasseen *et al.* 1993), it is nonetheless remarkable that this intensity estimate is comparable to those obtained from the theoretical galaxy evolution models and the direct UV galaxy counts.

### 3.3 Extrapolation of Ultraviolet Galaxy Counts

The most convincing and direct demonstration that galaxies must provide a significant contribution to the extragalactic UV background has recently been given by Armand, Milliard & Deharveng (1993). These authors base their analysis on observed UV ($\simeq 2000$ Å) galaxy counts obtained with a balloon borne UV telescope (Milliard *et al.* 1992). The balloon counts are complete down to a UV magnitude of $m \simeq 18.5$, which by itself yields an integrated galaxy background of $I_\lambda \simeq 30$ photons s$^{-1}$ cm$^{-2}$ sr$^{-1}$ Å$^{-1}$. Armand *et al.* extrapolate this resolved portion of the background to fainter magnitudes by use of the Guiderdoni & Rocca-Volmerange (1991) galaxy evolution models. Because of the various observational and theoretical uncertainties, the total background can only be predicted with certainty to lie in the range $I_\lambda \simeq 40\text{--}130$ photons s$^{-1}$ cm$^{-2}$ sr$^{-1}$ Å$^{-1}$. Nonetheless, this flux is in good agreement with those obtained through the other less direct approaches above, and clearly demonstrate that galaxies must be a significant source of extragalactic UV background radiation.

### 3.4 The Integrated UV Light of Quasars and AGNs

As opposed to the situation at higher X-ray energies, quasars and active galactic nuclei play only a marginal role in the case of the background in the UV. This conclusion follows implicitly from a point alluded to in Section 2.2, namely that current models for quasar and AGN evolution have difficulties explaining the level of meta-galactic ionizing background deduced from the proximity effect displayed by the Lyman forest absorption lines, and that this intensity is to begin with very faint compared to a nominal extragalactic UV background of $I_\lambda \approx 100$ photons s$^{-1}$ cm$^{-2}$ sr$^{-1}$ Å$^{-1}$.

Estimates of the quasar contribution to the UV background obtained through integration of models for quasar evolution are sensitive to the assumptions made concerning the average far- and extreme-UV quasar spectrum, the quasar turn-on epoch, and the intervening absorption. As an illustration, Figure 5 show the anticipated background spectra calculated for the three evolution models adopted by Bajtlik *et al.* (1988), and assuming an average $F_\nu \propto \nu^{-0.5}$ quasar continuum spectrum. If the intervening Lyman continuum absorption is ignored, the resulting integrated background spectrum is a power law with the same spectral index as that assumed for the quasars. Including the effects of absorption lowers the predicted fluxes further by a factor ∼ 3, and decreases the sensitivity to evolution effects by quenching the background contributions from higher redshifts emitted in the Lyman continuum. The intensity of the integrated quasar flux is predicted to be $I_\lambda \lesssim 10$ photons s$^{-1}$ cm$^{-2}$ sr$^{-1}$ Å$^{-1}$ throughout the far-UV. Very similar results are obtained for other quasar models (Martin & Bowyer 1989; Martin et al. 1991 and references therein). It follows that quasars probably at most contribute a few percent of the nominal extragalactic UV background flux.

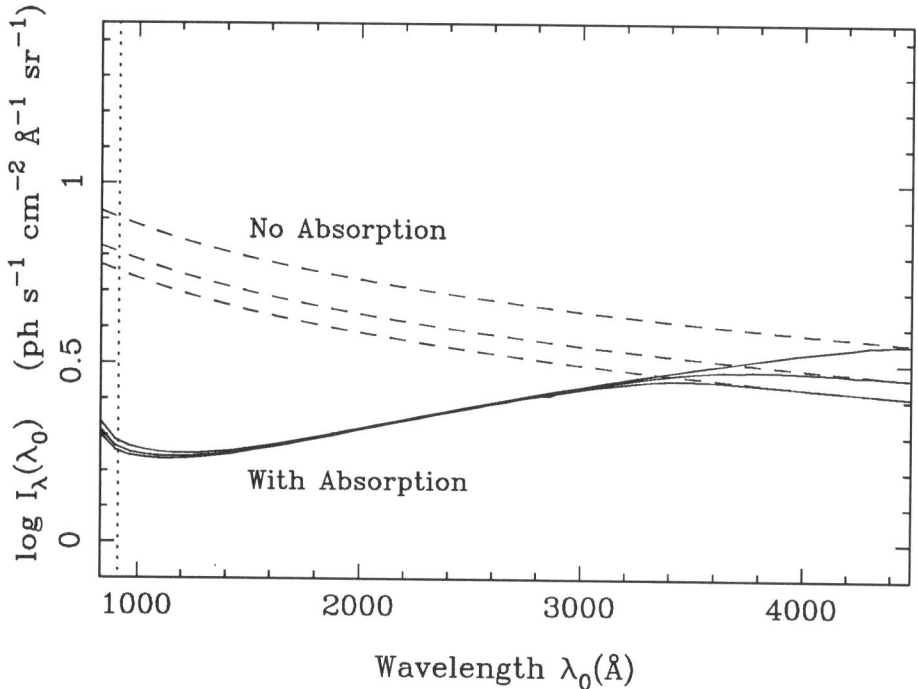

**Figure 5.** *The integrated UV background due to quasars calculated for the three evolution models adopted by Bajtlik et al. (1988) and used in Figure 2. An $F_\nu \propto \nu^{-0.5}$ quasar continuum spectrum was assumed. The dashed curves show the unabsorbed backgrounds and the full curves the same spectra when intervening Lyman continuum absorption is taken into account.*

As a parenthetical aside, the reason that quasars are a dominant background source in the X-rays, but not in the adjacent UV, can be traced to the fact that the nominal extragalactic far-UV background intensity of $I_\lambda \approx 100$ photons s$^{-1}$ cm$^{-2}$ sr$^{-1}$ Å$^{-1}$ is actually considerably brighter in terms of energy per octave in frequency than the extragalactic background flux in the X-rays.

## 4. RADIATIVE DECAY OF EXOTIC PARTICLES

A final more speculative potential source of diffuse extragalactic UV emission is the radiative decay of exotic particles of cosmological origin. Since Big Bang theory predicts the existence of a cosmological "sea" of neutrinos having a particle density similar to the photon density of the 2.7 K microwave background radiation, massive neutrinos and similar "inos" of various flavors have long been considered as candidate sources for missing matter. In particular, any type of exotic particle having a present day cosmological particle density of $n_\nu^0 \approx 100$ cm$^{-3}$ will be capable of closing the Universe if its mass is of order $m_\nu \approx 100h^2$ eV.

In some theories, such massive exotic particles are not stable, but decay into lighter particles under the emission of photons. It is easy to show that the energy of such decay photons is given by

$$E_\gamma = \frac{(m_H^2 - m_L^2)}{2m_H} \simeq \frac{m_H}{2} \tag{17}$$

where $m_H$ and $m_L$ are the masses of the heavy and light particles, and the last approximation is valid in the limit where $m_H \gg m_L$. From this equation it follows that any particle massive enough to provide closure density will have emit its decay radiation at $E_\gamma \simeq 50h^2$ eV, i.e. at ionizing extreme-UV wavelengths.

Depending on the rate of decay of the particles in question, the accumulated redshift-smeared emission from such particles could give rise to observable UV background radiation. Conversely, the UV background observations severely constrain the exponential decay time of any radiatively decaying cosmologically produced particle with a mass in the range $m_\nu \approx 10 - 100$ eV to be much larger than the age of the Universe, or $\tau_\nu > 10^{23}$ s (Kimble, Bowyer & Jakobsen 1981; Overduin, Wesson & Bowyer 1993).

The intensity and spectral shape of the decay background is easily calculated from equation (1) through insertion of the appropriate line emissivity

$$\epsilon_l(z) = \dot{n}_\nu (1+z)^3 \tag{18}$$

where $\dot{n}_\nu \simeq n_\nu/\tau_\nu$ is the particle density decay rate. This leads to

$$I_\lambda(\lambda_0) = \left[\frac{c}{H_0}\right] \frac{\dot{n}_\nu}{4\pi} \frac{1}{\lambda_l} \left(\frac{\lambda_0}{\lambda_l}\right)^2 \left[1 + \Omega\left(\frac{\lambda_0}{\lambda_l} - 1\right)\right]^{-\frac{1}{2}} \quad (\lambda_0 \geq \lambda_l) \tag{19}$$

where $\lambda_0$ and $\lambda_l$ are the observed and decay line wavelengths. The redshifted decay spectrum displays a characteristic jump at $\lambda_0 = \lambda_l$ and drops steeply toward the red as $I_\lambda \propto \lambda^{-2.5}$ ($\Omega \simeq 1$).

Following the suggestions of Cowsik & McClelland (1972) and De Rújula & Glashow (1980), much attention has in recent years focussed on the concept of decaying massive neutrinos as possible carriers of the missing mass. The astrophysical consequences of this idea have been investigated in some detail by Sciama (1990) who in a series of papers has argued that through suitable tuning of the parameters, massive decaying neutrinos are not only capable of explaining the missing mass problem, but also provide a convenient and omnipresent *in situ* source of ionizing radiation that is capable of explaining the ionization structure of both the interstellar medium of the galaxy (Sciama 1993) and the intergalactic Lyman forest clouds discussed in Section 2.1 (Sciama 1991). In order to accomplish all this, the neutrino properties need to be rather tightly constrained. Sciama's hypothetical neutrinos have a mass of $m_\nu \simeq 28$ eV and decay under emission of photons at a wavelength of $\lambda \simeq 890$ Å just below the Lyman limit. The neutrino exponential decay time is $\tau_\nu \simeq 1 - 2 \; 10^{23}$ s.

Figure 6 shows the resulting redshift smeared far-UV background calculated for these parameters. Although the edge of the decay spectrum is conveniently hidden by interstellar neutral hydrogen absorption, and therefore un-observable, both the absolute intensity ($I_\lambda \approx (100 - 600)h^{-1}$ photons s$^{-1}$ cm$^{-2}$ sr$^{-1}$ Å$^{-1}$) and the steep blue color of the decay spectrum are at best barely consistent with the existing observations of the UV background (cf. figure 9 of Martin et al. 1991), but can probably not be definitively excluded at this point given the uncertainties in the data (Overduin et al. 1993).

Other and more persuasive observations that point against the Sciama model have, however, been reported by Davidsen et al. (1991) and Dettmar & Schultz (1992) (and countered in Sciama 1993).

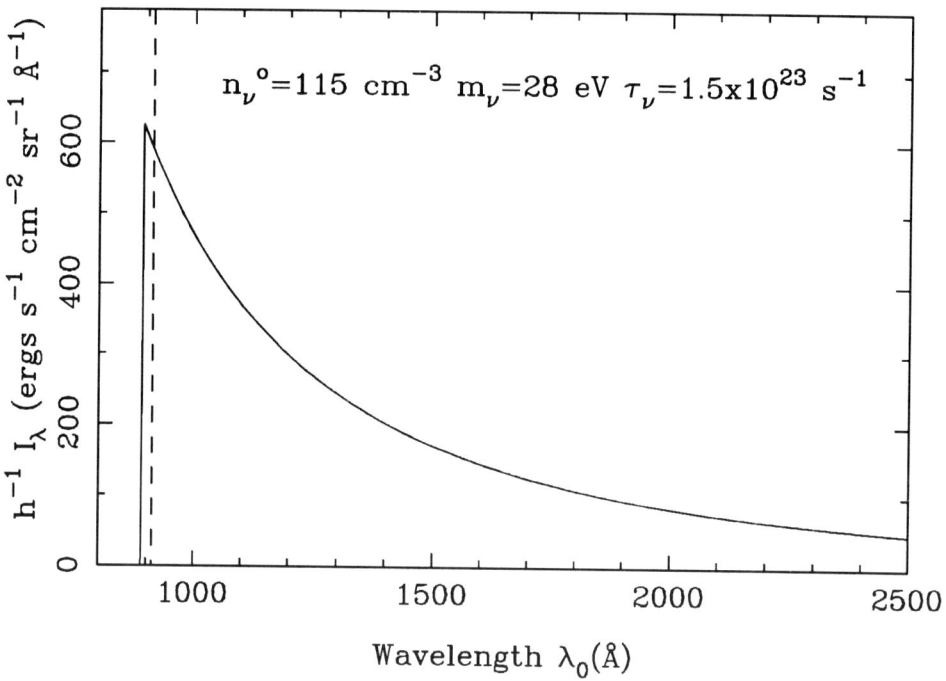

**Figure 6.** *The redshift-smeared UV background predicted for the decaying neutrino model of Sciama (1990). The step in the spectrum at the decay wavelength of λ890 Å falls below the Lyman limit (vertical dashed line) and is therefore not observable due to interstellar absorption.*

## 5. SUMMARY AND CONCLUSIONS

The three known potential sources of extragalactic UV background radiation have been discussed with an emphasis on evaluating their possible contributions to a nominal observed extragalatic UV background of intensity $I_\lambda \sim 100$ photons s$^{-1}$ cm$^{-2}$ sr$^{-1}$ Å$^{-1}$.

The most firm conclusions are possible in the case of the integrated UV light of galaxies. In particular, direct galaxy counts in the UV down to a limiting magnitude of $m \simeq 18.5$ have already "resolved" the background to an intensity of $I_\lambda \simeq 30$ photons s$^{-1}$ cm$^{-2}$ sr$^{-1}$ Å$^{-1}$, which extrapolated to fainter magnitudes by means of galaxy evolution models suggests that galaxies are probably the all-dominant source of extragalactic UV background radiation. There is also evidence that the UV background exhibits the small scale fluctuations expected from the accumulated light of galaxies.

The conclusion that thermal Ly$\alpha$ and HeII $\lambda$304 Å emission from the intergalactic medium, on the other hand, is a marginal contributor to the UV background is almost nearly as compelling, and is reached by drawing on results obtained through the study of quasar absorption lines. The intensity of the metagalactic ionizing background at high redshifts inferred from the so-called "proximity effect" displayed by the Lyman forest absorption lines suggests that the redshift-smeared backgrounds due to Ly$\alpha$ and HeII $\lambda$304Å recombination radiation from photoionized intergalactic gas are to be found at intensities far below current observational limits on the extragalactic background flux. The results of the Gunn-Peterson test at low redshifts derived from UV quasar spectra obtained with *IUE* constrain the intensity of redshifted collisionally excited Ly$\alpha$ emission from a shock-heated IGM component to an equally low intensity. The possibility that the IGM has gone through a phase of intense collisionally excited HeII $\lambda$304Å emission at $z \simeq 3 - 5$ cannot be completely ruled out until the HeII equivalent of the Gunn-Peterson test is carried out with *HST* or *Lyman/FUSE*. However, the statistics of quasar absorption lines imply that any far-UV background component stemming from $z \gtrsim 3$ will remain effectively hidden from view because of strong accumulated Lyman continuum absorption in the Lyman forest and Lyman limit types of quasar absorption systems.

The last source of extragalactic UV radiation considered, namely the decay from exotic cosmological particles, is by its nature more speculative and therefore less easy to rule out definitively. Nonetheless, in its latest incarnation as proposed by Sciama, the redshift smeared background predicted from this source has an intensity and spectral shape that strains the limits of existing observations.

## APPENDIX: DIFFUSE BACKGROUND FROM EXTRAGALACTIC SOURCES

Consider a source of diffuse background radiation having a continuum volume emissivity, $\epsilon(\lambda, t)$ as a function of wavelength, $\lambda$, and cosmic time, $t \leq t_0$. The background intensity due to this source at a given received wavelength, $\lambda_0$, results as the integration of the redshifted contributions from all epochs. The background generated during a time interval $dt$ is accumulated over the pathlength $cdt$ where $c$ is the speed of light. If $a(t)/a_0$ denotes the ratio of expansion of the Universe between the epoch of emission and today, the received intensity due to emission within $dt$ is therefore

$$dI_\lambda(\lambda_0) = \frac{1}{4\pi} \epsilon_\lambda(\lambda(t), t) \left[\frac{a(t)}{a_0}\right]^4 cdt \tag{A1}$$

where the emissivity is evaluated at the appropriate blue-shifted wavelength of emission, $\lambda(t) = \lambda_0 a(t)/a_0$. The factor in square brackets takes into account the dilution of the intensity due to the expansion of the Universe (as per Liouville's Theorem expressed in the adopted intensity units of photons s$^{-1}$ cm$^{-2}$ sr$^{-1}$ Å$^{-1}$). Integrating eq. (1) over the age of the Universe gives then for the total background intensity

$$I_\lambda(\lambda_0) = \frac{1}{4\pi} \int_0^{t_0} \epsilon_\lambda(\lambda(t), t) \left[\frac{a(t)}{a_0}\right]^4 cdt \tag{A2}$$

It is convenient to change variables from cosmic time to redshift, $(1 + z) = a_0/a(t)$, using the standard expression for a Friedman Universe

$$dt = \frac{1}{H_0} (1+z)^{-2} (1+\Omega z)^{-\frac{1}{2}} dz \qquad (A3)$$

where $H_0$ is the Hubble Constant. This leads to

$$I_\lambda(\lambda_0) = \frac{1}{4\pi} \left[\frac{c}{H_0}\right] \int_0^\infty \epsilon_\lambda(\lambda(z), z) (1+z)^{-6} (1+\Omega z)^{-\frac{1}{2}} dz \qquad (A4)$$

where $\lambda(z) = \lambda_0/(1+z)$.

Equation (A4) is the general expression for the accumulated background from any continuum emission source. The special case of redshift smeared monochromatic line emission is also of interest. In this case we have for the emissivity

$$\epsilon_\lambda(\lambda, z) = \epsilon_l(z)\delta(\lambda - \lambda_l) \qquad (A5)$$

Where $\lambda_l$ is the rest wavelength of the transition in question. Inserting (A5) into (A4) leads to the following expression for the intensity of the resulting continuum background at wavelengths $\lambda_0 \geq \lambda_l$

$$I_\lambda(\lambda_0) = \left[\frac{c}{H_0}\right] \frac{\epsilon_l(\tilde{z})}{4\pi} \frac{1}{\lambda_l} (1+\tilde{z})^{-5} (1+\Omega\tilde{z})^{-\frac{1}{2}} \qquad (A6)$$

where $\tilde{z} = \lambda_0/\lambda_l - 1$ is the redshift of emission.

## REFERENCES

Armand, C, Milliard, B., & Deharveng, J. M. 1993, A&A (in press)
Bahcall, J. N. et al. 1993, ApJS 87 (in press)
Bajtlik, S., Duncan, R. C., & Ostriker, J. P. 1988, ApJ 327, 570
Beaver, E. A, Burbidge, E. M., Cohen, R. D., Junkkarinen, V. T., Lyons, R. W., Rosenblatt, E. I., Hartig, G. F., Margon, B., & Davidsen, A. F. 1991, ApJ 377, L9
Bechtold, J., Green, R. F., Weymann, R. J., Schmidt, M., Estabrook, F. B., Sherman, R. D., Wahlquist, H. D., & Heckman, T. M. 1984, ApJ 281, 76
Bechtold, J., Weymann, R. J., Lin, Z., & Malkan, M. A. 1987, ApJ 315, 180
Blades, C., Turnshek, D., & Norman, C. (Eds) 1988, QSO Absorption Lines: Probing the Universe, (Cambridge: Cambridge University Press)
Bowyer, S. 1992, ARA&A 29, 59
Bruzual A. G. & Charlot, S. 1993, ApJ 405, 538
Bruzual A. G., Magris C. G., & Calvet, N. 1988, ApJ 333, 673
Burnstein, D., Bertola, F., Buson, L. M., Faber, S. M., & Lauer, T. R. 1988, ApJ 328, 440
Carswell, R. F. 1988, in QSO Absorption Lines: Probing the Universe, ed. C. Blades, D. Turnshek, & C. Norman, (Cambridge: Cambridge University Press), p. 91
Carswell, R. F., Whelan, J. A. J., Smith, M. G., Boksenberg, A., & Tytler, D. 1982, MNRAS, 198, 91
Cowsik, R., & McClelland, J. 1972, Phys Rev Let 29, 669

Davidsen, A., Bowyer, S., & Lampton, M. 1974, Nat 247, 513
Davidsen, A. F. *et al.* 1991, Nature 351, 128
De Rújula, A., & Glashow, S. L. 1980, Phys Rev Let 45, 942
Dettmar, R. J., & Schultz, H. 1992 A&A 254, L25
Green, R. F., Pier, J. R., Schmidt, M, Estabrook, F. B., Lane, A. L., & Wahlquist, H. D. 1980, ApJ 239, 483
Guiderdoni, B., & Rocca-Volmerange, B. 1991, A&A 252, 435
Gunn, J. E., & Peterson, B. A. 1965, ApJ 142, 1633
Henry, R. C. 1992, ARA&A 29, 89
Hunstead, R. W. 1988, in QSO Absorption Lines: Probing the Universe, ed. C. Blades, D. Turnshek, & C. Norman, (Cambridge: Cambridge University Press), p. 71
Hurwitz, M., Bowyer, S., & Martin, C. 1991, 372, 167
Ikeuchi, S., & Ostriker, J. P. 1986, ApJ 301, 522
Jakobsen, P., *et al.* 1993, ApJ (in press)
Kimble, R. A., Bowyer, S., & Jakobsen, P. 1981, Phys Rev Let 46, 80
King, I. R. *et al.* 1993, ApJ 397, L35
Kinney, A. L., Bohlin, R. C., Blades, J. C., & York, D. G. 1991, ApJS 75, 645
Koo, D. C., & Kron, R. C. 1992 ARAA 30, 613
Kulkarni, V. P., & Fall, M. S. 1993, ApJ 413, L63
Kurt, V. G., & Sunyaev, R. A. 1967, Cosmic Research, 5, 496
Kutyrev, A. S., & Reynolds, R. J. 1989, ApJ, 344, L9
Lampton, M., Deharveng, J. M., & Bowyer, S. 1990, in IAU Symp. No. 139, The Galactic and Extragalactic Background Radiation,eds. S. Bowyer & C. Leinert (Kluwer: Dordrecht)
Lanzetta, K. M. 1988, ApJ 332, 96
Lu, L., Wolfe, A. M., & Turnshek, D. A. 1991, ApJ 367, 19
Madau, P. 1992, ApJ 389, L1
Martin, C., & Bowyer, S. 1989, ApJ 338, 667
Martin, C., & Bowyer, S. 1990, ApJ 350, 242
Martin, C., Hurwitz, M.,& Bowyer, S. 1990, ApJ 345, 220
Martin, C., Hurwitz, M.,& Bowyer, S. 1991, ApJ 379, 549
Meiksin, A., & Madau, P. 1993, ApJ 412, 34
Milliard, B., Donas, J., Laget, M., Armaud, C., & Vuillemin, A. 1992, A&A 257, 24
Miralda-Escudé, J., & Ostriker, J. P. 1990, ApJ 350, 1
Miralda-Escudé, J., & Ostriker, J. P. 1992, ApJ 392, 15
Murdoch, H. S., Hunstead, R. W., Pettini, M., & Blades, J. C. 1986, ApJ 309, 19
Møller, P., & Jakobsen, P. 1990, A&A 228, 299
Osterbrock, D. E. 1989, Astrophysics of Gaseous Nebulae and Active Galactic Nuclei (Mill Valley: University Science Books), p. 23
Ostriker, J. P., & Ikeuchi, S. 1983, ApJ 268, L63
Overduin, J. M., Wesson, P. S., & Bowyer, S. 1993, ApJ 404, 460
Paresce, F., & Jakobsen, P. 1980, Nature 288, 119
Paresce, F., McKee, C. F., & Bowyer, S. 1980, ApJ 240, 387
Picard, A., & Jakobsen, P. 1993, A&A (in press)
Reimers, D., Clavel, J., Groote, D., Engels, D., Hagen, H. J., Naylor, T., Wamsteker, W., & Hopp, U. 1989, A&A 218, 71

Reimers, D., Vogel, S., Hagen, H. J., Engels, D., Groote, D., Wamsteker, W., Clavel, J., & Rosa, M. R. 1992, Nat 360, 561

Reynolds, R. J., Magee, K., Roesler, F. L., Scherb, F., & Harlander, J. 1986, ApJ 309, L9

Sargent, W. L. W., 1988, in QSO Absorption Lines, Probing the Universe, ed. J. C. Blades, D. Turnshek & C. A. Norman (Cambridge: Cambridge University Press), p. 1

Sargent, W. L. W., Steidel, C. C., & Boksenberg, A. 1989, ApJS 69, 703

Sargent, W. L. W., Young, P. J., Boksenberg, A., & Tytler, D. 1980, ApJS 42, 41

Sasseen, T. P., Bowyer, S., Wu, X., & Lampton, M. 1993, BAAS 25, 822

Sciama, D. W. 1990, ApJ 364, 549

Sciama, D. W. 1991, ApJ 367, L39

Sciama, D. W. 1993, ApJ 409, L25

Sherman, R. D. 1979, ApJ 232, 1

Sherman, R. D. 1982, ApJ 256, 370

Songaila, A., Cowie, L. L. & Lilly, S. J. 1990, ApJ 348, 371

Songaila, A., Bryant, W.,& Cowie, L. L. 1989, 345, L71

Steidel, C. C., & Sargent, W. L. W. 1987, ApJ 318, L11

Tinsley B. M. 1972, A&A 20, 383

Tinsley B. M. 1973, A&A 24, 89

Tinsley B. M. 1980, ApJ 241, 41

Tripp, T. M., Green, R. F., & Bechtold, J. 1990, ApJ 364, L29

Tytler, D. 1982, Nature 298, 427

Tytler, D. 1988, in QSO Absorption Lines: Probing the Universe, ed. C. Blades, D. Turnshek, & C. Norman, (Cambridge: Cambridge University Press), p. 176

Weymann, R. 1967, ApJ 147, 887

## DISCUSSION

**D. Henry**: Peter, I liked your remarks about the recombination radiation, although the predicted background contribution is low. Can we pump it up by having the gas clump over time?

**Jakobsen**: No, because it will never, no matter what you do, be brighter than the input flux you put in. The IGM is just acting as a photon slicer, right? So if you know how much is going in, that is it. It does not matter what you do with it, it will never be brighter than that. So if you believe in the proximity effect, it is not going to be a significant contributor.

**V. Khersonsky**: You classified the Lyman-$\alpha$ forest clouds as highly ionized. To me, the situation seems controversial; I mean, according to the recent discussion by Pettini and Carswell, Lyman $\alpha$ forest clouds could be cold, which might change completely the absorption models.

**Jakobsen**: The issue being raised here is that there are some observations indicating that some of the Lyman $\alpha$ forest systems have line widths more like 10 km/s, which is too narrow for photoionization temperatures. But that is very controversial.

**M. Rees**: In response to that, having had Hunstead, Carswell, and Pettini at Cambridge, I get the impression that there is really no very firm evidence for any $b$'s narrower than 50 km/s for any bonafide Lyman $\alpha$ system.

**Jakobsen**: I was talking to Max last week and he said they have now looked at the spectrum of Q1101, or whatever, the famous bright $z = 2.15$ quasar. And he has still got some systems at 10 km/s.

**Q.**: Could they be metal lines?

**Jakobsen**: Well, I guess that is the issue. They could be, but I am not an expert on this. Max should speak for himself.

**M. Rees**: The question I wanted to ask was about what you said about the helium at high redshifts. It seems to be quite possible that although the universe becomes an H II region for hydrogen at $z = 5$, it may be only at a lower redshift like, say, $z = 3$ that the helium becomes doubly ionized.

**Jakobsen**: Yes.

**M. Rees**: I would like to point out that if that is the case it will have an interesting consequence for the temperature of the Lyman $\alpha$ systems, because in a high redshift Lyman $\alpha$ system the heat input due to the helium is comparable to that of the hydrogen, and you would expect that, at the highest redshifts, the helium is singly ionized or doubly ionized. First of all, if you can see the helium line, it should be very strong; if you cannot see it, that is no good. But the $b$ value will be substantially less for the singly ionized helium than for the doubly ionized helium; therefore, you may get an indirect handle on whether the helium is doubly ionized or singly ionized by getting some evidence for the line widths and whether they are narrower in high redshift quasars. I mean other things being equal you would get a higher $b$ value when the helium is doubly ionized than when the helium is singly ionized.

**J. Peebles**: Could I ask a quick question? I liked very much your thoughts of 100 photons units as a contribution from galaxies. As you remark, you would not expect a very prominent signature in the spectrum. What does one know of the spectrum of the extragalactic background?

**Jakobsen**: There is no clear signature detected.

**J. Peebles**: As you say, because galaxies are very faint shortward of 912 Å, so if you go to longer wavelengths...

**Jakobsen**: Yes, then we should see a signature at 912 Å.

**J. Peebles**: Wouldn't you see, as you say, that the background is red?

**Jakobsen**: Sure, if we could measure the spectrum. I mean, one thing is getting the fluxes right, another is getting the exact spectrum of the extragalactic component.

**J. Peebles**: Must be exact. It is quite a big effect!

**Jakobsen**: To my knowledge you get a lot more UV. Galaxies turn out to have a somewhat flat UV spectrum in the integrated light because there is a lot more UV going back in time; the assumption here is that the star formation is decreasing. As a net effect, you get something that is sort of quasi flat. The reason is that there is an enormous increase in the look back time with increasing wavelength and this compensates for the blue galaxy spectrum. Let's remember that OB stars have Rayeigh-Jean-like spectra. The spectra just shoot up, but when you start redshifting them, they are compensated by the larger look back time. You get something that is a sort of flattish.

**A. Meiksin**: I just wanted to point out that the amount of UV radiation might be substantially higher than the estimates based on absorption from Lyman $\alpha$ clouds. Piero Madau looked at this, and there seems to be a window in the Lyman $\alpha$ forest around $10^{15}$ $cm^{-2}$. The number of systems begins to fall precipitously above $10^{15}$ $cm^{-2}$, based on some of the data from the Cambridge group (Carswell *et al.*). And the effect is to increase the estimate for the flux by about a factor of 3. This applies for galaxies as well.

**Jakobsen**: So you are changing the number density; you have different parameters from what we used.

**A. Meiksin**: Well, it is just that only the Lyman limit systems absorb the UV. The Lyman $\alpha$ forest itself might not absorb that much.

**Jakobsen**: No, it is like optical depth of unity or so. The Lyman-$\alpha$ forest, I mean.

**A. Meiksin**: Although I am not sure this would have much of an effect on the helium because the helium optical depth is higher.

**Jakobsen**: I would like to see the numbers you used and whether they are dramatically different from what we assumed.

**A. Meiksin**: In fact, there is a second effect which Mike Fall has looked at. At $z \simeq 3.5$ there can be some absorption by dust in damped Lyman $\alpha$ systems. Between $z = 3.5$ and 4, I mean. And you might get another factor of two.

**van der Kruit**: What is your opinion about future space experiment in the visible and near-ultraviolet and near-infrared observation? At high spectral or high spatial resolution and high sensitivity. What is your opinion?

**Jakobsen**: There has been no all-sky survey in the UV. My name was on Dick's [SMEX] proposal, and I obviously believe that we need it, both to map the diffuse background and also pick out the brightest point sources, such as UV bright quasars. But the big issue is, do you want to go for a deep point source survey or do you want to catch the diffuse background, as well? The latter is much easier observationally and instrumentally, which is why I think Dick's approach is the faster one. It would be wonderful to do a deep UV Schmidt survey of the sky, but it is more ambitious. Then, of course, there is spectroscopy and the HeII Gunn-Peterson test. The problem here is that nearly all the $z > 3$ quasars we have looked at with the HST-FOC are completely absorbed out at $\lambda 304$ Å. Out of 25 carefully selected objects, which basically represents most of the known $z > 3$ quasars worth looking at, there is only one that has any trace of UV flux that might extend to $\lambda 304$ Å, but its flux level is too low for a follow-up with a grating spectrograph. Maybe, with STIS we might be able to do this, but for now the quasar is just too faint for HST. Secondly, it would be wonderful to go to shorter wavelengths, below Lyman $\alpha$, both for the diffuse sky survey and for the spectroscopy, but this is more difficult, because you have to have windowless detectors and everything that goes with that. The helium Gunn-Peterson problem is an order of magnitude easier at redshift 2 than at redshift 3, because the opacity is smaller and quasars are brighter; these two things together give you a factor of 10 relief in required sensitivity, which means that with modest instrumentation you could actually do some significant quasar absorption spectroscopy at, say, 1000 Å. Those are the things I would like to see happening in the UV. And, indeed, some of them are at some level.

**B. Espey**: Just since Avery brought up the question about corrections to the canonical value of $j(\nu) = 10^{-21}$ erg/cm$^2$/s/sr/Hz. There is a poster that I have presented where it is shown that you can actually force down, through various systematic corrections, the estimated value of $j(\nu)$. That is actually going to cross over with the predicted value at some point. I think you may actually find a few additional corrections which will actually make the QSO contribution more important.

**Jakobsen**: Well, one question is what are quasars doing intrinsically at extreme ultraviolet rest wavelengths. We have no clue. But there I will say that the few moderately absorbed objects we have seen with HST at longer wavelengths indicate that quasars are intrinsically very bright in the extreme UV. It looks like an extrapolation of the visible power spectrum is not wrong to within a factor of 2 or 3.

**B. Espey**: I have actually done the calculations not just using Cloudy but using the same spectrum that Piero Madau used for his simulations and also using the optically thick background he calculated. And you can actually bring QSOs to within a factor of a few of what is needed.

**Q.**: You can try and put limits on the present ionizing background by looking for H$\alpha$ emission from galaxies that have HI and no local sources of ionization. Are those interesting?

**Jakobsen**: Yes, they are actually. Len Cowie and Esther Hu have done this and Reynolds has done this. The idea is look at some cloud at high latitude or outside our galaxy, measure the H$\alpha$ flux and then work backwards to figure out what is the ionizing input flux required to produce the H$\alpha$ you see. The derived intensity of the ionizing background at low redshift is substantially smaller than the $10^{-21}$ erg/s/cm$^2$/sr/Hz you get from the proximity effect.

**M. Fall**: You can also measure the ionizing radiation at low redshifts from the proximity effect, with the HST key project survey. Varsha Kulkarni, a student of mine, has just done that, in fact. We get enough lines, because the background is lower. The value we get is about $10^{-23}$ at redshift of 0.5 in the same units in which it is $10^{-21}$ at a redshift of 3. So it is down by about 2 orders of magnitude. It is where you would expect it to be.

**Jakobsen**: Based on quasar evolution models.

**M. Fall**: It is still a small sample and there are uncertainties of factors of 3 or 4 or so. But it is at least an order of magnitude down from what you get at high redshift.

**Jakobsen**: I would have thought there were too few lines, but of course, ... the contrast is much higher actually.

**M. Rees**: Can you say anything about the helium at low redshift? As you would expect, the quasar UV drops more precipitously than the stellar UV; and in the quasar UV the helium may be relatively more important than the hydrogen. So I wonder if one could rule out the possibility that helium has recombined between $z \simeq 2$ and the present.

**Jakobsen**: Not directly, because you cannot observe it. I mean 304 Å at redshift $z = 2$ are received at 912 Å. So, for anything below $z = 2$ you have to see out of the galaxy in the Lyman continuum. This is not possible.

**M. Rees**: I wonder whether there are indirect arguments.

**Jakobsen**: Not that I know of. There is no way you are going to do the He 304 Å Gunn-Peterson test at $z = 1$ unless you have an intergalactic satellite or something similar.

**M. Rees**: But what about the singly ionized helium absorption lines? You can see those at $z < 2$.

**Jakobsen**: Yes.

**M. Rees**: They would be much stronger than Lyman $\alpha$ in the Lyman $\alpha$ clouds.

**Jakobsen**: They are not. There is a famous quasar called HS1700+6416, discovered by Reimers et al., that is the brightest object in the UV ever seen so far. It has got a redshift of 2.78, which is just too small to get to HeII 304 Å with HST, but Reimers et al. have looked at it with the FOS. Their gorgeous spectrum shows the Lyman valley absorption with all the little edges and everything. There are a few cases in there where they claim to have seen the HeI 584 Å lines. It is a paper in *Nature*. (**360**, 561 (1992), Eds.)

**M. Rees**: Yes, but in those cases it was weaker than the hydrogen line.

**Jakobsen**: Yes, it was. The HeI column density was indeed down by something like $10^{-2}$ with respect to HI.

**M. Rees**: At a ratio consistent with helium being doubly ionized?

**Jakobsen**: Absolutely. But I think those systems, this is something I wanted to check, have relatively large column density in general and they would not really be classified as Lyman forest systems. The main thing is that they have seen weak HeI 584 Å, as expected. It is the first time anybody has seen helium in quasar absorbers.

# THE OPTICAL EXTRAGALACTIC BACKGROUND RADIATION

J. A. Tyson
AT&T Bell Laboratories
Murray Hill, NJ 07974

**Abstract.** The extragalactic night sky is filled with the light from stellar formation and heavy element production integrated back to the epoch of maximum quasar density. This is a review of the status of our understanding of the faint blue galaxies and the extragalactic background light (EBL) from 0.15 to 2.3 $\mu$m. New data on faint galaxies and the EBL are also presented. Ultra deep CCD imaging surveys have revealed a population of faint blue galaxies covering the sky. To a surface brightness of 29 B mag arcsec$^{-2}$ there are about 100 background galaxies per square arcminute anywhere in the sky. These 25–27 magnitude galaxies are apparently distributed over a broad redshift range of 1–3. A gravitational lens test of the redshift distribution of these galaxies also implies that most of them lie beyond redshift 0.7. Although measured redshifts of galaxies grow monotonically with magnitude, there may be several populations of galaxies at redshifts greater than 0.5; there is evidence that the mix of types is different at high and low redshift. The data do *not require* galaxy number non-conservation, and a variety of models of stellar population evolution are consistent with the data. Studies of the angular correlations of faint galaxies show a red/blue effect and very little clustering in faint blue galaxies. The seeing-deconvolved half light radii of these galaxies approach 0.8 arcsecond at the faint end.

## 1. INTRODUCTION

Optical images of the extragalactic sky show a variety of galaxies with various luminosities and distances, all seen in projection. During the past decade ultra-deep optical imaging using CCDs over the wavelength range 0.3–1 $\mu$m have revealed a high surface density of faint blue galaxies. At a flux level corresponding to 1 photon/pixel/minute collected in a 4-meter telescope there are about 300,000 galaxies per square degree on the sky. These galaxies have apparent magnitudes between 25 and 28 B mag. Although too faint for spectroscopic redshift determination, several tests indicate that the redshift of these galaxies extends between 0.7 and 3. Observations suggest that we may be seeing many of these galaxies at an epoch of formation of much of their stellar content. The resulting UV-bright spectrum, when redshifted to redshifts of 1–3, could produce the observed blue spectral shape.

The extragalactic background radiation in the optical part of the spectrum originates in young stellar populations in galaxies at moderate to high redshift. The subject has a long history, but only recently have CCD surveys achieved the required sensitivity to capture most of the extragalactic light of the night sky. Perhaps one of the more intriguing questions is the nature of the 25–27 B magnitude so-called "faint blue galaxies" (FBGs). While there are an increasing number of clues to the physical nature of the FBGs (colors, numbers, correlations, morphological scale lengths, and rough redshift limits), I think it is fair to say that we don't even fully understand the type mix and star formation activity of the more nearby $0.1 < z < 0.4$ galaxies. For example, the possibility of a redshift-dependent luminosity function is still an open question. The situation for the FBGs, being a superposition of galaxies over a greater span of look-back time, is bound to be more complicated. Happily, the summed extragalactic background light at some wavelength may be calculated directly from the galaxy number counts. Of course this does not affect our state of ignorance regarding the faintest sources of this radiation, particularly at blue and UV wavelengths.

The intensity of the extragalactic background light (EBL) is influenced strongly by galaxy evolution (stellar lifetimes and population evolution), and to a much lesser extent by cosmology (Harrison 1964). Number-magnitude counts which rise with magnitude like dex (0.4 mag) or faster continue to add to the EBL at the faintest magnitudes, and if this slope never falls below 0.4 the surface brightness of the extragalactic sky would approach that of a typical galaxy. These considerations are related to Olbers' paradox. The finite luminous lifetime of stellar populations offers a way out of both problems.

## 1.1 Theoretical Estimates of the EBL

Whitrow & Yallop (1965) first gave the form of the bolometric flux in an arbitrary cosmology with coeval galaxies starting at some formation redshift. Partridge & Peebles (1967) estimated the light from primeval galaxies in a model in which star formation began at high redshift giving the EBL a 30 $\mu$m IR excess and large angular sizes. The optical EBL is probably due to evolving spirals: Wyse (1985) estimated that less than 18% of the EBL light is from ellipticals, and that the elliptical light is red. Models of mild continuous star formation in galaxies tend to predict EBL similar to that observed in the faint galaxy population. Yoshii & Takahara (1988) calculated a no-evolution EBL flux of $\lambda F_\lambda = 1.1 \times 10^{-6}$ erg cm$^{-2}$ sec$^{-1}$ sr$^{-1}$ at 0.36 $\mu$m wavelength. For evolving galaxies with formation redshifts ranging from 3 to 5 (and nearly independent of deceleration parameter) they estimate an EBL about two times higher than this, close to the observed value, but the shape of their EBL spectrum is very different that that observed (see below). EBL from faint galaxies is very sensitive to evolution and relatively insensitive to cosmology, particularly at wavelengths where much of the redshifted UV from early star formation appears. Differential galaxy counts in the blue part of the spectrum are relatively more sensitive to luminosity evolution. The sum total EBL, being an integral over the flux times number counts at each wave band, is even less sensitive to cosmology. Generally, small changes in the galaxy luminosity function can compensate for changes in $\Omega$.

## 1.2 Upper Limits to the Diffuse EBL

It is useful to derive the diffuse EBL for comparison with other observations of diffuse background light at various wavelengths. Upper limits to the EBL have been set by all-sky photometry (Roach & Smith 1968; Mattila 1976; Dube et al. 1977, 1979).

Progress in this field was reviewed by Toller (1983). The optical background is dominated by foreground emission (atmosphere, zodiacal light, galactic cirrus, aurora) which makes any absolute EBL measurement difficult; this makes optical EBL more elusive than its counterparts at some other wavelengths (*e.g.*, sub-mm, radio, X-ray) where direct absolute measurements of the diffuse background emission are possible. As a result, discussion of the optical EBL is necessarily focussed on the contribution to the EBL from discrete sources to to about 1 arcminute diameter. Total diffuse EBL chopping experiments have suffered from systematic errors larger than the EBL itself. For example, in chopping between a Lynds dark nebula and nearby blank sky the Lynds nebulae were unfortunately *brighter* than neighboring blank sky because of the back scattered Galactic starlight from nebular dust. One would also need to know where the "dark" nebula is in relation to the edge of the galaxy. These studies are very important since they are the only ones that attempt a direct measurement of the diffuse optical EBL on angular scales larger than one arcminute. A series of observations currently being undertaken by R. Bernstein, B. Madore, & W. Freedman using the Carnegie telescopes in a clever modern variation of Dube's experiment may yield a more accurate determination of the diffuse optical EBL.

A convenient but unfortunate unit often used to express the light of the night sky is $S_{10}(\lambda)$, which is the equivalent bolometric surface brightness of a 10th mag star of specified spectral type measured at wavelength $\lambda$, if its light is uniformly spread over a square degree of sky. One gets a feeling for the possibilities for systematic error in all-sky EBL photometry by reviewing the relative intensity of the competing sources: in units of $S_{10,V}$, airglow = 40 (atomic lines) plus 50 (bands and continuum), zodiacal light away from the zodiac = 100, stars fainter than 6th V mag = 30 (galactic poles) or 95 (mean sky) or 320 (mean in galactic plane), diffuse galactic light and cirrus = 20, summing to a mean sky zenith background of 290 $S_{10,V}$ at low geomagnetic latitudes with no moon. By comparison, the detected EBL from discrete galaxies less than 1 arcmin in size is 0.5 $S_{10,V}$, a mere $1.7 \times 10^{-3}$ of the other diffuse optical backgrounds. CCDs can, however, be calibrated to a reproducible accuracy of $10^{-5}$ so that accurate subtraction of diffuse backgrounds on angular scales larger than the CCD is possible in principle. Since measurements are always made at some wavelength, and the spectrum of the EBL is not known a-priori, a more physical unit would be $\lambda F_\lambda = \nu F_\nu$ in W m$^{-2}$ sr$^{-1}$, which gives equal weight to the energy contribution at each wavelength.

All sky photometry has yielded upper limits around 1 $S_{10}(V)$, where 1 $S_{10}(V)$ = $1.2 \times 10^{-9}$ erg cm$^{-2}$ sec$^{-1}$ sr$^{-1}$ Å$^{-1}$ = $5.3 \times 10^{-9}$ W m$^{-2}$ sr$^{-1}$, at 4400 Å. Dube et al. (1979) observed an EBL at 5115 Å of $1 \pm 1.2$ $S_{10}$ (5115 Å) corresponding to $\lambda F_\lambda < 10^{-5}$ erg cm$^{-2}$ sec$^{-1}$ sr$^{-1}$ at 5100 Å, about three times higher than the EBL due to the discrete faint blue galaxies (FBGs). If the objects dominating the EBL appear smaller than 5 arcsec, as the data imply, then the direct CCD imaging sensitivity is over 100 times the sensitivity of the integrated sky chopping techniques. The diffuse EBL on scales of 10–30 arcsec at K-band is currently better constrained: assuming Gaussian statistics Boughn et al. (1986) obtained an upper limit for the total diffuse K-band EBL of $7 \times 10^{-20}$ erg cm$^{-2}$ Hz$^{-1}$ sr$^{-1}$ = $9 \times 10^{-6}$ erg cm$^{-2}$ s$^{-1}$ sr$^{-1}$ at 2.2 $\mu$m for fluctuations on scales of > 10 arcsec, only slightly above the summed flux from discrete objects in K-band (see below). Gunn (1965) & Shectman (1974) considered statistically the fluctuations in the EBL due to galaxy clustering. This is directly measurable in deep wide-field CCD imaging data, but involves assuming a distribution function. This may soon be possible, since some ultra-deep CCD fields have been reimaged many times, permitting extraction of the fluctuation distribution function separately for the detector + sky noise and the diffuse EBL.

## 2. FAINT SURFACE BRIGHTNESS SURVEYS

Whether the numerous "field" galaxies form a single population has been a long-standing question. Photographic surveys to a surface brightness of 26 magnitude arcsec$^{-2}$ showed as many as 17,000 faint galaxies per square degree at 24th magnitude (Jarvis & Tyson 1981; Kron 1982). However, photographic surveys of galaxies suffer from a selection bias due to the achievable limiting surface brightness for small objects occupying less than a few square arcseconds. Galaxies, particularly at high redshift, have peak surface brightness which can approach the plate limit. Figure 1 shows the effect on the apparent redshift distribution of introducing such a limit in a mild evolution model. The number of galaxies found at high redshift is reduced by the surface brightness cut typical of photographic limit plates, along with whatever angular diameter threshold is used for object detection. Seeing reduces central surface brightness and can move an object below the surface brightness or angular diameter thresholds. Slit spectrographs impose their own diameter limits.

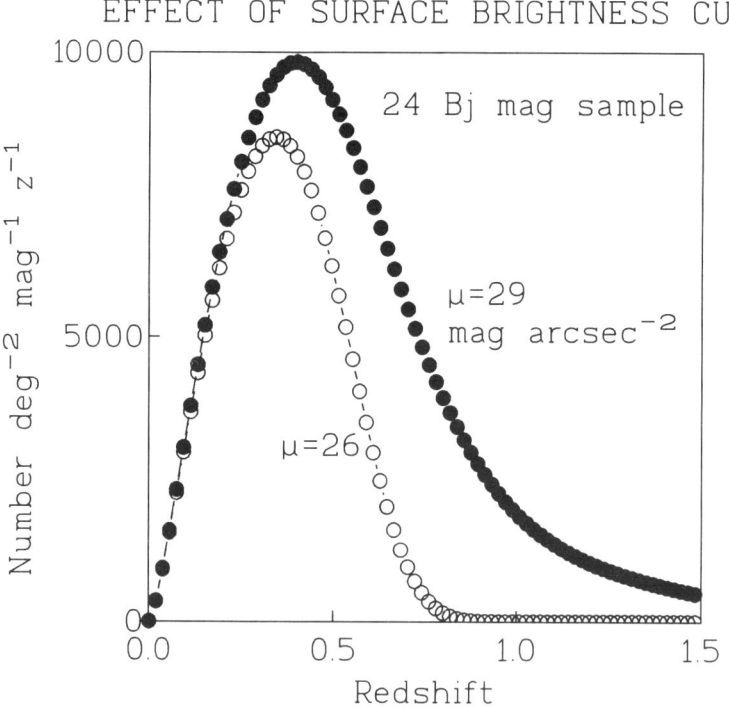

**Figure 1.** *The effect of a surface brightness limit is shown in a model number-redshift plot. A Bruzual model with mild luminosity evolution is run with surface brightness cuts at 26 B mag arcsec$^{-2}$ [appropriate for photographic surveys] and at 29th mag arcsec$^{-2}$ [typical of 3-4 hour CCD imaging]. The selection effects on a redshift survey which uses photographically identified galaxies combine this selection with that of the spectrograph.*

## 2.1 CCD Surveys

Deep CCD imaging survey of 7000 faint galaxies were carried out in 12 random high galactic latitude fields (140 arcmin$^2$ area) to 29 $B_j$, 28 R, and 26 I mag arcsec$^{-2}$ limiting surface brightness (Tyson 1988; Tyson & Seitzer 1988), and in a survey of 100 galaxies as deep in B, V and I in three 1.3 arcmin$^2$ fields (Lilly et al. 1991). Ultra-deep CCD imaging to 30 mag arcsec$^{-2}$ has been obtained in three fields (Guhathakurta et al. 1990), and a large area CCD survey is now under way. This involves as many as 100 disregistered exposures of one field in a given filter. This 3-d stack of disregistered exposures contains all the information about the systematic errors in the CCD, telescope and sky, and the true object luminosity distribution, in separable form. Data processing (see Tyson 1990) produces a catalog of object multi-band photometric data. There are over 200,000 FBGs per square degree in a one magnitude bin at 27th B magnitude. Figure 2 shows a monochrome image of part of one of the old CCD survey fields, 3 arcminutes across.

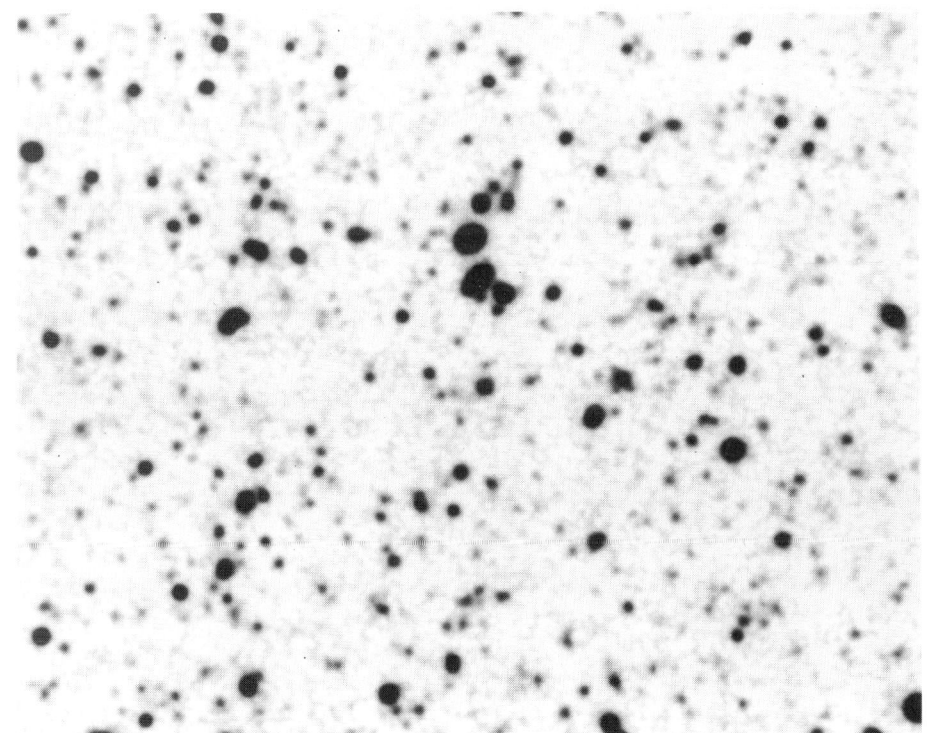

**Figure 2.** *Part of a deep image of a high latitude survey field. This part of the field measures 3 arcmin across. This field is blank on a Schmidt survey plate.*

Between 20 and 26 B magnitude the average number of galaxies per magnitude per square degree is given approximately by Log N = 0.45$B_j$ - 6.55, where the blue $B_j$ passband (close to the photographic IIIaJ+GG385 passband) is effectively 3700–5100 Å (Gullixson et al. 1993). However, for $B_j$ magnitudes fainter than 23 the N($B_j$) data and models show significant departures from this simple relation.

Figure 3 summarizes faint galaxy count data for five wavelength bands. The U counts are from Guhathakurta *et al.* 1990, Majewski 1993, and Koo 1986; the $B_j$, R, and I counts are from our most recent data on 15 fields; and the K counts are from Gardner *et al.* 1993. The N(m) counts in the $B_j$ band have average slope 0.45. Note that this slope of 0.45 is super-critical in the sense that the contribution to the EBL is monotonically increasing with magnitude. The N(m) slope in the U-band is even steeper. The slope of the N(m) counts for longer wavelengths is sub-critical, going to a "no-evolution" slope of 0.3 at 0.9 $\mu$m wavelength. While all counts have been corrected for lost area due to objects, only the $B_j$, R, and I counts have had full Monte Carlo recovery efficiency corrections applied. Only the U counts at the last three points require this additional correction. We estimate this to be upward by dex 0.2.

The enhancement of the blue and U band counts is apparently coming from objects which are UV-bright in their rest frame. Tinsley (1980) was the first to point out that a high star formation rate at a given epoch would produce a bump in the number-magnitude counts. The small but highly significant bump seen in the blue counts around 25 $B_j$ mag, and more prominently in the U counts, is possibly an effect of this kind. A more accurate representation of the blue counts over the range 15–27 $B_j$ mag, averaged over the survey fields, is given by the following relation:

$$\log N = 1.43\ B_j - 0.14\ B_j^{3/2} - 13.59, \qquad (1)$$

$$15 < B_j < 27 \text{ mag},$$

where N has units of deg$^{-2}$ mag$^{-1}$. Equation 1 follows the data from relatively local galaxies at 15th magnitude with Euclidean log $N$—mag slope of 0.6 to the beginning of the turn-over at $B_j = 26$–27 mag. The error bars at the faint end for the average counts arise primarily from statistical fluctuations in the crowding correction Monte Carlo simulations which were performed on the real data images. Fluctuations in galaxy number count from field to field in CCD surveys are inversely correlated with limiting magnitude because of Poisson noise in the bright count statistics and because some brighter galaxies are more clustered on the sky. These field-to-field fluctuations range from 50% at 20th mag to 10% at 27th mag. Finally, for most of the bands the data at the bright end are a mix of CCD and various photographic surveys with correspondingly disparate error bars as a function of magnitude. At a surface brightness of 29 B magnitude arcsec$^{-2}$, the sky is about 15% covered with FBGs. The number counts rise less steeply with magnitude at longer wavelengths, with the slope of the log number-magnitude relation decreasing from 0.5 at 0.4 $\mu$m wavelength, to 0.3 at 0.9 $\mu$m, to less than 0.28 at 2.2 $\mu$m.

I have resisted the temptation to draw families of lines in Figure 3 corresponding to various evolution and no-evolution models, since they show wide scatter at the faint end. As discussed below, there are examples of both kinds of models which can be made to agree with the faint counts in several bands; the details of the model SFR, luminosity functions, and selection effects dominate cosmology. However, one robust property of the counts in the blue band is that they exceed simple no-evolution models using a locally normalized Schechter luminosity function for a $(\Omega, \Lambda) = (1,0)$ cosmology by as much as a factor of ten at the faint end. The longer look-back times and larger volumes in open cosmologies make it relatively easy to build models with larger numbers of faint galaxies.

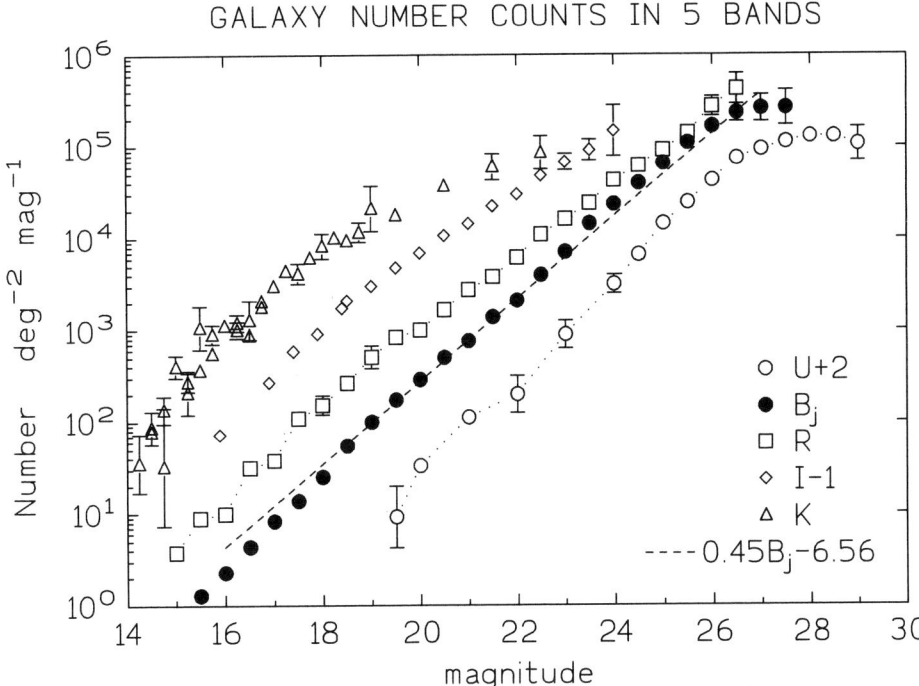

**Figure 3.** *Faint galaxy differential number counts dN(m) as a function of magnitude in five color bands are plotted to their completeness limits. Data are from a number of surveys (see text). Note the decrease in slope $d\log N/dm$ with wavelength of the faint part of the number counts. The bump in the counts at the faint end becomes more pronounced at short wavelengths.*

## 2.2 Faint and Blue

The factor of two increase in Log N - magnitude slope from red to blue wavelengths translates into a blue trend at faint magnitudes. Indeed, even 2.2 µm-selected faint galaxies have extreme blue $B_j$-R colors (Cowie et al. 1991). There was some evidence of a bluing trend from 18 to 22 B mag for field galaxies in the photographic work of Kron (1980, 1982). This trend continues to fainter magnitudes; the deep multi-band CCD surveys detect a twenty fold increase in surface number density and have much higher photometric precision than plates. While the mean $B_j$-R color of zero redshift galaxies is about 1.5–2 mag at 21 $B_j$ mag, many FBGs at 26–28 $B_j$ mag have $B_j - R < -0.2$. The 1.5 mag of bluing in B-R of these FBGs to 27 B magnitude is shown in Figure 4 (see also Tyson 1988; Guhathakurta 1989). Some of these galaxies are as blue as O stars. Color–magnitude plots of stars and galaxies for wide fields show the FBGs going bluer than the stellar blue subdwarf limit in the Galaxy. In the context of stellar evolution these extreme blue colors are usually taken as evidence that one is seeing starburst galaxies at redshifts of 1–2, so that the UV excess is redshifted into the $B_j$ passband. Unfortunately, there are insufficient data on starburst galaxy UV spectral energy distributions shortward of 1500 Å to construct reliable models for high z.

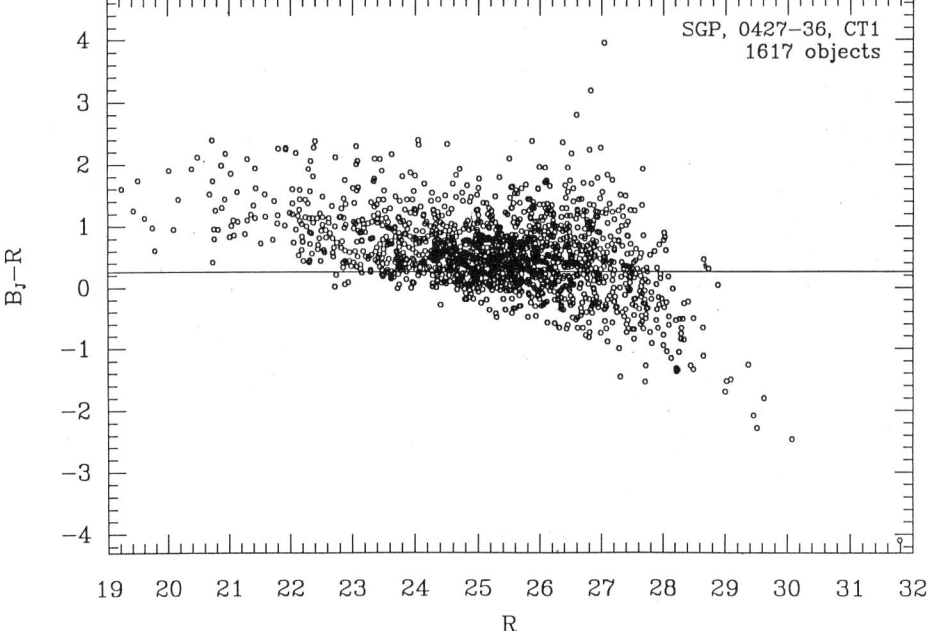

**Figure 4.** *Galaxy $B_j$-R color vs total R magnitude (Guhathakurta et al. 1990). Identical isophotes, determined from the sum image, were used in the two bands in order to avoid systematics. The bluing trend continues to the completeness limit.*

## 3. THE NATURE OF THE FAINT BLUE GALAXIES

What are the FBGs? A population of extreme blue galaxies is expected from metallicity production arguments: an estimate by Cowie, *et al.* (1988) suggested that star-forming galaxies would produce as much as 40% of the observed metals. Several lines of evidence imply that most of these blue galaxies, between 24 and 27 B magnitude, are distributed smoothly over a range of redshifts between 0.7 and 3. Babul & Rees (1992) argued that the majority of the FBGs may be dwarf ellipticals undergoing starburst at redshift about 1, most fading below detectability at the present epoch via supernova wind dispersal. Generally, models which introduce a new population of galaxies at high redshift have a sufficient number of additional adjustable parameters to fit the observed number counts.

Based on estimates of the local galaxy luminosity function Loveday, *et al.* (1992) conclude that the surface density of galaxies exceeds a no-evolution estimate by a factor of two by $B_j$ = 20–21 mag; Colless *et al.* (1990) claim a factor of two excess by 22.5 $B_j$ magnitude. Naturally these comparisons depend critically on the details of the models. Some models with evolution (Yoshii & Takahara 1988; Guiderdoni & Rocca-Volmerange 1989) underproduce the blue number counts in a flat universe. Other models, both with and without evolution, have claimed consistency with the observations (see below). A good review of this luminosity function problem is given by Eales (1993). There have also been claims, based on undefined evolution models, for a significant deficit of high reshift galaxies in the faintest redshift surveys. However, a mild evolution model shown in Figure 5 agrees well with the redshift data, but is not unique. Spectroscopy of a large sample of galaxies to 25th mag in many fields is required for the next step.

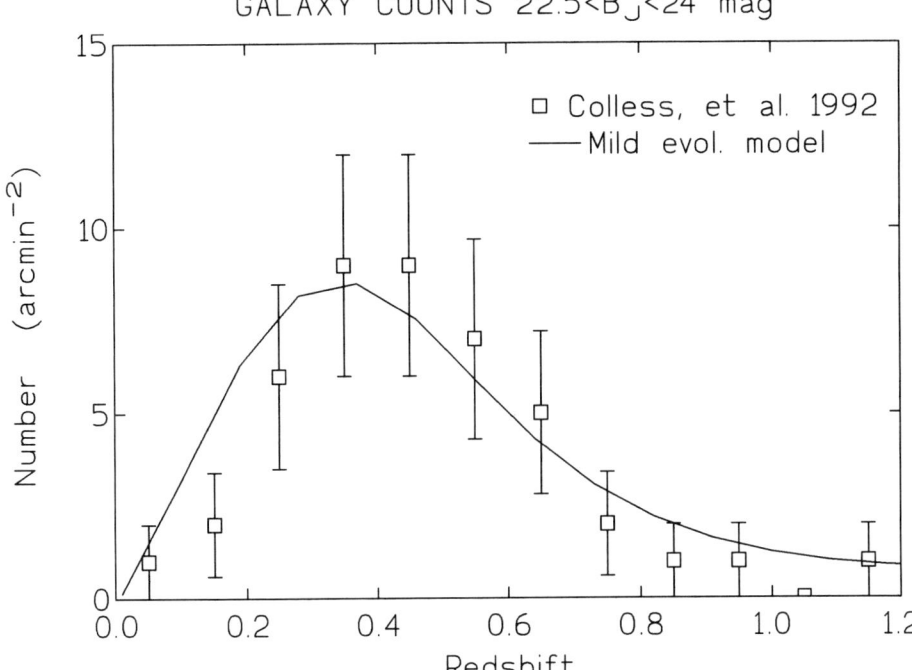

**Figure 5.** *Number-redshift relation from the survey of Colless, et al. 1993, together with a mild-evolution Bruzual model prediction including the surface brightness sample selection and lost-light corrections. No significant deficit of high redshift galaxies is seen.*

Not surprisingly each of these models confirms that a redshift survey to 23rd magnitude does not sample a significant fraction of the population of faint galaxies to 26th magnitude. There has been some confusion in the literature where, on occasion, galaxies between 21–23 B mag are called FBGs (mostly in reports of spectroscopic work, where it is difficult or impossible to go to fainter magnitudes). What should we expect for the redshift of a randomly chosen 24th to 26th magnitude galaxy? Figure 6 shows differential number-redshift counts at 24–26 B magnitude in a variety of recent models.

The expected number of galaxies per log redshift interval dN/dLog z ($\Delta$Log z = .04) for the $\Omega = 0.1$ C model of Guiderdoni & Rocca Volmerange (1989), which does not include a lost-light correction, are shown, together with one of our mild evolution models for $\Omega = 0.2$ and $H_0 = 50$. The $(\Omega, \Lambda) = (.2, .8), (.2, 0)$ and $(1, 0)$ models of Yoshii et al. (1993) are also shown. The $(.2,.8)$ model yields 24–26 mag counts which are too high. These models use slightly different luminosity functions. Models corrected for observational selection should be compared with raw data, while the others should be compared with corrected data. See also Carlberg & Charlot (1992).

A broad range of FBG redshifts is a feature of all these models and is less dependent on model parameters than the number-redshift relation at brighter magnitudes. Redshifts of 23 mag field galaxies in most models peak around $z = 0.4$, consistent with existing galaxy spectroscopic data. On the other hand, the bulk of the FBG population is over 16 times fainter ($m > 26$ mag) and samples redshifts up to 3. It is interesting that, given the very different parameters in these models, most FBGs are predicted to

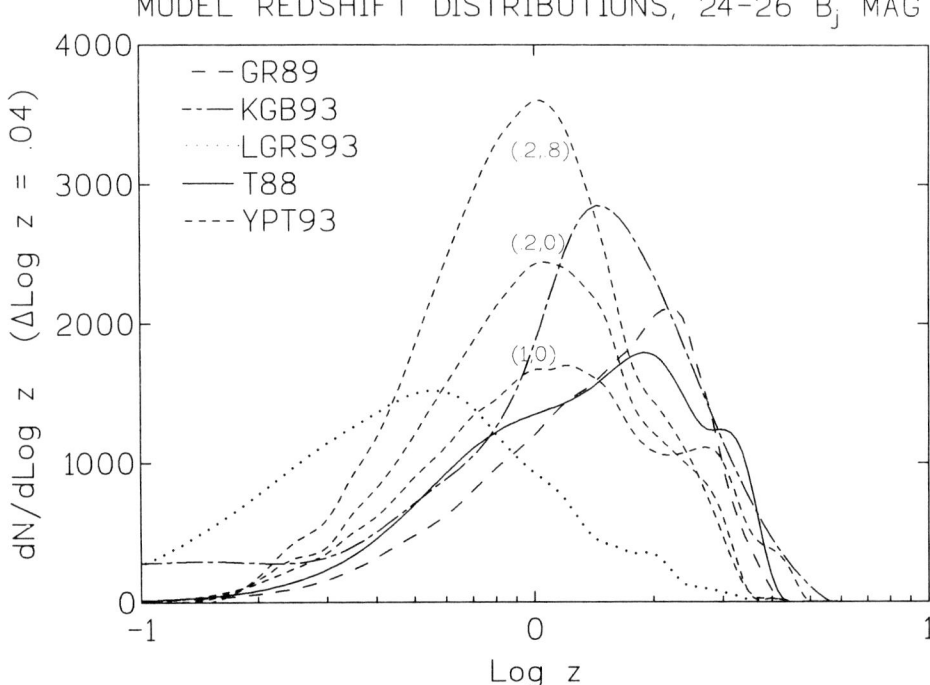

**Figure 6.** *Predicted differential number-redshift relation $dN(z)/d\mathrm{Log}\,z$ at magnitudes 24–26 $B_j$ ($\Delta \mathrm{Log}\,z = 0.04$), for galaxies in several models which are roughly consistent with the observed $N(z, m < 22)$. The models cover a wide range of evolution and cosmology. The models GR89 (Guiderdoni & Rocca-Volmerange 1989) and LGRS93 (Lacey et al. 1993) do not include observational selection corrections. Also shown are the mild evolution models T88 (Tyson 1988) and YPT93 (Yoshii et al. 1993), both with observational selection corrections at 29 $B_j$ mag arcsec$^{-2}$ and minimum area 2 arcsec$^2$, and the no-evolution heuristic model KGB93 (Koo et al. 1993).*

lie at $z > 1$. The non-evolution KGB93 model has the bulk of 24–25 B mag FBGs at redshifts above one.

### 3.1 Observed Distribution in Redshift

Redshift data will provide the only hope of disentangling the trends of galaxy mix, luminosity functions, evolution scenarios, and cosmology. Figure 7 shows the sum of all redshift surveys of field galaxies, excluding AGNs and QSOs, to date. The recent LDSS-2 (Glazebrook *et al.* 1993) survey at 23-24 mag is shown as filled circles. Overplotted are mean redshift lines from a mild evolution model, for three different galaxy types.

The trend to redshift $\approx 1$ at 25th $B_j$ magnitude is clear. Note that a typical galaxy seen at z=1 may be a 0.1 $L^\star$ galaxy and that at 25th magnitude there may be a wide range in redshifts extending from 1–3. Ten meter and larger telescopes will be required to obtain redshifts of non-lensed galaxies fainter than 24th magnitude. The Colless *et al.* and Broadhurst, *et al.* redshift surveys imply that the comoving number density of galaxies brighter than about 0.1 $L^\star$ was roughly three times higher at z=0.3 than

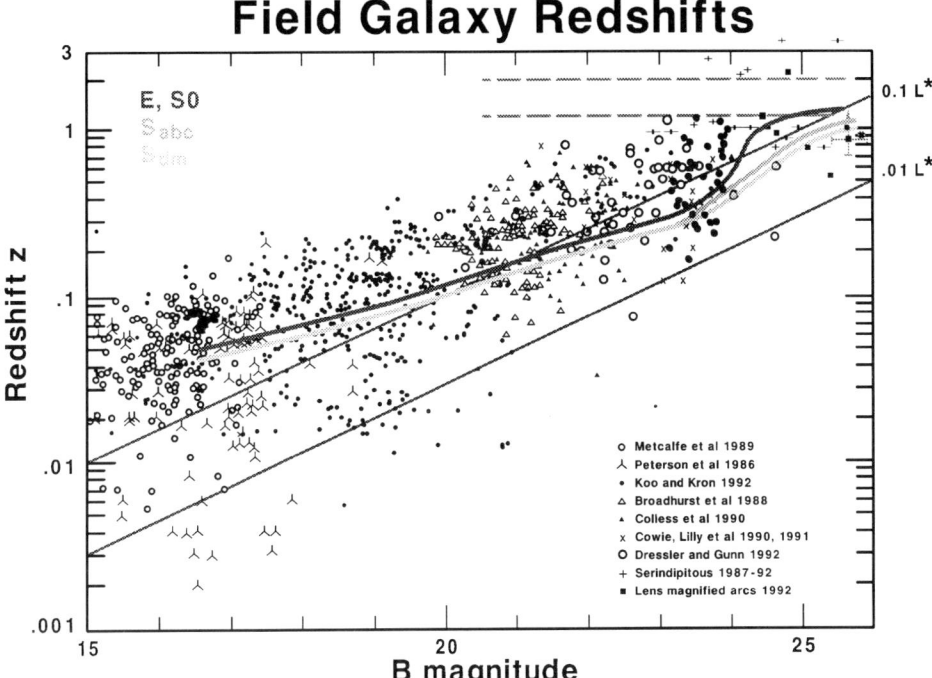

**Figure 7.** *A plot of field galaxy redshift vs $B_j$ magnitude, from all the redshift surveys to date (adapted from Koo & Kron 1992). New survey data, including Ly-$\alpha$ surveys and lens arc surveys, have been added. Note the absence of a dominant population of extreme dwarf galaxies at the faint end. The band (dashed lines) between redshifts 1.2 and 1.8 is excluded from ground based surveys due to inaccessibility of Ly-$\alpha$ or oxygen 3727 Å emission lines.*

now (Eales 1993; Lonsdale & Chokshi 1993). In the absence of merging, many FBGs must evolve to dwarf luminosities at the current epoch.

Although redshifts of individual galaxies fainter than 24th magnitude are generally not known, rough statistical limits have been set via gravitational lens image distortion in the case of several foreground massive clusters with independently determined masses. Thus, a lower redshift limit for these faint galaxies may be derived from their response to a gravitational lens placed in the foreground. A good test is provided by the dark matter lens in the well studied cluster 0024+16 (z=0.39). Roughly the same number of FBGs per square arcminute are seen behind these red clusters, but their images are stretched by the gravitational lens and are aligned orthogonal to the vector to the cluster center. We take this as evidence that most of the background galaxies are at redshifts larger than 0.7.

An upper redshift limit for these FBGs may be determined from its spectral energy distribution over the wavelength range 0.3–1 $\mu$m, and the discontinuity in the spectrum of stars and galaxies at the Lyman break of hydrogen at 912 Å (Guhathakurta et al. 1990). Most of these faint galaxies appear to have a redshift less than three. If a significant fraction of these galaxies were at redshift greater than 2.8, the Lyman break would be shifted through the 0.32–0.36 $\mu$m "U" passband, causing these galaxies to either drop out in this U band or to have a drop in flux between $B_j$ (0.37–0.5 $\mu$m) and

U wavelengths. Lyman breaks from most stars in these primeval galaxies are probably at least a factor of two. In addition, these galaxies would have even more hydrogen than present galaxies, which would absorb all the Lyman continuum photons, causing these galaxies to be black in the U-band for redshifts larger than 3. This Lyman break has recently been seen by Dickinson & Spinrad in the spectrum of a bright galaxy associated with a radio source at redshift 3.8.

## 3.2 Galaxy Morphology Trends

When surface brightness effects are taken into account, detected galaxies at high redshift are expected to be sub-$L^\star$ in all models. When combined with the additional expected magnitude-size correlation, model scale lengths decrease monotonically with apparent magnitude. Galaxies at 26–27th magnitude are observed to have half-light diameters approaching twice the stellar point-spread function. Thus, it is possible to get some statistical information on the average radial profile of the FBGs, at a given magnitude, by making use of the observed stellar point spread function obtained from brighter stars in the same image. After seeing deconvolution (Tyson et al. 1993) galaxies fainter than 26 $B_j$ mag are found to have average exponential scale length 0.2–0.5 arcsec. The raw intensity profile of an average of 100 25th B mag galaxies is shown in Figure 8 along with the stellar PSF.

What fraction of faint galaxies are low surface brightness galaxies? A search of the deep CCD catalogs for objects classed as "diffuse" turns up less than 5% of all galaxies in the magnitude range 23–26 $B_j$. Monte Carlo tests in which we insert low surface brightness galaxies finds them efficiently, so that these kind of galaxies are probably not missed in the deep CCD surveys, although nearby versions of them would be missed because their angular size would be larger than the chopping angle in the shift-and-stare observations. Galaxies of 26th $B_j$ mag appear to be somewhat more extended in comoving metric size than brighter galaxies at low redshift, roughly like (1+z).

Galaxy angular scale data from a number of sources are shown in Figure 9, in which the best fit exponential scale radii are plotted vs total $B_j$ magnitude. Two methods of PSF deconvolution were used: (1) PSF quadrature subtraction using the cataloged first radial moments of the intensity profiles of stars and galaxies, and (2) maximum likelihood solution of a Bayesian integral PSF convolution with an exponential profile. Monte Carlo seeing simulations were used to test these methods. Using 120,000 galaxies to 25th mag and 15,000 to 27th Bj mag we find, to surface brightness 29 Bj mag arcsec$^{-2}$, an average "half-light" (half total luminosity) radius of $0.86''$ for Bj=23–24, and $0.7''$ for Bj=25–26. The former is not too different from preliminary results from the HST Medium Deep Survey parallel observations (Griffiths et al. 1993).

Plotting a mean value for $R_s$ is perhaps misleading since, in a given magnitude bin each survey field shows a large range of $R_s$ values significantly in excess of noise. In addition to the mix of all types in such an average, which tends to lower the best fit scale lengths below their value for a spiral-only sample, there is the question of disturbed morphology. While some galaxies may exhibit a disturbed morphology with multiple peaks, the average scales shown in Figure 9 represent the composite scale of the galaxies. The lines plotted in Figure 9 show schematically the behavior of the scale length in simple single population mild luminosity evolution models for several types of galaxies. Model comoving sizes are multiplied by 1+z to account roughly for collapse of the star forming region. A tidally triggered galaxy formation model of Lacey et al. 1993 is also shown; this is not significantly different from the scale lengths of spiral galaxies

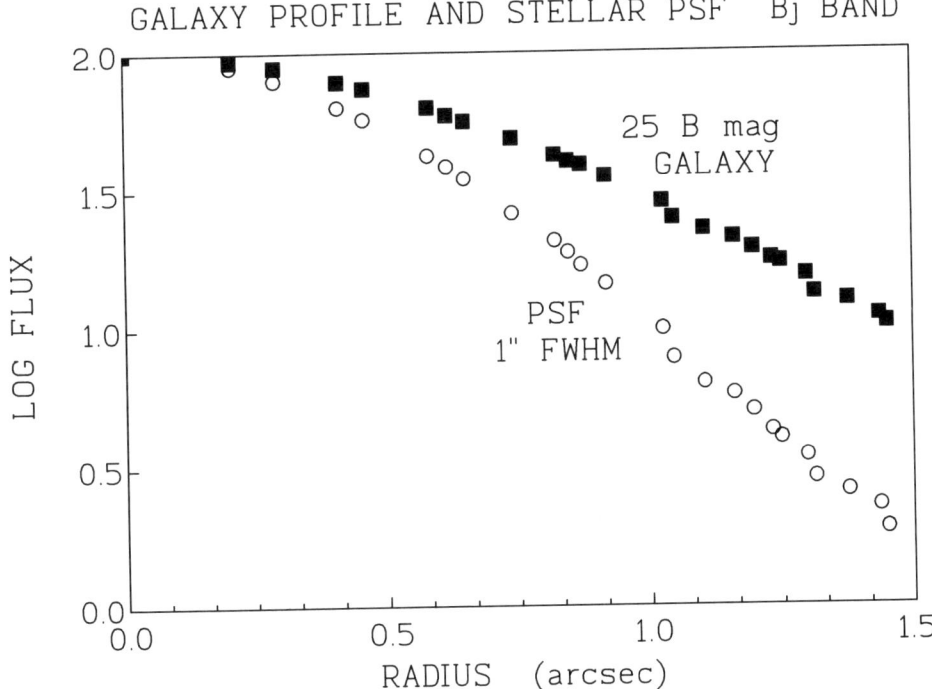

**Figure 8.** *The raw profile of an average 25th B mag galaxy is shown, along with the stellar PSF. In each magnitude bin an integral deconvolution technique minimizes sensitivity to noise in solving for the deconvolved exponential scale.*

in the mild evolution model, but their colors in this magnitude range are more than 1 mag redder than the data shown in Figure 4. The bump in $R_s$ for the mild evolution model at 24–26 mag is due to higher SFR at z=1–2 (see model curves in Figure 7), which also produces the bump in the counts similar to that seen in Figure 3.

Broadhurst et al. (1988), and more recently Cowie et al. (1991) and Babul & Rees (1992), suggested that the FBGs may be dwarfs undergoing starburst which have faded below detection in recent times. Starbursting dwarf galaxies at high redshift tend to have exponential scale radii less than 0.1 arcsec (see Carlberg & Charlot 1992), so that the FBGs are not likely to be these unless star formation in dwarf galaxies at z=1 takes place over a very much larger comoving volume than in nearby dwarfs. In place of the observed local LF, Koo et al. (1993) have proposed an $\Omega = 0.1$ no-evolution model which choses from 154 combinations of absolute magnitude and color class, attempting a match to N(m) and color data; this leads them to unusual LFs. They effectively increase the numbers of Sdm type at moderate redshift (.5–1) and thus would predict very small sizes for the majority of faint galaxies (bottom curve in Figure 9).

Parametrizing a typical high redshift galaxy by a single scale length may be misleading if these objects have disturbed morphology with multiple peaks within 1–2 arcsec. Such objects would have large apparent average scale lengths. Baron & White 1987 showed that in CDM theory the properties of the FBGs are consistent with their being primeval galaxies at z = 2–3, with formation continuing to z = 1–2. While their large scale lengths are borne out by the data shown in Figure 9, their model counts fall below the observed counts. The morphology of these model forming galaxies is diffuse and

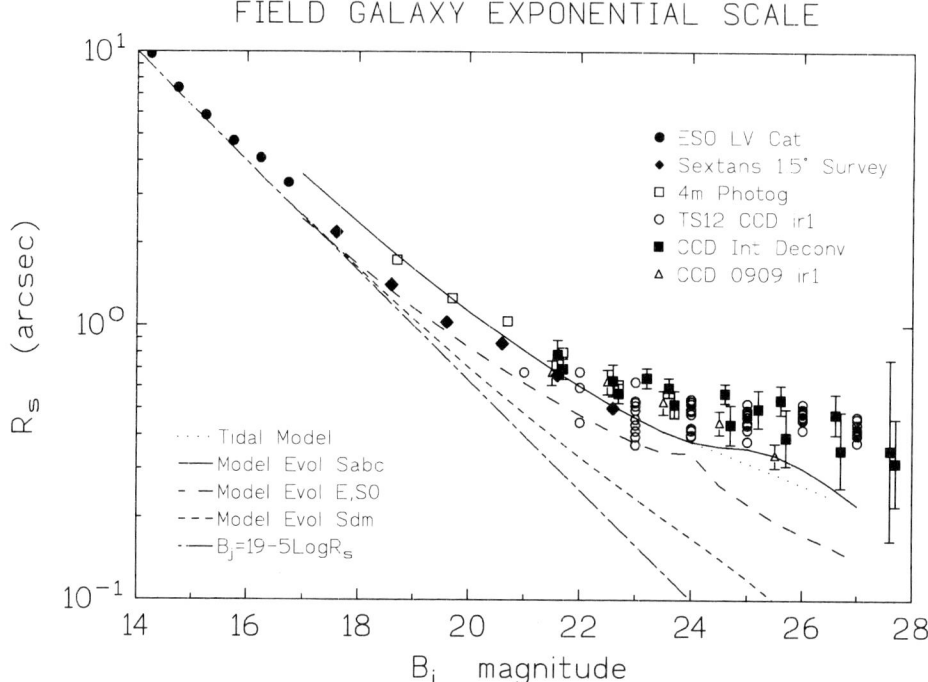

**Figure 9.** *Data and simple models for the seeing-deconvolved angular exponential scale radius as a function of apparent $B_j$ magnitude. The angular scale length does not show the classic minimum, even as a function of redshift, due to surface brightness dimming. There is no evidence for any color-band dependence of $R_s$.*

complex by comparison to nearby galaxies, a result which is not in disagreement with the data. Recent HST WFC deep parallel data show galaxies at 23–24 mag with a higher incidence of disturbed morphology than in brighter samples (Windhorst et al. 1993). But a large fraction of FBGs would have to have a disturbed morphology to bias the deconvolved scales shown in Figure 9. It will be interesting some day to obtain imaging of individual FBGs with $0.1''$ resolution at 29 mag arcsec$^{-2}$ surface brightness.

Are there trends in the seeing-deconvolved principal axis ratio of galaxies with magnitude? We have studied the faint galaxy ellipticity distribution for field and cluster galaxies. The FBGs have considerably higher ellipticities than equally faint galaxies in clusters. Only about 10% of the faint field galaxies have observed (uncorrected for PSF) ellipticities less than 0.2. When corrected for seeing, the axis ratios increase even further. Nevertheless, the observed ellipticity distribution of the FBGs after PSF deconvolution is not significantly different than that of the nearby population of field galaxies, which is dominated by spirals. This may suggest that disks have been forming over a wide range of look-back times. Finally, it must be mentioned that the large average galaxy scale lengths shown in Figure 9 are easily accommodated in a cosmology with positive cosmological constant, a model which produces very large counts and is also consistent with correlation data (Efstathiou et al. 1991; Yoshii et al. 1993).

## 3.3 Galaxy-Galaxy Correlation

The nature of the FBG population(s) will also be constrained by their angular and spatial correlations $w(\theta, mag)$ and $\xi(r, z)$. Clustering is type dependent; nearby late-type spirals have nearly a factor of four lower clustering amplitude than early type galaxies (Giovanelli et al. 1986). Like the clustering at the current epoch, the red galaxies at z=0.3 appear to be more strongly clustered than the blue galaxies (Bernstein et al. 1993); this red/blue difference in clustering is shown in Figure 10 for a large sample of 20–23 mag galaxies for which the redshift distribution is known.

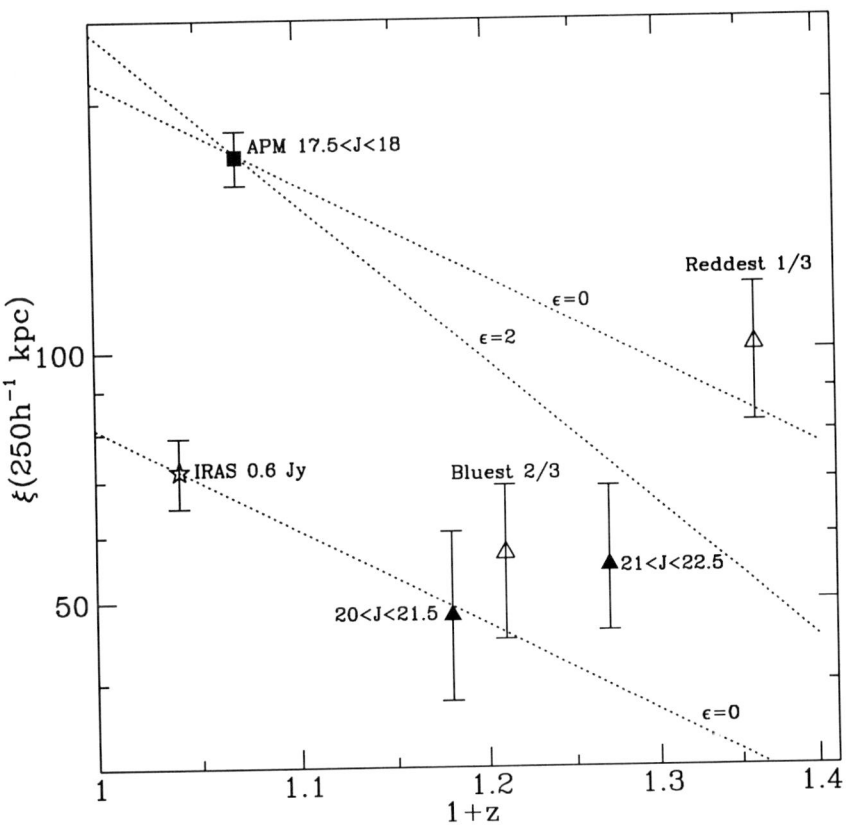

**Figure 10.** The two-point correlation function vs redshift for galaxies in various surveys (from Bernstein et al. 1993). Pairs of lines labeled $\epsilon = 0$ (clustering fixed in proper coordinates) and 2 (much faster growth of clustering) chart the trend. Note the stronger clustering of red galaxies. The clustering of the FBGs [see Figure 11] is so low that a maximal redshift distribution must be used (Efstathiou et al. 1991).

The clustering at z=0.3 is about half the $\xi(r = 250h^{-1}$ kpc) seen in nearby galaxies, yet quite similar to the clustering of nearby IRAS galaxies. This stronger correlation at redder wavelengths was also observed in our very faint data (Efstathiou et al. 1991); this implies, not surprisingly, that redder types which are at lower redshift are being selected in the red band. Perhaps the FBGs are progenitors of the IRAS-selected population. The amplitude of the FBG correlations fall below predictions of theories of linear growth of structure at the faint high-redshift end. By 26th $B_j$ magnitude the amplitude of $w(30\text{ arcsec})$ is ten times lower than linear CDM, even if most of the 26 mag galaxies are put at redshifts approaching 3 (Efstathiou et al. 1991). But this may not necessarily imply an additional population of fading unclustered FBGs.

Yoshii et al. (1993) have compared the correlation data on faint galaxies with N-body simulations which are allowed to go non-linear. Although is it always impossible to identify the present epoch in such simulations, they find good agreement with observations in a simulation with low bias. Both highly biased CDM and open models disagree significantly with the clustering observations. CDM-like models with power on large scales may be consistent with these data; then merging enhances early nonlinear growth of correlations (Melott 1992), corresponding to a value of the clustering growth parameter $\epsilon$ larger than 0; i.e. more clustering than fixed in proper coordinates. To $B_j < 22.5$ mag the evolution in clustering of blue galaxies appears to be due to an increasing fraction of late type or star-forming galaxies at the faint end, combined with weaker small-scale clustering for late types. These blue galaxies appear to be clustered more like scaled predecessors of the IRAS galaxies. This similarity may suggest ongoing starburst activity. Merging scenarios, even if self-similar, do not solve both the low correlation and high density problems. Since low redshift dwarf galaxies are highly correlated with low redshift $L^\star$ galaxies, which are themselves highly correlated, this very low correlation for the FBGs rules out the possibility that a substantial fraction of the FBGs are nearby dwarfs.

Deep imaging of a 1.5 degree field with 120,000 detected galaxies to 26 $B_j$ magnitude lets one follow $w(\theta)$ over a wide range (Postman et al. 1993). Figure 11 shows the amplitude of $w(1\text{ arcmin})$ as a function of $B_j$ magnitude from 18th to 26th mag. The monotonic decrease to the small correlations at 25.5 mag reported in Efstathiou et al. (1991) for several small CCD fields is clearly seen.

Also shown in Figure 11 are data from Maddox et al. (1990) [open triangles] and Koo & Szalay (1984) [open squares]. The curves are from N-body simulations by Yoshii et al. (1993): $(\Omega, \text{bias}, \Lambda) = (1,1,0)$ dotted, $(1,1.5,0)$ dashed, $(.2,0,.8)$ solid. The $w(\theta, m)$ data alone cannot distinguish between these three models, but the observed amplitude is much lower than either highly biased CDM models or open models with zero cosmological constant.

Most models of structure formation show some increase in correlation amplitude at the faint end (Carlberg & Charlot 1992) whereas the observations at the faint end follow a simple power law. Introducing interaction-induced starburst in these models makes no significant change in the correlation amplitude for FBGs. Perhaps many FBGs are being seen during initial starburst activity at an early epoch, before nonlinear clustering. In any case, models for FBGs which attempt to account for the high surface number density and very low correlation are forced into a maximal redshift distribution extending at least to $z = 3$.

## 3.4 Discussion

Is the sky similarly crowded with faint galaxies at K band? It is conceivable that there was some early stellar burning at redshifts higher than 6 (so as not to exceed the observed 0.9 $\mu$m galaxy counts) substantially raising the K-band counts and total flux. If there were a bright phase of stellar burning at redshift about 10–20 (population 3 stars), they would be seen in a deep K-band survey. The new HgCdTe area imagers may well create as much of a revolution in K-band imaging as the silicon CCDs have in

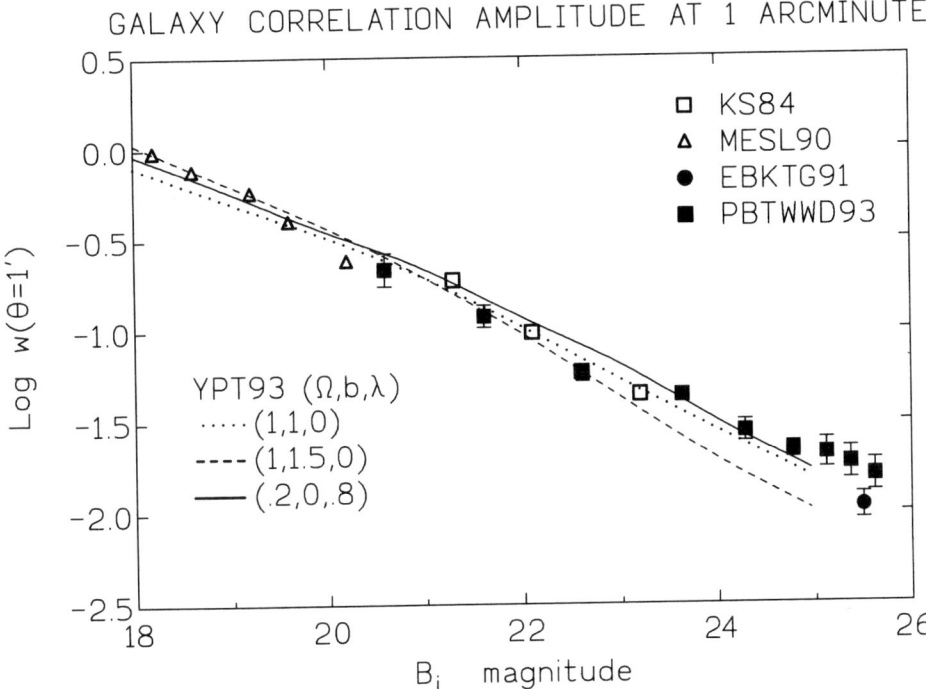

**Figure 11.** Observations of the amplitude of the galaxy-galaxy angular correlation at one arcminute is shown vs $B_j$ magnitude. The solid triangles are data from a 1.5 square degree field containing 120,000 galaxies. The various theoretical curves are discussed in the text.

the optical, but deep surveys to date using these devices have covered an area of only one square arcminute at 22nd K magnitude (Gardner et al. 1993).

In going from the optical wavelengths to the near-IR to obtain galaxy counts, the initial hope was that one would be rid of the uncertain K-corrections from the unknown far UV portion of galaxy spectra that might affect our understanding of the blue counts. However, even K-selected galaxies turn out not to be at terribly high redshifts. For example, for a typical galaxy at a redshift $z = 0.5$ the rest wavelength corresponding to the 2.2 $\mu$m K-band is about 1.5 $\mu$m, and little is known about the current-epoch luminosity function of galaxies at 1.5 $\mu$m. So that even no-evolution predictions are difficult to make for the K-band. Dusty primevals may be mixed with the large numbers of FBGs. In this connection, it is perhaps significant that we do not see many very red objects at faint I or K magnitudes. Either there was negligible galaxy formation (proto-ellipticals and proto-bulges) for $z > 5$, or there was adequate dust from a very much earlier star formation epoch to hide that luminous phase. But that much dust would seriously lessen the likelihood of finding $z = 4-5$ QSOs. The counts at the faint end of the K-band data are surprisingly low, given the I-K color of somewhat brighter galaxies within the combined completeness limits. Given the extreme blue colors of K-selected galaxies, it would follow that B-selected galaxies would have even bluer colors, due to color selection bias; the FBGs dominate the bluer passbands. Yoshii et al. (1993) point out that standard models of luminosity evolution of a single population of evolving galaxies cannot reproduce both the K-band counts and the I-band counts with the same model parameters, independent of assumed cosmology.

## 3.5 Models Revisited

Will it be possible to constrain cosmological models with number-magnitude N(m) counts of these faint galaxies? Speculation on the nature of these FBGs has ranged from low redshift dwarf galaxies undergoing star formation, to high redshift primeval galaxies forming the majority of their initial stellar population. There now exist a variety of models consistent with most observations with different evolution scenarios, including various stellar formation histories, merging, galaxy luminosity functions, stellar initial mass functions, and cosmological parameters. Without number-redshift data little if any cosmological information can be uniquely determined from the number-magnitude-color counts alone. Under restricted model scenarios in which the shape of the galaxy luminosity function is assumed invariant, stellar formation begins at some universal epoch (different for spirals and ellipticals), and the arbitrary assumptions of no merging and no episodic star formation, various authors have remarked that it is difficult to fit the observed galaxy counts with a high value of the cosmological density $\Omega$, the number-magnitude counts exceeding evolution model counts for $\Omega = 1$ at 27th magnitude by a factor of ten (Bruzual 1987; Arimoto & Yoshii 1987; Tyson 1988; Yoshii & Takahara 1988).

Through the effect of volume, changes in cosmology affect the numbers of galaxies seen at some redshift N(z). In general, galaxy number-magnitude counts N(m) are relatively more sensitive to evolution, and number-redshift counts are more sensitive to cosmology. The unavoidable result is that we will never be able to conclude much of interest cosmologically from galaxy number-magnitude counts without also having good number-redshift information complete to the same surface brightness. Even then, we will have to know something about the shape of the galaxy luminosity function vs redshift. It is illuminating to play with evolution models, relaxing the assumption of invariant luminosity function, introducing appropriate observational selection, episodic SFR, and/or merging. While it is possible to fit the observed N(m,$\lambda$) with low $\Omega$ and the traditional models, it is also possible to get a good fit by varying the faint tail of the luminosity function vs look-back time or introducing galaxy number non-conservation.

Merging models (galaxy number non-conservation) have been proposed by White 1989, Cowie et al. 1991, and Broadhurst et al. 1992. But in a merging galaxy model which properly includes observational selection effects, Yoshii (1993) finds too few counts at 24–26 mag by a factor of 3–4 in a $(\Omega,\Lambda) = (1,0)$ model. There is also a problem getting rid of enough galaxies by merging: Dalcanton (1993) has shown that only 5–15% of the mass in galaxies at z=0.35 can have merged into the local population of ellipticals and spirals. To search for merging at z = 0.3 very long exposure HST imaging will be helpful, particularly if redshifts of the galaxies can be obtained. It is perhaps risky to argue that irregular 2-dimensional morphology implies merging since there are several other expected mechanisms for this morphology in deep images.

At this time, we know even less about the nature of galaxies at z=1 than we do at z=0.3, where model calculations already extrapolate galaxy luminosity functions eight magnitudes below $L^\star$. General galaxy evolution models have more adjustable parameters than necessary for a good fit to the N(m,$\lambda$) data. This situation changes when we have observational constraints on the N(z) distribution, and it will perhaps be possible in coming years to obtain corresponding rough constraints on $\Omega$ using these

N(z) constraints together with some limits on the luminosity evolution. Koo, et al. (1993) have proposed a no-evolution model which instead changes the mix of galaxy types as a function of redshift. Lacey et al. (1993) have a tidal interaction model. Models which have been successful in explaining some data have been less compatible with other kinds of data on the FBGs. My Bruzual model (1988) predicts evolving bright ellipticals which probably should have been seen in redshift surveys of selected objects at 24th mag; faint galaxies in the heuristic Koo et al. model are too blue and probably too small; the Lacey et al. model gives colors which are far too red. Moreover, most models assume a single time-invariant luminosity function. In any case, the total EBL is an integral over redshift of all sources of luminosity and is less sensitive to cosmology.

## 4. THE OPTICAL EBL FROM DISCRETE OBJECTS

Providing the observations go to sufficiently faint surface brightness to catch most of the light from these objects, the EBL from all discrete objects can be calculated directly from the galaxy count data by summing flux × count products. Deep imaging in several wavelength bands can be used in this way to construct a rough spectral distribution for this EBL. Since discrete objects are detected, one is not sensitive to any uniform diffuse component in the EBL, so that the EBL one derives by adding up flux of discrete objects must be considered a lower limit if there were a smooth extragalactic visible component uniform on greater than 1 arcminute scales.

### 4.1 EBL From Integral Flux Counts

Galaxy counts vs isophotal magnitude may be used to calculate directly the EBL, down to the surface brightness threshold for detection. The CCD data covering the range 0.3–1 $\mu$m can be used to calculate the EBL down to 30 $B_j$ mag arcsec$^{-2}$. The observed counts flatten fainter than 27 $B_j$ mag, with most of the EBL flux originating from galaxies around 24 B mag. Integrating the flux times the differential number counts, including the uncertainties at the faint end of the counts, we arrive at the following EBL at five wavelengths covering the octave $\lambda = 3200$–24000 Å:

$$\lambda F_\lambda = 2.5(+.07 - .04) \times 10^{-6} \text{ erg cm}^{-2} \text{ sec}^{-1} \text{ sr}^{-1} \text{ at 3600 Å.}$$
$$\lambda F_\lambda = 2.9(+.09 - .05) \times 10^{-6} \text{ erg cm}^{-2} \text{ sec}^{-1} \text{ sr}^{-1} \text{ at 4500 Å.}$$
$$\lambda F_\lambda = 2.9(+.09 - .05) \times 10^{-6} \text{ erg cm}^{-2} \text{ sec}^{-1} \text{ sr}^{-1} \text{ at 6500 Å.} \quad (2)$$
$$\lambda F_\lambda = 2.6(+.3 - .2) \times 10^{-6} \text{ erg cm}^{-2} \text{ sec}^{-1} \text{ sr}^{-1} \text{ at 9000 Å.}$$
$$\lambda F_\lambda = 7.2(\pm 1) \times 10^{-6} \text{ erg cm}^{-2} \text{ sec}^{-1} \text{ sr}^{-1} \text{ at 2.2} \mu\text{m.}$$

In the blue this is 30 times the surface flux threshold of these observations and is equivalent to a diffuse EBL of $0.53(+.02 -.01)$ $S_{10,G2V}(V)$. The flattening of the N(m) counts originally seen in the $B_j$ band have been seen now more prominently in the U band, and thus represents the outer shell of magnitude contributing significantly to the EBL. Very faint diffuse objects larger than 30 arcsec are not counted in the EBL sum in equation 2, and wide-area searches for nearby examples of such low surface brightness galaxies will be worthwhile (Dalcanton 1994). Given that flattened profile low surface brightness galaxies with radial scales larger than 6 arcseconds comprise less than a few percent of the observed number counts in the magnitude range 23–27 B mag, additional EBL from these low surface brightness galaxies appears to be negligible.

Galaxies fainter than 20 $B_j$ mag contribute about 75 percent of the EBL at 4500 Å. The redshift shell which dominates the EBL is dependent on the galaxy luminosity evolution $L^\star(z)$. The corresponding magnitude shell contributing most to the EBL at 4500 Å is that magnitude beyond which the slope dN/dm drops below 0.4: 26 $B_j$ mag. From what little we know of the redshifts of these galaxies, the outer redshift shell contributing to the EBL ranges over redshifts 1–3. Interestingly, the $B_j$-R color of K-selected faint galaxies is very blue and indistinguishable from the FBGs; no population of K-bright galaxies was found, suggesting that starburst activity takes place mainly at redshifts less than 5. A starburst phase at redshift greater than 20 would not be excluded, however. Due to the shallow number-magnitude slope in the K band, most of the 2 $\mu$m EBL is contributed by 16–18 K magnitude galaxies at comparatively low redshift. Thus, the nature of the FBGs at 2 $\mu$m is separated from the issue of the EBL at long wavelengths; the B-K color of the EBL is unrelated to the colors of the FBGs. Some galaxies at K band have low surface brightness and small scale lengths, and these may be the FBGs. The K band surveys do not yet go faint enough to adequately sample the 27th B magnitude FBGs. Thus, the $B_j$-K color of the EBL can be redder than the color of a typical FBG because giant stars in bright galaxies contribute so heavily to the K luminosity. The $B_j$-R color of the FBGs is near zero, like O stars, so that most of their energy in the 0.3–2 $\mu$m region is in the blue.

## 4.2 Spectral Distribution in the Optical

The spectrum of the EBL is related to the integrated star formation rates in galaxies over all relevant redshifts, including K-correction. As such, it can be a diagnostic of galaxy formation scenarios. The U, $B_j$, R, and I surface flux integrals for the EBL are plotted in Figure 12 vs wavelength. Note the short wavelength rise. The error bars on the EBL flux are $3\sigma$, and are mostly due to field-to-field fluctuations in numbers of bright galaxies. The spectral distribution of the optical EBL is slightly redder than flat in $F_\nu$ (see Guhathakurta et al. 1990). This SED is more UV-bright than nearby galaxies and is probably due to star formation at redshifts up to 2–3 in the FBGs. A theoretical prediction by Guiderdoni & Rocca-Volmerange (1990), for two cosmologies and different luminosity evolution is shown in Figure 12 as dashed and dotted lines. This large formation redshift model EBL is not very different from previous models by Yoshii & Takahara (1988) and Partridge & Peebles (1967). Extending continuous galaxy formation down to smaller redshift would have the effect of raising the blue flux, more in agreement with the data.

Note that the diffuse component of the EBL must be a small fraction of this discrete component, since the 10–300 arcsec upper limit for large diffuse sources at 2.2 $\mu$m (Boughn et al. 1984) is only slightly above the measured EBL from discrete sources.

## 4.3 Extension to the UV

A review of non-imaging EBL surveys in the UV was given by Paresce & Jakobsen (1980). Many models for the FBGs predict a source spectrum rising in the UV to Lyman-$\alpha$. More sensitive UV and near-UV surveys could determine if the galaxy counts at a given flux level are significantly below that seen at 3200–4500 Å, as might be expected in these models due to the redshifted Lyman continuum. If spiral formation continues to redshifts near one, then the new UV sensitive CCDs could be used to detect the bright galaxies expected in the 2400–3000 Å band from an orbiting telescope. The Hubble Space Telescope, if repaired, might detect the bright nuclei of some of these blue galaxies. Galaxies fainter than 24 B mag contribute about 30% of the EBL in the blue. Since the redshift of this faint population is mainly larger than 0.7, their expected contribution to 10 arcsec scale fluctuations of the EBL would be negligible for wavelengths shorter than 1700 Å if these galaxies are "black" in their Lyman continuum. Sasseen et al. (1994) have recently improved on the measurement by Martin & Bowyer (1988) at 1600 Å: $< 50$ photon cm$^{-2}$ sec$^{-1}$ sr$^{-1}$. This corresponds to a flux of $< 6 \times 10^{-10}$ erg cm$^{-2}$ sec$^{-1}$ Å$^{-1}$ sr$^{-1}$, or $< 1$ $nW$ $m^{-2}$ $sr^{-1}$. This upper limit is plotted along with the optical-IR data in Figure 12. A testable prediction of luminosity evolution models is a strong UV excess for low redshift galaxies undergoing starburst. It would be useful to have this direct imaging UV observational test; these galaxies should be visible in a sensitive UV imager with sub-arcsecond resolution.

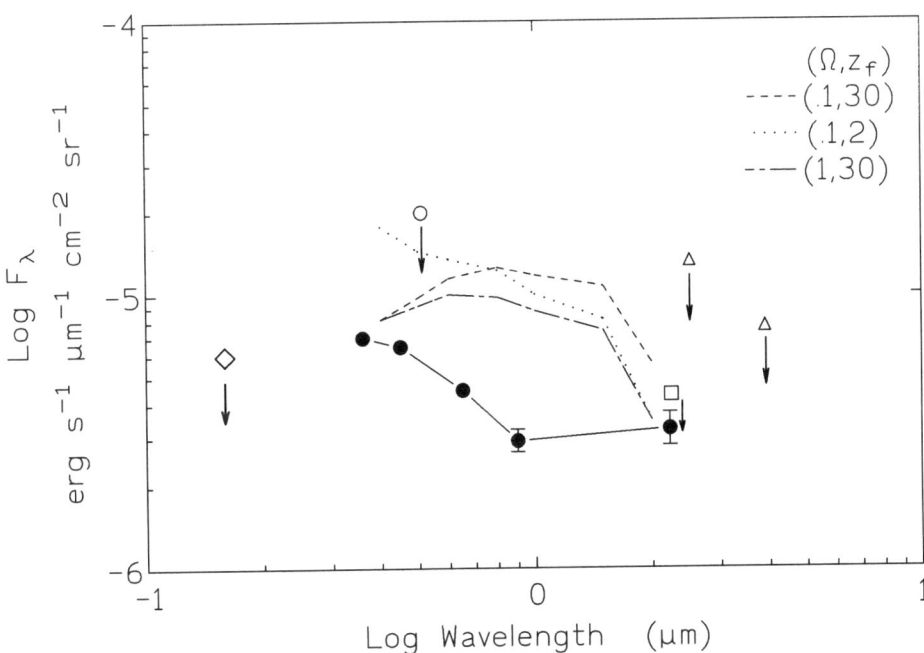

**Figure 12.** *The sum extragalactic background flux from the faint galaxy population as a function of wavelength from 1600 Å to 22000 Å. The error bars are $3\sigma$, arising mostly from field-to-field fluctuations in the brighter galaxy numbers. Note the blue excess of the EBL spectral energy distribution. Several diffuse EBL upper limits are plotted. A recent theoretical prediction is also shown (dashed and dotted lines).*

### 4.4 Isotropy

The faint galaxy counts in 15 randomly chosen high galactic latitude fields and 33 other fields are isotropic on the angular scales between fields (10–180 degrees) to about 10% ($2\sigma$). Since these galaxies are at redshifts of order 1 and above, the galaxies in different fields are out of causal communication (reminiscent of the cosmic microwave background). On angular scales smaller than 30 arcsec we find occasional voids in each of the survey fields. These "dark lanes" are probably not intergalactic dust clouds (with implications for $\Omega_{baryon}$) but rather open channels in the 3-dimensional galaxy distribution.

### 4.5 Surface Brightness

Even with mild luminosity evolution the surface brightness of galaxies at $z > 1$ is expected to be considerably fainter than nearby galaxies. For example, in 1 arcsec seeing an Sb galaxy at a redshift of 1.5 is expected to have an apparent central surface brightness fainter than 27 $B_j$ mag arcsec$^{-2}$, falling by one mag at a radius of one arcsec, and reaching 30 $B_j$ mag arcsec$^{-2}$ at 2 arcsec. The total EBL from discrete objects is equivalent to a diffuse surface brightness of 28.8 $B_j$ mag arcsec$^{-2}$. However it is instructive to plot the diffuse surface brightness of FBGs, binned by magnitude, as a function of magnitude. Figure 13 shows this surface brightness in the $B_j$ band for the light from FBGs averaged over the sky, in one magnitude bins (solid line).

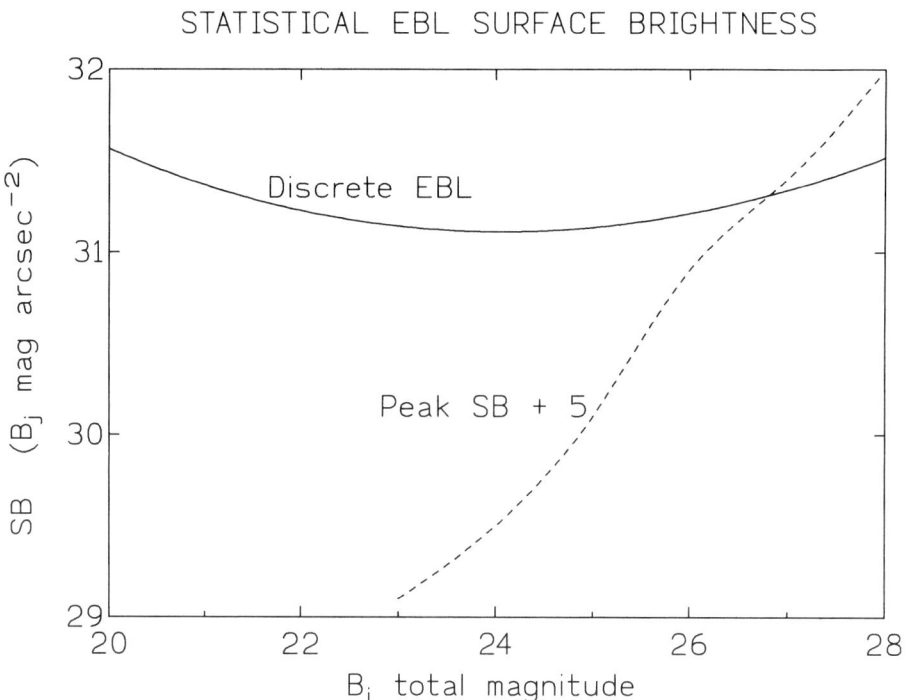

**Figure 13.** *The summed FBG extragalactic background surface brightness in the $B_j$ band (in 1 mag bins), along with the average seeing-deconvolved peak surface brightness of the FBGs, is plotted vs magnitude.*

This EBL surface brightness is calculated from the observed magnitude and differential number counts Also shown is the seeing deconvolved central surface brightness of FBGs vs magnitude, obtained from the deconvolved mean scale length and magnitude (dashed line). A cautionary note: the deconvolved scale lengths exhibit wide scatter, and we would expect the same for the peak surface brightness. Also, the dashed curve in Figure 13 is the extrapolated central surface brightness not including any bulge (see Bosma & Freeman 1993). The curves in Figure 13 impliy that if there is a confusion limit, it is at fainter surface brightness than 30 $B_j$ mag arcsec$^{-2}$, although some galaxies in the deepest images have overlapping outer isophotes. Recently, we have examined a field in which 20 hours of cumulative exposures were obtained. Most of the galaxies appear resolved in 1.1 arcsec FWHM seeing, many with average surface brightness of 28 $B_j$ mag arcsec$^{-2}$ inside a 30 $B_j$ mag arcsec$^{-2}$ isophote.

## 5. SUMMARY

There is an isotropically distributed population of UV-excess galaxies. Their number density is high: about 300,000 per square degree. There appears to be a leveling off of the N(m) counts in the blue and U wavelength bands, giving most of the contribution to the extragalactic background light at magnitudes between 24 and 26 U mag. The nature of the faint blue galaxies remains unclear, although there is evidence that the visible mix of galaxy type or population changes with redshift, and the FBGs are distributed broadly in redshift. Their high surface number density and low correlation implies a maximal redshift distribution extending up to 3. Not many FBGs can have redshifts above 3, due to the lack of Lyman continuum absorbed FBGs in ultra-deep U band imaging. Their average angular size is maximal and their correlations are minimal by comparison to theories.

Because of their large volume, models with open cosmologies can more easily accommodate the count excess at 26 mag. Models with flat cosmologies must adjust one or more of the other available parameters: LF, SFR senarios, merging, cosmological constant, etc. Galaxy number non-conservation is not required, even in flat cosmologies. Multiple poulations of galaxies and a changing type or luminosity mix as a function of redshift is perhaps the most natural explanation of all the observations taken together. Among the single population models of FBGs those with low spatial correlation and open cosmology, or vacuum dominated flat models are in best agreement with the data. It is very possible that there was a population of fading galaxies at moderately high redshift, and future studies to ultra faint flux levels at wavelengths from radio to x-ray would be required for their characterization (see Treyer & Silk 1993). If the FBGs have faded below detection at $z = 0 - 0.1$ in photographic surveys, it is puzzling why they are not seen in the large volume now covered by 4 magnitude deeper CCD surveys at $z = 0.2 - 0.5$.

The optical EBL from summing over the FBGs is close to that required for production of all metallicity, with little room left for an added diffuse EBL. Taken together with the correlation data, the blue color of this EBL and the absence of a comparable density of faint red galaxies at K band implies that we are seeing galaxy formation over a range of redshifts from roughly 1 to 3. Thus, FBGs will be valuable for statistical cosmology; because of their large angular sizes and redshifts these FBGs are a useful tool in statistical gravitational lens studies of foreground mass distributions.

I would like to acknowledge my collaborators in this FBG research over the past decade: Gary Bernstein, Pat Boeshaar, Carol Christian, Raja Guhathakurta, Craig Gullixson, Neal Hartsough, John Jarvis, Greg Kochanski, Steve Majewski, Phil Fischer, Pat Seitzer, Frank Valdes, Ethan Vishniac, Pat Waddell, and Rick Wenk. The staff and telescope operators at CFHT, CTIO and KPNO have been very helpful, sometimes beyond the call of duty. Bill Baum, Art Hoag, and the staff at Lowell Observatory greatly aided the calibration of the $B_j RI$ filter system. I have also had helpful discussions recently with Richard Ellis, Mike Fitchett, Jim Gunn, Richard Kron, Carol Lonsdale, Adrian Melott, Jordi Miralda-Escudé, Gus Oemler, Marc Postman, Chris Pritchet, and Simon White.

## REFERENCES

Arimoto, N. & Yoshii, Y. 1987, A&A, **173**, 23;
Babul, A. & Rees, M. J. 1992, MNRAS, **255**, 346.
Baron, E., & White, S. D. M. 1987, ApJ, **322**, 585.
Bernstein, G. M., Tyson, J. A., Brown, W. R., & Jarvis, J. F. 1993, preprint.
Bosma, A., & Freeman, K. C. 1993, AJ, **106**, 1394.
Boughn, S. P., Saulson, P. R., & Uson, J. 1986, ApJ, **301**, 17.
Broadhurst, T. J., Ellis, R. S., & Shanks, T. 1988, MNRAS, **235**, 827.
Broadhurst, T. J., Ellis, R. S., & Glazebrook, K. 1992, Nature, **355**, 55.
Bruzual, G. 1987, ApJS, **53**, 497; private communication (1987).
Carlberg, R. G., & Charlot, S. 1992, ApJ, **397**, 5.
Colless, M. M., Ellis, R. S., Broadhurst, T. J., Taylor, K., & Peterson, B. A. 1993, MNRAS, **261** 19.
Couch, W. J., Jurcevic, J. S. & Boyle, B. J. 1993, MNRAS, **260** 241.
Cowie, L. L., Lilly, S. J., & Gardner, J. 1988, ApJ, **332**, L29.
Cowie, L. L., Songaila, A., & Hu, E. M. 1991, Nature, **354**, 460.
Cowie, L. L., Songaila, A. & Hu, E. M. 1991, Nature, **354** 460.
Dalcanton, J. J. 1993, ApJ, **415**, L87.
Dalcanton, J. J. 1994, in preparation.
Dube, R. R., Wickes, W. C., & Wilkinson, D. T. 1977, ApJ, **215**, L51.
Dube, R. R., Wickes, W. C., & Wilkinson, D. T. 1979, ApJ, **232**, 333.
Eales, S. 1993, ApJ, **404** 51.
Efstathiou, G., Bernstein, G., Katz, N., Tyson, J. A. & Guhathakurta, P. 1991, ApJ, **380** L47.
Gardner, J. P., Cowie, L. L., & Wainscoat, R. J. 1993, ApJ, **415**, L9.
Giovanelli, R., Haynes, M. P. & Chincarini, G. L. 1986, ApJ, **300** 77.
Glazebrook, K. et al. 1993, in press.
Griffiths, R. E. et al. 1993, ApJ, submitted.
Gullixson, C. A., Boeshaar, P. C., Tyson, J. A., & Seitzer, P. 1993, preprint.
Gunn, J. E. 1965, Ph.D. Thesis, Cal. Tech.
Guhathakurta, P., Tyson, J. A., & Majewski, S. 1990, ApJ, **357**, L19.
Guhathakurta, P. 1989, PhD. Thesis, Princeton University.
Guiderdoni, B., & Rocca-Volmerange, B. 1989, A&A, **227**, 362.
Guiderdoni, B., & Rocca-Volmerange, B. 1990, in Galactic & Extragalactic Background Radiation, IAU Symposium 139, Eds. S. Bowyer & C. Leinert (Kluwer, New York) p. 365.
Harrison, E. R. 1964, Nature, **204**, 271.
Jarvis, J. F., & Tyson, J. A. 1981, A. J., , **86**, 476.

Jones, L. R., Shanks, T. & Fong, R. 1988, in Bergeron, J. et al., eds., *High Redshift and Primeval Galaxies* (Gif-sur-Yvette: Editions Frontieres) p. 29.
Koo, D. & Kron, R. 1992, ARAA, **30** 613.
Koo, D. C. & Szalay, A. S. 1984, ApJ, **282** 390.
Koo, D. C. 1986, ApJ, **311**, 651.
Koo, D. C., Gronwall, C. & Bruzual, G. A. 1993, ApJ, **415** L21.
Kron, R. G. 1980, ApJS, **43**, 305.
Kron, R. G. 1982, Vistas in Astronomy, **26**, 37.
Lacy, C. & Cole, S. 1993, MNRAS, **262** 627.
Lacey, C., Guiderdoni, B., Rocca-Volmerange, B., & Silk, J. 1993, ApJ, **402**, 15.
Lilly, S. J., Cowie, L. L., & Gardner, J. P. 1991, ApJ, **369**, 79.
Lonsdale, C. J. & Chokshi, A. 1993, AJ, **105** 1333.
Loveday, J., Peterson, B. A., Efstathiou, G. & Maddox, S. J. 1992, ApJ, **390** 338.
Maddox, S. J., Efstathiou, G., Sutherland, W. J. & Loveday, J. 1990a, MNRAS, **242** 43.
Maddox, S. J., Sutherland, W. J., Efstathiou, G., Loveday, J. and Peterson, B. A. 1990b, MNRAS, **247** 1.
Majewski, S. 1993, private communication.
Martin, C. & Bowyer, S. 1989, ApJ, **338**, 677.
Mattila, K. 1976, A&A, **47**, 77.
Melott, A. L. 1992, ApJ, **393**, 45.
Paresce, F. & Jakobsen, P. 1980, Nature, **288**, 119.
Partridge, R. B., & Peebles, P. J. E. 1967, ApJ, **148**, 377.
Postman, M., Bernstein, G. M., Tyson, J. A., Wenk, R. A., Windhorst, R. A., & Dressler, A. 1993, in preparation.
Roach, F. E. and Smith, L. L. 1968, Geophys. J.R.A.S., **15**, 227.
Sasseen, T. P., Lampton, M., Bowyer, S., and Wu, X. 1994, ApJ, in press.
Shectman, S. A. 1974, ApJ, **188**, 233.
Tinsley, B. M. 1980, ApJ, **241**, 41.
Treyer, M-A. & Silk, J. 1993, ApJ, **408**, L1.
Tyson, J. A., 1988, AJ, **96** 1.
Tyson, J. A. 1990, in *CCDs in Astronomy*, G. Jacoby, Ed. (Astron. Soc. Pacific), p. 1.
Tyson, J. A. & Seitzer, P. 1988, ApJ, **335**, 552.
Tyson, J. A., Wenk, R. A., Guhathakurta, P., & Bernstein, G. M. 1993, in preparation.
Tyson, J. A. & Jarvis, J. F. 1979, ApJ, **230** L153.
White, S. D. M. 1989, in The Epoch of Galaxy Formation, ed. C. S. Frenk, R. S. Ellis, T. Shanks, A. F. Heavens, & J. A. Peacock (Dordrecht: Kluwer), 15.
Whitrow, G. J. & Yallop, B. D. 1965, MNRAS, **130**, 31.
Yoshii, Y. 1993, ApJ, **403**, 552.
Yoshii, Y. & Takahara, F. 1988, ApJ, **326**, 1.
Yoshii, Y., Peterson, B. A., & Takahara, F. 1993, ApJ, **414** 431.
Windhorst, R. A., et al. 1993, ApJ, submitted.
Wyse, R. F. G. 1985, ApJ, **299**, 593.

## DISCUSSION

**S. White**: May I ask you about the angular two-point correlation function?

**Tyson**: At 25 mag and fainter the observed $w(\theta)$ is smaller than expected (Efstathiou, et al.). Let me try making a connection with brighter galaxies by showing you what Gary Burnstein and several of us have got for $w(\theta)$ by analyzing some old deep 4m photographic data. There are some old data and some new CCD data, but they are primarily photographic data from the 70s, showing a very nice agreement with Pritchet and Infante going from 0.1 arcminute out to 10 arcminutes or so. This is at 22nd magnitude. Now there is a problem if you take this data together with the $n(z)$ measurements of Broadhurst and Colless et al., and then try to calculate $\xi(r)$. I think this may be related to the problem that Eales and others have pointed out with regard to the luminosity function normalization. If you take the redshift data for even a 22nd magnitude slice together with the number – magnitude counts, and try to calculate what the normalization of the luminosity function is then relative to now, assuming that there is no shift horizontally in terms of luminosity evolution, then you are a factor of 3 higher in normalization. You have to jack up the numbers by a factor of 3 at a redshift of 0.3 to explain everything. However, instead of moving the curve up you can move it to the right just a little, mimicking luminosity evolution. You do not run into any problem with that, and you can also explain the data that way. In any case, because of this great excess of galaxies, models about disappearing galaxies encounter some difficulties.

We find that for a reasonable range of $\epsilon$ the two-point correlation $\xi(r)$ for this sample of 22nd magnitude galaxies appears to be in fair agreement with the IRAS galaxy clustering, but not the clustering in the APM data.

**J. Peebles**: The summary is that you take the redshift distribution, and the present correlation function, and do not correct the predicted $n\phi$.

**Tyson**: Yes. There is a problem. Marc Postman and Gary Bernstein have just calculated the two-point correlation for a field with 200,000 galaxies. This is from the $1.5^o$ photographic field. I have plotted the simple linear CDM prediction for a range of possible redshifts for the faint blue galaxies, from Efstathiou et al. I think by the time you get to 26th magnitude there are some serious problems with this plot, and I think they would reveal themselves in the two-point function.

**S. White**: Tony, is the amplitude you have here consistent with the one Efstathiou uses, which was in the first paper on your very faint correlation functions? There were some theoretical predictions in there for the predicted amplitude based on models where the correlation function scaled as a power law in redshift. Are those the models you have used to get the blue point here?

**Tyson**: Yes, this is that same range of $\epsilon$'s and range of redshifts; $\epsilon$ goes from 0 to $-1.2$.

**S. White**: Well, in fact, if you actually take the simulations and you measure the $\epsilon$'s from well outside that range, for instance, you take $\epsilon$s large and positive, from 1.2

to 1.5, there is no conflict with the data points. But you still need the wide redshift distribution, so it is still necessary that the galaxies are spread over a very wide range in redshift.

**Tyson**: I wonder what physics are we learning from that. We were forced into using this maximal distribution of redshifts in order to get plausibly close to the observed data.

**S. White**: The reason why the simple power law models did not work is because you are measuring the correlation function on very small scales, about 200 kpc true separation, where it is very non-linear. On those scales and over this redshift range, it grows much faster than predicted by the simple scaling model.

**J. Peebles**: Could I ask, Tony, whether if you went to a low density cosmological model, so you would have a greater path length and, so, a greater diminution in the angular correlation function, you would get away with a narrower range in redshift to look more like what you see in the redshift samples?

**Tyson**: Quite possibly. We did not try that in the Efstathiou *et al.* paper, but I am sure that would work. So there are a number of lines of evidence pointing towards low $\Omega$. You can introduce scenarios in which you have evolving galaxies or disappearing galaxies and still fit the data.

**J. Peebles**: Except that if the galaxies disappear, they have to be eaten in such a way that they were not pre-clustered. There was this thought that the high blue counts at the faint end were due to the fact that galaxies are observed in pieces. You might then have expected that the pieces were clustered, since they were coming together to form a single galaxy and you might then have expected a behavior more like the blue line in your plot, is that fair?

**Tyson**: Well, it depends on the angular scale of the data because, generally, we cannot see subclumps that are closer than a few arcseconds.

**J. Peebles**: If you do not see them, you count them as one galaxy and this is the excess counts of blue galaxies that you see. And if you believe that these are fragments of galaxies that are coming together to form single galaxies at a lower redshift then you might have anticipated that the two-point correlation function would be higher at that apparent magnitude.

**S. White**: Well, I thought that was the case, but if you actually use their limits and you integrate out to the scales that are relevant, let's say, 300 or 400 kpc, you are allowed to have 4 or 5 fragments close to each individual fragment, so that you could just get away with having them merge together in 4's or 5's.

**Tyson**: Could Richard Griffiths show us a picture of what is seen in the HST Medium Deep Survey? One of the things that the MDS is very useful for is trying to see if galaxies are multiple at high angular resolutions, but, necessarily at much higher surface brightness.

**R. Griffiths**: This is the deepest survey we have had from the so-called Medium

Deep Survey. It is 15 orbits in the I-band with HST, roughly 30 minutes or so per orbit. There is quite a high fraction of apparently close interacting systems, this upper object here for instance is double, and it is not unusual to find that as much as 30% of all the objects appear to be multiple or interacting systems. We only have redshifts for the brightest objects so far ($z \simeq 0.3$).

**Tyson**: On the other hand, to take another point of view, it is also true that if you look at our groundbased data, which are at a different angular scale, the eye convinces you that there is an excess of pairs. But when we actually to measured it, we found it was surprisingly small.

**J. Peebles**: There are lots of close pairs of galaxies at low redshift. I wonder if the incidence of these close pairs is what you would expect if there had been no evolution in the clustering of galaxies. I mean, have you estimated the two-point correlation function?

**R. Griffiths**: We have not yet. The trouble is also that we see differences from field to field. This is the deepest one we have, with the exception of the data that are just been coming in for the past six weeks. The first field we got near 3C273 did not seem to show as great an incidence of multiple systems. Anyway, Lyman Neuschafer is presently working on this problem. Lyman, do you want to comment on the two-point correlation?

**L. Neuschafer**: We do not have information on the correlation function at small angular scales. The survey probably extends out to 24th magnitude. In my thesis, I measured the slope of the correlation function to flatten with the apparent magnitude. If one makes the assumption that the observed flattening is real, then the amplitude is predicted to decline more strongly than the models shown here and it can be made very consistent with the observations with $\epsilon = 0$.

**M. Rees**: Just a speculative comment about the interpretation of the weak correlations. It probably is not a problem, as Simon mentioned, but if one believes that star formation in a starburst is affected by the UV background, it is worth noting that the UV from the neighboring galaxy 200 kpc away would dominate the UV background and one could imagine that, if what you are seeing is a short-lived starburst, there may be some inhibition of two neighbors bursting simultaneously. If an effect like that were important, it would give you spuriously small correlations.

**S. White**: I have a question about something else. The last time I heard Richard Ellis talk about the work he had been doing with a student, they seemed to have opposite conclusions from the ones you came to. Do you understand why you have different answers?

**Tyson**: I think so. This refers to a test of the lower part of the redshift distribution of the faint galaxies by observing the number of galaxies distorted into arcs by a foreground massive cluster. Part of the difference is that they have chosen clusters for which they know very little independently about the mass distribution inside them. There are at most a few galaxies in these clusters whose velocity has been measured, not enough to measure the velocity dispersion. So I think that until we get these difficult measurements for a full sample of galaxy clusters going all the way from modest redshifts up to perhaps z=0.6, with dynamically determined mass distributions, we cannot safely play this game.

**S. White**: So, you would say the reason they did not see any lensed galaxies was because there was no lens.

**Tyson**: Well, in a sense, that is what I am saying. But, in fact, they did see some lensed galaxies (arcs). The point is that their galaxy colors and numbers did not easily fit the model; but they were relatively bright and low redshift.

**T. Heckman**: I have a comment. There is a constraint on very low surface brightness galaxies, which Len Cowie has pointed out, set by the nucleosynthetic products. You cannot hide an arbitrarily large amount of luminous energy in the universe, because you know how much metals and helium have been produced by stars over the history of the universe.

**Tyson**: Thank you for reminding me, I should have mentioned that. It was first pointed out by Len Cowie that for a population of galaxies that is approximately flat in $f(\nu)$, you lose photometric color sensitivity to redshift; but that there is, in principle, a proportionality between the surface brightness of the night sky from all of those evolving galaxies and the total metallicity production up until present. The value of the constant of proportionality probably involves dirty physics; the answer will depend on whether it was embedded in dust and how much was re-emitted in nebulae and things of this sort. It is a messy calculation, but there is a rough proportionality. Estimates suggest that the observed faint blue galaxy numbers account for a significant fraction of the metallicity production. So you can have maybe twice the measured extragalactic background made up by faint blue galaxies and understand our current metallicity.

**R. Giacconi**: Did you mention the number of faint blue galaxies per square degree?

**Tyson**: The integrated number is nearly 1 million per square degree, integrated to 28th B or R magnitude.

**J. Peebles**: Tony, you made the striking remark that the galaxies detected in the deep K-band counts are detectable in the optical. I take it has to follow that there are many galaxies you detect in the optical that are not detected in the deepest K-band.

**Tyson**: Yes. If you take deep K-band data in the same field where you take deep optical data, you find both of them out to the limit the K-band survey. That is usually the limiting factor. It is very difficult to do faint K-band photometry to the equivalent faintness of the deep optical data.

**J. Peebles**: Am I right in assuming that you have detected in the the optical band galaxies that are not picked up in the K-band?

**Tyson**: Yes, the optical data goes fainter than one can currently go in the K-band.

**J. Peebles**: And you get a lot bigger number density. And I would be curious to know whether the galaxies that are detected in the K-band turn out ever to have blue apparent colors in the optical.

**Tyson**: They do. It is striking that K-selected galaxies have very blue B-R colors.

**J. Peebles**: These galaxies look like, in the K-band, ordinary $L^*$ giant galaxies. And this does not go with the dwarf galaxy properties in the visible.

**Tyson**: Starburst activity could produce this. It is useful to plot the summed galaxy flux vs wavelength. If we average the K-band number counts and plot it in Jy/sr, and do the same at various wavelengths from UV to IR, stellar populations at different stages of evolution will dominate different wavelengths. The UV is dominated by recent starbursts. The UV background is within perhaps a factor of two of what is required to ionize the intergalactic medium. The K-band counts are integrated down to $23^m$. Fainter than 18 $K^m$ you stop gaining in contribution to the extragalactic background. The peak in the K-band where most of the extragalactic background is coming from is around $K=18^m$.

**B. Wang**: I see. So are you saying that up to $K=23^m$ every galaxy you find in the K-band can be found in the B-band.

**Tyson**: Essentially yes.

**B. Wang**: I have another question. You said there is a correlation between the equivalent width of the [OII] 3727 Å emission lines and the B magnitude but there is no correlation between the redshift and the equivalent width. Doesn't that mean you have a selection effect, that is, you would tend to select galaxies with higher star formation rates, or miss galaxies?

**Tyson**: Yes, if star formation is responsible for the luminosity as you go to the faint limit you are selecting galaxies with larger luminosity at higher redshift and presumably larger equivalent widths in oxygen.

**B. Wang**: So, again, in principle, the redshift distribution is biased, is that true?

**Tyson**: Yes. That effect may be seen in the redshift-magnitude plots. At the faint magnitude limit of each one of those redshift surveys there is an absence of points at low luminosity. The point that Koo and Kron made in presenting that plot is that, if you just overlay all the redshift surveys, then you get a redshift trend with magnitude. And I am sure we are experiencing the same bias at the faint end of the gravitational lens surveys, too.

**R. Griffiths**: You mentioned that some of the faint blue galaxies could be dwarfs, but the sizes that you showed seem to indicate that many of those objects are, indeed, dwarfs.

**Tyson**: Well, about 1%, if you believe our numbers. Based on a recent survey of 125,000 faint galaxies, low surface brightness "diffuse" galaxies are not in the majority, their numbers running around 1%.

**I. Morgan**: For the faintest EBL, what is the number count slope of the blue band at the faintest magnitude? Is it still above 0.4?

**Tyson**: Good question. No. It may be rolling off, giving a sort of Tinsley-like bump. I should have pointed out that it better roll off at some faint apparent magnitude, because if it keeps going like 0.4 steeper, the extragalactic night sky background would diverge. One can use Rodger Dube's experiment to try to give a constraint; Dube's upper limit to the EBL is 1 $S_{10,V}$, and then you must have an abrupt cutoff around 32nd magnitude. And if the summed EBL is 0.5 $S_{10,V}$, then one might expect a smooth roll-off of the sort we appear to see. It is interesting to ask what the mean surface brightness of the faint blue galaxies is. Just take the faint blue galaxies by themselves and spread them out over a square degree and ask what contribution those make in terms of surface brightness: 28.8 in the blue band per square arcsecond.

**S. White**: You say that when we look in the K-band we find substantial blue emission for nearly every object that is detected.

**Tyson**: That is what Cowie and collaborators have found.

**S. White**: What has happened to the bright elliptical galaxies?

**Tyson**: Well, they are there, but they are not dominating the counts at the redshifts sampled.

**S. White**: But if you see blue emission, does that mean you have no bright elliptical galaxies or the bright elliptical galaxies are forming stars?

**Tyson**: Possibly the latter, and the main question is what redshift is it?

**I. Morgan**: So, there is a problem. There seems to be a lack of objects not forming stars.

**Tyson**: There are red objects. A few are red dwarf stars in our disk and halo. A few might be ellipticals at redshift less than perhaps 1.5 or 2, depending on whose luminosity evolution model you believe. Something like 10 or 20% of the objects have non-flat spectra, I think.

# INFRARED BACKGROUND (OBSERVATIONS)

Michael G. Hauser
NASA Goddard Space Flight Center
Greenbelt, MD 20771

**Abstract.** The Diffuse Infrared Background Experiment (DIRBE) on the Cosmic Background Explorer (*COBE*) satellite is designed to conduct a sensitive search for isotropic cosmic infrared background radiation over the spectral range from 1.25 to 240 $\mu$m. The cumulative emissions of pregalactic, protogalactic, and evolving galactic systems are expected to be recorded in this background. The DIRBE instrument, a 10-spectral band absolute photometer with an $0.7° \times 0.7°$ field of view, has mapped the full sky with high redundancy at solar elongation angles ranging from $64°$ to $124°$ to facilitate separation of interplanetary, Galactic, and extragalactic sources of emission. Initial sky maps show the expected character of the foreground emissions, with relative minima at wavelengths of 3.5 $\mu$m and longward of 100 $\mu$m. Conservative limits on the isotropic infrared background are given by the minimum observed sky brightness in each DIRBE spectral band. Extensive modeling of the foregrounds is required to isolate or strongly limit an extragalactic infrared component. The *COBE* Far Infrared Absolute Spectrophotometer (FIRAS) experiment has already established strong upper limits to any isotropic background from 500 $\mu$m to 5000 $\mu$m wavelength in excess of the 2.726 K blackbody radiation of the cosmic microwave background.

## 1. INTRODUCTION

The search for cosmic infrared background radiation (CIBR) is a relatively new field of observational cosmology. Measurement of this distinct radiative background, expected to arise from the cumulative emissions of pregalactic, protogalactic, and galactic systems, would provide new insight into the cosmic 'dark ages' following the decoupling of matter from the cosmic microwave background radiation (see, for example, early papers by Partridge & Peebles 1967; Low & Tucker 1968; Peebles 1969; Harwit 1970; Kaufman 1976; and more recent discussions by Bond, Carr, & Hogan 1986, 1991). Observationally, there have been no corroborated detections of the CIBR, though possible evidence for an isotropic infrared background in data from rocket experiments has been reported (Matsumoto et al. 1988; Matsumoto 1990; Noda et al. 1992). The Diffuse Infrared Background Experiment (DIRBE) on the *COBE* spacecraft is the first satellite instrument designed specifically to carry out a systematic search for the CIBR.

The FIRAS instrument on the *COBE* spacecraft, designed primarily to make a precise measurement of the spectrum of the cosmic microwave background radiation, is also a powerful instrument for the CIBR search. As described by Mather at this Symposium, the FIRAS measures the absolute sky brightness with a 7°-diameter beam from a wavelength of 100 $\mu$m to 1 cm.

The search for the CIBR is the most exploratory of the *COBE* objectives. Even from a spaceborne instrument, this cosmic fossil is far more difficult to observe than the cosmic microwave background radiation (CMBR). Whereas the CMBR is the dominant diffuse celestial radiation at millimeter wavelengths, the local infrared foregrounds from interplanetary dust and the Galaxy are far brighter than expected CIBR levels. Preliminary conservative limits on the CIBR based on the DIRBE data at the south ecliptic pole, and a comparison of DIRBE measurements with those from instruments on sounding rockets and the *Infrared Astronomical Satellite* (*IRAS*), were presented by Hauser et al. (1991). The DIRBE investigation is still in its data collection and data reduction phase. In this talk, I report updated CIBR limits in the DIRBE spectral range based upon the faintest sky brightness observed at each wavelength and a more recent calibration of the DIRBE data than that used by Hauser et al. (1991), and note the new stringent limits at submillimeter wavelengths implied by the FIRAS data.

## 2. THE SEARCH FOR THE COSMIC INFRARED BACKGROUND

The strong cosmological motivations for measuring the CIBR have been extensively discussed in the references cited above, and are discussed further by Lonsdale at this Symposium. I therefore restrict my comments to providing some perspective on the observational search. In the past three decades while we have measured the properties of the CMBR in detail, we have only slowly learned something about what the CIBR isn't, but we still can't say what it is.

At the time of the discovery of the CMBR in the mid-1960's, there were no direct observational limits on cosmic electromagnetic radiation in the more than three decades from optical to millimeter wavelengths (for example, see Figure 6.3 in Peebles 1993). In particular, we could not by direct observations of the infrared sky brightness exclude the possibility that a radiant energy density as large as that required to close the Universe might be present in the spectral range from 1 $\mu$m to 1 cm! When the *COBE* proposals were written in 1974, a few pioneering rocket measurements in the infrared had limited that unlikely possibility, but one could not rule out a CIBR with an order of magnitude more integrated energy density than the CMBR, the dominant known contribution to the cosmic electromagnetic radiation. At the time of the *COBE* launch in 1989, though there had been more infrared instruments on rockets and the *IRAS* mission, there were still no definitive detections of a CIBR at any wavelength. Ressell & Turner (1990) compiled a comprehensive accounting of the diffuse photon background from radio to $\gamma$-ray energies. If the CIBR were in fact comparable to the measurements or upper limits compiled by Ressell & Turner, it was still possible that the CIBR might contain more cosmic energy density than any other part of the electromagnetic spectrum. Clearly, determining the cosmic infrared background is important for sharpening our knowledge of the radiant energy content of the Universe, as well as for providing clues to past and present astrophysical processes.

## 3. THE OBSERVATIONAL CHALLENGE

The search for the CIBR is impeded by two fundamental challenges: there is no unique spectral signature of such a background, and there are many local contributors to the infrared sky brightness at all wavelengths, often quite bright. The lack of distinct spectral signature arises in part because so many different sources of primordial luminosity are possible (see, *e.g.*, Bond, Carr, & Hogan 1986), and in part because the primary emissions are then shifted into the infrared by the cosmic red-shift and dust absorption and re-emission. Hence, the present spectrum depends in a complex way on the characteristics of the luminosity sources, on their cosmic history, and on the dust formation history of the Universe.

Setting aside the difficult possibility of recognizing the CIBR by its angular fluctuation spectrum (Bond, Carr, & Hogan 1991), the only identifying CIBR characteristic for which one can search is an isotropic glow. One must solve the formidable observational problem of making absolute brightness measurements in the infrared. One must then discriminate and remove the strong signals from foregrounds arising from one's instrument or observing environment, the terrestrial atmosphere, the solar system, and the Galaxy. Particular attention must be given, of course, to possible isotropic contributions from any of these foreground sources. Finally, if an extragalactic isotropic residual remains, one must evaluate the contribution from galaxies over their luminous lifetime to distinguish their light from that of pregalactic or protogalactic sources.

## 4. THE COBE DIFFUSE INFRARED BACKGROUND EXPERIMENT

The *COBE* mission and FIRAS instrument have been described by Mather at this Symposium. I therefore focus on the Diffuse Infrared Background Experiment (DIRBE). The primary aim of the DIRBE is to conduct a definitive search for an isotropic CIBR, within the constraints imposed by the local astrophysical foregrounds. The experimental approach is to obtain absolute brightness maps of the full sky in 10 photometric bands at 1.25, 2.2, 3.5, 4.9, 12, 25, 60, 100, 140, and 240 $\mu$m. In order to facilitate discrimination of the bright foreground contribution from interplanetary dust, linear polarization is also measured at 1.25, 2.2, and 3.5 $\mu$m. Because of the Earth's motion within the interplanetary dust cloud, the diffuse infrared brightness of the entire sky varies over the course of a year. To monitor this variation and use it to help discriminate the signal contribution from the interplanetary dust cloud, the DIRBE field of view was offset from the *COBE* spacecraft spin axis by 30°. This produces a helical scan of the sky during each *COBE* orbit, modulating the signal from any spherically symmetric, Sun-centered cloud. Over the course of six months this yields observations of all celestial directions hundreds of times at all accessible solar elongation angles (depending upon ecliptic latitude) in the range 64° to 124°. All spectral bands view the same instantaneous field-of-view, and the helical scan allows the DIRBE to sample fully 50% of the celestial sphere each day. The instrument is designed to achieve a sensitivity at each wavelength for each $0.7° \times 0.7°$ field of view of $\nu I_\nu = 1$ nW m$^{-2}$ sr$^{-1}$ (1 $\sigma$, 1 year). This is well below the actual sky brightness, and below many of the predictions for the CIBR brightness, at all wavelengths.

The DIRBE instrument is an absolute radiometer, utilizing an off-axis folded Gregorian telescope with a 19-cm diameter primary mirror. The optical configuration (Magner 1987) is carefully designed for strong rejection of stray light from the Sun, Earth limb, Moon or other off-axis celestial radiation, or parts of the *COBE* payload

(Evans 1983). The instrument, which is maintained at a temperature below 2 K within the *COBE* superfluid helium dewar, measures absolute brightness by chopping between the sky signal and a zero-flux internal reference at 32 Hz. Instrumental offsets are measured by closing a cold shutter located at the prime focus. A radiative offset signal in the long wavelength detectors arising from JFETs (operating at about 70 K) used to amplify the detector signals was identified and measured in this fashion. Internal radiative reference sources are used to stimulate all detectors when the shutter is closed to monitor the stability and linearity of the instrument response. The highly redundant sky sampling and frequent response checks provide precise photometric closure over the sky and reproducible photometry to $\sim 1\%$ or better for the duration of the mission. Calibration of the photometric scale is obtained from observations of isolated bright celestial sources.

The primary DIRBE survey of the sky was carried out from December 11, 1989 until depletion of the liquid helium on September 21, 1990. The interior of the dewar subsequently warmed to about 50 K. Though the detectors at wavelengths longer than 4.9 $\mu$m no longer provide useful data, the 1.25 to 4.9 $\mu$m detectors continue to provide usable data at about 20% of the original sensitivity. The present plan is to operate the *COBE* through the end of 1993. A more detailed description of the *COBE* mission has been given by Boggess *et al.* (1992), and the DIRBE instrument has been described by Silverberg *et al.* (1993).

The DIRBE sky maps show the dominant anticipated features of Galactic starlight and zodiacal light at short wavelengths, and emission from the interplanetary and interstellar media at long wavelengths. Four all-sky false color images each containing data from three adjacent DIRBE bands ranging from the near-infrared to the submillimeter are given by Hauser (1993). These images dramatically illustrate the challenge of distinguishing the CIBR from signals arising in our local cosmic environment. The DIRBE data are clearly a valuable new resource for studies of the interplanetary medium and the Galaxy.

## 5. CURRENT LIMITS ON THE COSMIC INFRARED BACKGROUND RADIATION

Until the various foreground cosmic sources of diffuse infrared radiation are properly discriminated and subtracted, an effort of the DIRBE team currently under way, the most credible direct observational limits on the CIBR are the minimum observed sky brightnesses. The preliminary DIRBE spectrum toward the south ecliptic pole, one of the darkest directions in the sky at many wavelengths (Hauser *et al.* 1991, Table 2), showed that the faintest levels of the foreground emissions occur at 3.4 $\mu$m ($\nu I_\nu = 150 \pm 60$ nW m$^{-2}$ sr$^{-1}$) and near 241 $\mu$m ($\nu I_\nu = 70 \pm 40$ nW m$^{-2}$ sr$^{-1}$), confirming these as the most sensitive spectral windows for the CIBR search. The sensitivity of the *COBE* DIRBE and FIRAS measurements in each of their respective fields-of-view is generally well below these observed sky brightnesses toward the ecliptic pole (Hauser *et al.* 1991). With the *COBE* data in hand, discrimination of foreground emission, rather than measurement sensitivity, is clearly the major challenge in searching for the CIBR.

Since publication of the spectrum of the south ecliptic pole, which was based on a quick-look reduction and initial calibration of the DIRBE data, the complete data set acquired while the DIRBE was cryogenically cooled has been processed with an improved calibration. Table 1 lists, for each DIRBE wavelength, a representative darkest sky value observed. At wavelengths where interplanetary dust scattering or emission is strong, the sky is darkest near the ecliptic poles. At wavelengths where the interplanetary cloud signal is rather weak (*i.e.*, at 3.5 µm and longward of 100 µm), the sky is darkest near the galactic poles or in minima of HI column density. A 20% error is shown for each value, representative of the present absolute calibration uncertainties. These darkest sky brightnesses are the current DIRBE upper limits to any isotropic infrared background.

As described by Mather at this Symposium (see also Mather et al. 1994), the CMBR spectrum in the wavelength range 0.5–5 mm deviates from a 2.726 K blackbody shape by less than 0.03% of the peak intensity. Taking this as an upper limit to an additional cosmic infrared background implies $\nu I_\nu < 340/\lambda(\mu m)$ nW m$^{-2}$ sr$^{-1}$. This limit does not make allowance for systematic errors in separating the Galactic signal from the FIRAS CMBR signal. A complete systematic error analysis is currently in progress; Table 1 shows twice the above limit as an estimate of the proper magnitude. Analysis of the FIRAS high frequency data, combined with foreground modeling, will limit or provide measurements of the CIBR in the 100 µm to 500 µm range. It should be noted that the *COBE* DIRBE and FIRAS sky brightness measurements have been compared at 140 and 240 µm as a check on the calibrations of both instruments. The two sets of measurements were found to be consistent within the present calibration uncertainties.

Table 1.
Upper Limits on the Cosmic Infrared Background

| $\lambda$ µm | $\nu I_\nu$ nW m$^{-2}$ sr$^{-1}$ | Reference |
|---|---|---|
| 1.25 | 480 ± 96 | DIRBE dark sky |
| 2.2 | 190 ± 38 | " |
| 3.5 | 83 ± 17 | " |
| 4.9 | 260 ± 52 | " |
| 12 | 2800 ± 560 | " |
| 25 | 2400 ± 480 | " |
| 60 | 300 ± 60 | " |
| 100 | 90 ± 18 | " |
| 140 | 140 ± 28 | " |
| 240 | 28 ± 6 | " |
| 500–5000 | $680/\lambda(\mu m)$ | Mather et al. (1994) |

There have been reports of upper limits on, or possible detections of, isotropic residuals in the infrared sky brightness from several rocket experiments (Matsumoto et al. 1988; Matsumoto 1990; Noda et al. 1992). These investigators have arrived at these limits after attempting to discriminate the various foreground components of emission contributing to their measurements. Because of the limited sky and spectral coverage available to discriminate contributions to the measured signal and the brief time to check possible systematic measurement errors, conclusions from the rocket experiments require confirmation. Where the DIRBE and rocket observations have been compared (Hauser et al. 1991; Noda et al. 1992), the actual sky brightness measurements have been generally similar.

As a complement to the CIBR upper limits which can be set by diffuse infrared background measurements, measurements of galaxies in the infrared allow estimation of lower limits to the total extragalactic infrared background. For example, Cowie et al. (1990) have estimated the integrated contribution of galaxies at 2.2 $\mu$m to be $\nu I_\nu = 5$ nW m$^{-2}$ sr$^{-1}$ on the basis of deep galaxy counts. Hacking & Soifer (1991) have used galaxy luminosity functions derived from $IRAS$ data to predict minimum diffuse backgrounds (integrated to z=3) at 25, 60, and 100 $\mu$m of 1, 2, and 4 nW m$^{-2}$ sr$^{-1}$ respectively. Beichman & Helou (1991) have used synthesized galaxy spectra, also based largely on $IRAS$ data, to estimate the diffuse infrared background due to galaxies. At 300 $\mu$m, their minimum estimated brightness (integrated to z=3) is 2 nW m$^{-2}$ sr$^{-1}$. The integrated galaxy far-infared background contribution may exceed these estimates substantially if there has been evolution in galaxy luminosity or space density: deeper counts from future space infrared observatories such as $ISO$ and $SIRTF$ will improve these estimates. Even these minimum extragalactic background contributions should be detectable if the foreground contributions to the $COBE$ measurements can be modeled to about the 1% level, a difficult but perhaps achievable goal based on our initial studies.

## 6. CONCLUSION

The theoretical implications of the current FIRAS limits on the cosmic infrared background are discussed by Mather at this Symposium and by Wright et al. (1994). The implications of the more general infrared background limits are discussed by Lonsdale at this Symposium. To put the present observational status into the historical context introduced at the beginning of this talk, the upper limits of Table 1 yield an upper limit to the integrated infrared background from 1 to 300 $\mu$m of 5 $\mu$W m$^{-2}$ sr$^{-1}$, still a factor of 5 larger than the integrated CMBR. Table 1 shows that this limit on the CIBR integral is dominated by the thermal emission peak from the interplanetary dust at 12 and 25 $\mu$m. Careful modeling of the sky brightness contribution from the interplanetary dust cloud will substantially reduce the limit on the integrated infrared background. However, limits on the CIBR will clearly remain relatively weak in the spectral decade from about 6 to 60 $\mu$m until measurements can be made from outside the interplanetary dust cloud.

Measurement of the cosmic infrared background radiation will enhance our understanding of the epoch between decoupling and galaxy formation. The high quality and extensive new measurements of the absolute infrared sky brightness obtained with the DIRBE and FIRAS instruments on the $COBE$ mission are allowing a definitive search for this elusive background, limited primarily by the difficulty of distinguishing it from bright astrophysical foregrounds.

The author gratefully acknowledges the contributions to this report by the many participants in the $COBE$ Project, to his colleagues on the $COBE$ Science Working Group, and especially the many scientists, analysts, and programmers engaged in the DIRBE investigation. The National Aeronautics and Space Administration/Goddard Space Flight Center (NASA/GSFC) is responsible for the design, development, and operation of the $COBE$. Scientific guidance is provided by the $COBE$ Science Working Group. GSFC is also responsible for the development of the analysis software and for the production of the mission data sets.

## REFERENCES

Beichman, C. A. & Helou, G., 1991, ApJ, 370, L1.
Boggess, N. et al. , 1992, ApJ, 397, 420.
Bond, J. R., Carr, B. J., & Hogan, C. J., 1986, ApJ, 306, 428.
Bond, J. R., Carr, B. J., & Hogan, C. J., 1991, ApJ, 367, 420.
Cowie, L. L., et al., 1990, ApJ, 360, L1.
Evans, D. C., 1983, SPIE Proc. 384, 82.
Hacking, P. B. & Soifer, B. T., 1991, ApJ, 367, L49.
Harwit, M., 1970, Rivista del Nuovo Cimento Vol. II, 253.
Hauser, M. G., 1993, in *Back to the Galaxy*, AIP Conf. Proc. 278, ed. S. S. Holt & F. Verter (New York: AIP), 201.
Hauser, M. G., et al., 1991, in *After the First Three Minutes*, AIP Conf. Proc. 222, ed. S. Holt, C. L. Bennett, & V. Trimble (New York: AIP), 161.
Kaufman, M., 1976, Ap. Sp. Sci, 40, 369.
Low, F. J. & Tucker, W. H., 1968, Phys. Rev. Lett, 22, 1538.
Magner, T. J., 1987, Opt. Eng., 26, 264.
Mather, J. C., et al., 1994, ApJ, 420, 439.
Matsumoto, T., 1990, in *The Galactic and Extragalactic Background Radiation*, IAU Symposium 139, ed. S. Bowyer and C. Leinert (Dordrecht: Kluwer), 317.
Matsumoto, T., Akiba, M., & Murakami, H., 1988, ApJ, 332, 575.
Noda, M. et al., 1992, ApJ, 391, 456.
Partridge, R. B. & Peebles, P. J. E., 1967, ApJ, 148, 377.
Peebles, P. J. E., 1969, Phil. Trans. Royal Soc. London, A, 264, 279.
Peebles, P. J. E., 1993, *Principles of Physical Cosmology*, (Princeton: Princeton University Press).
Ressell, M. T. & Turner, M. S., 1990, Comm. on Astrophys., 14, 323.
Silverberg, R. F. et al., 1993, Proc. SPIE Conf. 2019, *Infrared Spaceborne Remote Sensing*, ed. M. S. Scholl (Bellingham: SPIE), 180.
Wright, E. L., et al., 1994, ApJ, 420, 450.

## DISCUSSION

**D. Henry:** Mike, I do not think you should be bashful about your idea of sending something up beyond the zodiacal cloud. My memory says that the zodiacal cloud disappears rather abruptly, just somewhere beyond the orbit of Mars. The real challenge would be preserving the cryogen for a year or two or whatever it would take to go up there. But your wonderful data have shown that this is something that is definitely going to be neat. I think it is a great idea!

**Hauser:** As a matter of fact, it was not a totally facetious remark. We have ideas of doing that. We even have a name for the experiment and would like to do some technical studies, not only just for cryogens. We think, in fact, that with radiative cooling one could do a great deal. Maybe we will get an opportunity for a ride in deep space. I agree with you, it is only a few A.U. that are needed. The Pioneer missions clearly showed us that; they measured the optical zodiacal light and showed it to die rather dramatically once you get out a few more A.U. from the sun than we are. So I think it is technically a very doable experiment. There is a question of polical will and resources, of course.

**F. Stecker**: I would like to mention a technique that I think is a potentially powerful way of actually measuring the intergalactic radiation field without having to worry about the galactic foreground. That is, to use the high energy gamma-rays spectra from high redshift quasars to determine the intergalactic infrared field. Infrared photons pair-produce with the gamma-rays, and you can look for the absorption feature in the gamma-ray spectrum of these objects.

**Hauser**: I agree. In fact, I would like to see that analysis although I do not particularly have time to do that myself.

**J. Peebles**: Mike, may I congratulate you on this remarkable progress in turning a theoretical subject into one that is empirical? As one semi-empirical remark, I would like to point out that if you want to turn roughly an $\Omega h^2$ in baryons equal to 0.02 into heavy elements abundances of 3%, and if you do it at a redshift z=1, I compute the $\nu I(\nu)$ you need is, in your units (W cm$^{-2}$ sr$^{-1}$), of the order $10^{-11}$, the depth of your little trough at about 3 $\mu$m.

**Hauser**: So, we are excited. If you guys will give us something that bright, I think we will find it! If you give us a value of $10^{-13}$, we are going to have a tough time of it.

**J. Peebles**: If you make the metals at a modest redshift, as is indicated by the relatively low metallicity in the damped Lyman$\alpha$ systems, that stuff has got to be there. There is no hiding, as you remark.

**B. Partridge**: Mike, could you say something about the emissivity indices of the interplanetary and the interstellar dust, and how you derived those?

**Hauser**: Let me start with the interplanetary dust. If we just take the DIRBE data alone out to 100 $\mu$m where it is quite obvious, the energy distributions are fit as well by $n = 0$, that is to say, a black particle distribution, as any other. I do not have numbers for corresponding spots on the sky at 140 and 240 $\mu$m that we can add to that. I know that my FIRAS colleagues, in particular Dale Fixsen, have recently been studying evidence for interplanetary emission in the FIRAS data going out to longer wavelengths. There is quite clearly a signal there, and I think there is evidence that $n > 0$ is more consistent with the data at those wavelengths than $n = 0$, but this value is not well determined, because the signal-to-noise ratio is pretty terrible. Furthermore, at those long wavelengths we are dominated by the Galactic part of the signal.

The interstellar dust is an interesting question. The FIRAS group published (Wright E.L. et al. 1991, ApJ 381, 200) the galactic continuum spectrum. If you fit it with a single temperature dust model, you get a spectral index $n \simeq 1.6$. If you fit it with the two temperature component model, which allows for the cold dust component as well as the one at a few tens of Kelvin, a combination of $n = 2$ spectra is found to fit quite nicely. So, basically one is looking around for any kind of real handle that can discriminate these possibilities. Single temperature dust is obviously a gross approximation to reality; but empirically it fits the observed distribution pretty well. Theoretically, one expects that at these long wavelengths it ought to be $n = 2$. I am just not sure there is enough information in the measurements to really distinguish these possibilities.

**Q.:** You described how you set your upper limits by saying you went in the database, and at each wavelength you selected the measurement which was the lowest in time and position. I wonder if that is not giving you too strong constraints for your upper limits because, if you have a whole bunch of measurements of the same thing, then you would pick the one which was the largest fluctuation down from the true value.

**Hauser:** We average the brightness of these pixels over some finite amount of time, so we are not just picking a statistical fluctuation that is low. In fact, we have done this dark patch thing in several different ways, picking different places on the sky, so the numbers that I have shown you are sort of representative rather than literally a negative fluctuation of the statistics on the brightness on any point. I do not think we are making that kind of error here. And certainly the signal-to-noise ratio in the numbers that we have is quite appreciable, so we cannot be off by very much on those numbers.

**Q.:** Mike, the intensity picture you showed seems sort of flat-topped to me.

**Hauser:** Oh, you mean the square edges near the ecliptic plane? Actually this is one of the major IRAS discoveries. There is a rather hard edge to the blue band along the ecliptical plane. When you do a scan across that, you do see a few percent enhancement above a smooth profile at those points. The IRAS team investigated this and recognized that it was interesting. If you take away a smooth shape, then you get an enhancement and those are called the interplanetary dust bands now. Their apparent ecliptic latitude allows them to be associated with dust debris from named families of asteroids with orbits of particular inclinations. I will show you one other Figure since you have asked this question. On the left is a profile at 25 $\mu$m from ecliptic pole to ecliptic pole. There is a solid line on that curve, which is a cosecant $\beta$ law at high ecliptic latitudes, and which is smooth down towards the plane. If you look at the residuals from that empirical function near the ecliptic plane, you see sort of three prominent peaks which are, in fact, your hard edges, and these are the interplanetary dust bands. Their brightness is a tiny fraction of the total brightness. If you think about the geometry of this cloud, you will see pairs of enhancements where the absolute value of the latitude equals the inclination of the orbits of these dust particles; so $\pm 10°$ turns out to be a pair and there is another pair actually at $\pm 1°$. With the DIRBE resolution the one-degree pair tends to clump up right around the plane. The bands in the IRAS data have been extensively studied and people are busy modelling the dynamics. You can actually determine the orbital parameters of these particles. Peolpe are doing self-consistent models of this cloud. For example, Stan Dermott (Univ. of Florida), is doing that sort of game. An interesting new thing in the DIRBE data is that we see those bands not only in emission but also in the scattered light from the interplanetary dust as well. We showed this at the AAS meeting in Atlanta about a year and a half ago.

**J. Peebles**: So, if you want to avoid the zodiacal light why don't you fly normal to the plane?

**Hauser**: It takes too much energy, Jim. We now have the first out-of-the-ecliptic mission in flight. The Ulysses is going to fly over the solar poles and it has taken many years to get out to Jupiter and then swing out of the ecliptic plane by gravitational assist; and unfortunately it did not carry a DIRBE along.

# THE INFRARED BACKGROUND (THEORY)

Carol J. Lonsdale
Infrared Processing and Analysis Center
California Institute of Technology
and Jet Propulsion Laboratory
Pasadena, CA 91125
USA

## 1. INTRODUCTION

The infrared region of the spectrum is a very interesting and important one for the study of extragalactic backgrounds, both galaxian and pre-galactic. This review concentrates on backgrounds from protogalaxies and galaxies. Detailed discussions of many possible pre-galactic backgrounds have recently been presented by Bond, Carr and Hogan (1986, 1991); see Carr (1992) for a recent review of these and Wright et al. (1993) for a summary of how well some of them have fared in light of the most recent COBE observational results. I will concentrate on the approximate wavelength range 10 to 500$\mu$m since both the near-infrared and the submillimeter range are discussed by others.

The infrared spectral region is a good place to look for detectable backgrounds from the integrated light of galaxies and protogalaxies for several reasons. First, young galaxies and protogalaxies may have been relatively much more luminous at far-infrared wavelengths, compared to the optical-UV, than galaxies at the present epoch. Metallicity can increase rapidly during the early evolution of galactic systems, and if dust formation follows suit the dust optical depth and the far-infrared luminosity can rise dramatically at the expense of the obscured optical-UV luminosity (Wang 1991a,b; Lonsdale 1992; Mazzei, De Zotti and Xu 1993).

Figure 1 illustrates a second reason for the importance of the infrared region to background studies of galaxies and protogalaxies: the prominent "windows" between the various foregrounds and the cosmic microwave background radiation (CMB). This figure is similar to several which have been shown already at this meeting, depicting the intensity $\nu I_\nu$ in W/cm$^2$/sr. The main foregrounds in the infrared spectral region shown in Figure 1 are the zodiacal light which peaks near 1$\mu$m and falls into the far-infrared, the interplanetary dust (IPD) emission peaking near 10$\mu$m, and the interstellar dust (ISD) peaking beyond 100$\mu$m. There are two main infrared "windows": one near 3$\mu$m and the second at about 300$\mu$m.

It is not simply the existence of these windows that marks their importance, but also the fortunate chance that they happen to coincide very nicely with the two prominent peaks in the spectral energy distributions of moderate-to-high redshift galaxies: the stellar spectral energy distribution of nearby galaxies peaks near 1$\mu$m, thus moves into the 3$\mu$m window with increasing redshift, while the dust re-emission peak of ISM-rich galaxies peaks near 60 to 100$\mu$m, moving into the 300$\mu$m window with increasing redshift. Thus there is a rich hunting ground for the integrated stellar light of galaxies in the near-infrared window, and another one for the dust emission of galaxies in the far-infrared window. Conversely it will be difficult to ever measure the integrated light of galaxies or protogalaxies in the 5 to 30$\mu$m region unless spacecraft can be sent to the more distant reaches of the solar system where the interplanetary dust emission is much reduced.

Another reason why the infrared spectral region is one of the most valuable for studying the background light due to galaxies is that there is a strong *positive* far-infrared K-correction with redshift. Unlike the situation in the UV through near-infrared spectral region, the energy distributions of galaxies at $\lambda \gtrsim 80\mu$m fall with increasing wavelength with a very steep dependence on $\lambda$, therefore at longer wavelengths than this the K-correction can almost counter the cosmological effects of luminosity distance and surface brightness dimming so that the apparent flux density at a fixed observing frequency has little dependence on distance. The same effect holds to a more limited extent in the $3 - 20\mu$m region.

Finally, it is possible that intervening galaxies may produce sufficient obscuration to eliminate optically-selected background quasars from flux-limited samples (Ostriker and Heisler 1984; Wright 1990; Fall and Pei 1993), especially if there has been strong evolution of the dust optical depth (Wang 1991a,b; Mazzei et al. 1993). It is therefore also possible that such an effect will obscure background young galaxies and protogalaxies. At far-infrared wavelengths, not only will the obscuration be low enough to be insignificant, but the dust which is responsible for extinguishing the optical-UV light will re-emit this light in the far-infrared and submillimeter.

Mike Hauser has given us an excellent summary of the observational results on the infrared backgrounds. For the purposes of my discussion of the theoretical backgrounds expected from galaxies and protogalaxies, I summarise on Figure 1 the most recent observational limits from DIRBE and FIRAS on COBE. As Mike has described, the DIRBE data do not yet include any foreground subtraction, pending the very difficult task of modelling the galactic emission in detail. The spectacular FIRAS results of Mather et al. (1993) do include a galactic foreground subtraction, and have a maximum deviation from the CMB blackbody spectrum in the $2 - 20$ cm$^{-1}$ region of 0.03% of the peak of the CMB spectrum. However, as Wright et al. (1993) have noted, this foreground subtraction is not appropriate for modelling the cosmological backgrounds due to the integrated light of galaxies because the background itself is expected to have a spectral shape similar to that of the galaxy, thus the "galactic foreground" that has been subtracted could include some cosmological background. Wright et al. used a csc|$b$| method of galactic foreground subtraction to avoid this problem, and from their integrated galaxy light model fits I estimate a maximum deviation of about twice that inferred by Mather et al. This is illustrated by the upper of the two heavy solid lines in Figure 1.

## 2. INTEGRATED INFRARED EMISSION OF GALAXIES

IRAS has shown us that the far-infrared emission of galaxies in the local universe constitutes a large fraction, $\sim 30\%$, of the total energy output of normal galaxies, and

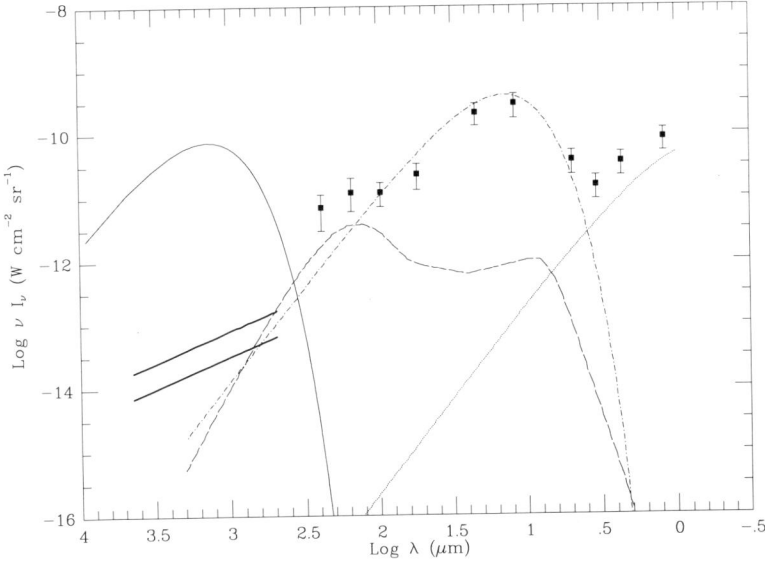

**Figure 1.** *The infrared foregrounds and COBE limits. The solid curve is the cosmic microwave background (CMB) radiation, and the other three curves are various foregrounds, as derived by Beichman and Helou (1991; C. Beichman, private communication): the dotted curve is the reflected solar zodiacal emission, the dot-dash curve the thermal emission from interplanetary dust, and the dashed curve the thermal emission from interstellar dust scaled to a brightness of $I_\nu(100\mu m) = 1$ MJy/sr, which is representative of the typical sky brightness in regions of the weakest cirrus emission at high galactic latitude. The lower short heavy solid line illustrates the maximum deviation from the CMB measured by FIRAS (Mather et al. 1993), $3.4 \times 10^{-8}$ erg cm$^{-2}$ s$^{-1}$ sr$^{-1}$ cm, which is 0.03% of the peak of the CMB spectrum, while the upper line represents an estimate of this maximum deviation adopting a more conservative galactic foreground subtraction (see text). The solid squares depict the DIRBE limits at the north ecliptic pole, as discussed by M. Hauser—note that these do not include any subtraction of the galactic or zodiacal foregrounds.*

that for some starburst and active galaxies the far-infrared fraction can be much higher (see Lonsdale 1992 and references therein). It follows by analogy to derivations of the expected integrated light in the optical band (e.g., Partridge and Peebles 1967) that the integrated light of galaxies in the infrared region can be expected to be at least a few percent of the peak of the CMB. Hacking, Condon and Houck (1987) found evidence for strong luminosity and/or density evolution of IRAS galaxies, at a rate comparable to that of quasars, which further enhances the expected integrated background radiation (see also Franceschini et al. 1988b; Lonsdale and Hacking 1989; Lonsdale et al. 1990; Saunders et al. 1990; Hacking and Soifer 1991).

The simplest kind of model for the integrated emission of galaxies involves taking a local luminosity function (LF) and a functional form for the evolutionary law, and extrapolating backwards in time to match the observed flux and redshift distributions of deep galaxy samples in a particular wavelength band. The model distributions are then integrated over flux and redshift, to some maximum redshift, $z_{max}$, to derive a background intensity. In many models the entire luminosity function is fixed in shape and it translates *en masse* in luminosity and/or density at the given rate. Such models are called translational models, and they were originally proposed for the evolution of radio sources and quasars. The physical interpretation of such evolution involved an increasing luminosity of AGN with lookback time for luminosity evolution, or an increasing fraction of galaxies possessing AGN at earlier times for density evolution. In the context of far-infrared bright galaxies, power law translational evolution would describe an increasing luminosity of starbursts and/or AGN with lookback time, or an increasing fraction of galaxies undergoing starburst episodes and/or AGN events with lookback time. Thus this kind of evolution is not well suited for describing the recently popular merging scenarios, in which smaller galaxies merge to form larger ones—such scenarios implicitly involve evolution of the shape of the luminosity function.

A summary of the parametric models of the far-infrared background that have been published based on translational evolution is given in the upper panel of Table 1. The wavelengths or wavelength range modelled, the cosmology, and $z_{max}$ adopted by these various authors is given. The parameters $\alpha$, $\beta$, $\kappa$ and Q define the parametric form of the assumed evolution behavior of the LF with redshift, as follows:

$$L(t) = L(t_0)(1+z)^{\alpha} \qquad (1)$$

$$\rho_{co}(t) = \rho_{co}(t_0)(1+z)^{\beta} \qquad (2)$$

describing power law luminosity and density evolution, respectively, where $\rho_{co}$ is the co-moving density, and:

$$L(t) = L(t_0) e^{[\kappa(1-(t(z)/t_0)]} \qquad (3)$$

for exponential luminosity evolution. For $\Omega = 1$, $(t(z)/t_0) = (1+z)^{-3/2}$ defining the relation used by Oliver et al. (1992): $L(t) = L(t_0) \exp\{2/3Q[1-(1+z)^{-3/2}]\}$.

For the first four models in Table 1 all galaxies comprising the LF are allowed evolve at the same rate, while the model of Franceschini et al. (1991) allows for three galaxies types to evolve at different rates. The model of Treyer and Silk is different in nature from the others in that they do not allow the luminosity function of "normal" galaxies to evolve with time, but add a new population of dwarf galaxies whose characteristic space density, $\phi^*$ (Mpc$^{-3}$), is the parameter that evolves:

$$\phi^*_{dw} = 6.0 \times 10^{-2} (\frac{0.7}{z} + 1)^{-1} h^3 \qquad (4)$$

where $h = H_0$ in units of 100 km/s/Mpc. This model is designed to explain the steepness of the observed blue number counts of galaxies with a population of dwarf galaxies which is present at $z = 0.7$ but has faded to invisibility by the current epoch. Treyer and Silk also investigate a model based on the number density evolution of dark matter halos in a cold dark matter scenario, which predicts a large abundance of low-mass halos. The integrated background light produced by this model is very similar to that of the blue dwarf model described by equation (4).

A different kind of model assumes that the evolution with lookback time of galaxies in the far-infrared is principally due to the natural evolution of their stellar populations and interstellar medium, without necessarily invoking dramatic starburst or AGN events. The most sophisticated kind of model involves population synthesis using stellar photometric and chemical evolution prescriptions to model in detail the spectral energy distributions of various galaxy types as a function of time. The evolutionary behavior of

## Table 1. Far-Infrared Evolutionary Models

| Model | $\lambda$ ($\mu$m) | $H_0$ | $\Omega$ | $z_{max}$ | $\alpha,\beta,\kappa,Q$ | Figure |
|---|---|---|---|---|---|---|
| Weedman 1990 | 100 | 75 | 0.1 | 4 | $\alpha = 2.5$ | 4 |
| Hacking and Soifer 1991 | 25, 60, 100 | 75 | 1 | 0.5, 3 | $\beta = 4$ $\alpha = 2,3$ | 4 |
| Beichman and Helou 1991 | 10–1000 | 40–100 | 0–1 | 1–5 | $\beta = 2,4$ | 4 |
| Oliver et al. 1992 | 4–1500 | 50 | 1 | 1–7 | $\alpha = 3.15$ $\beta = 6.7$ $Q=3.2$ | 5 |
| Franceschini et al. 1991 | 12–1000 | 50 | 0.1 | 5 | normal galaxies: $\kappa = 0$ starbursts: $\kappa = 3.2$ AGN: $\kappa = 2.5$ | 4 |
| Treyer and Silk 1993 | 12–550 | 50 | 1 | | see text | 4 |
| Wang 1991b | 20–1000 | 50 | 1 | 2,5,20 | | 6 |
| Franceschini et al. 1993 | 20–1000 | 50 | 1 | 4.5 | | 6 |

models of this type is usually dictated by an assumed dependence of the star formation rate on some power of the gas mass or density. As for the purely parametric LF translational models, the synthesis models are matched to the flux and redshift distributions of deep galaxy samples and then integrated to derive backgrounds.

Population synthesis modelling of evolving galaxies is a very active field in the UV through near-infrared bands, but it is less well developed at far-infrared wavelengths because the dust re-emission of stellar photons is an extra dimension which most modelers have yet to tackle. The population synthesis models of Mazzei, Xu and de Zotti (1992) and Mazzei, De Zotti and Xu (1993) are the first to fully incorporate the re-emission of starlight by dust within galaxies. These closed box models incorporate chemical evolution, thus at early epochs the dust content grows with time as the metallicity increases with the return to the interstellar medium of enriched gas from evolved stars. The dust optical depth reaches a maximum and some point in time and then declines again as star formation gradually uses up the ISM.

The rate of change of the dust opacity is a function of the initial star formation rate, which is largest for the earliest type galaxies and smallest for the latest types. The behavior of $\tau$ with time is illustrated for several galaxy types in Figure 2: a much more rapid relative far-infrared/optical-UV evolution is expected with lookback time for early type than for late-type systems, because the latest type systems are still reaching their peak optical depth at the present epoch. Late-type systems in the local universe do indeed emit similar amounts of energy in the optical-UV and the far-infrared, whereas ellipticals now emit only a very small fraction of their energy in the far-infrared.

In Figure 3, I reproduce a figure from Mazzei and de Zotti (1993), which shows model spectral energy distributions for a range of ages for elliptical galaxies, compared to the observed data for the high redshift IRAS galaxy F10214+4724 (see Section 3.1 for a discussion of this object). The oldest of these models represents a present day

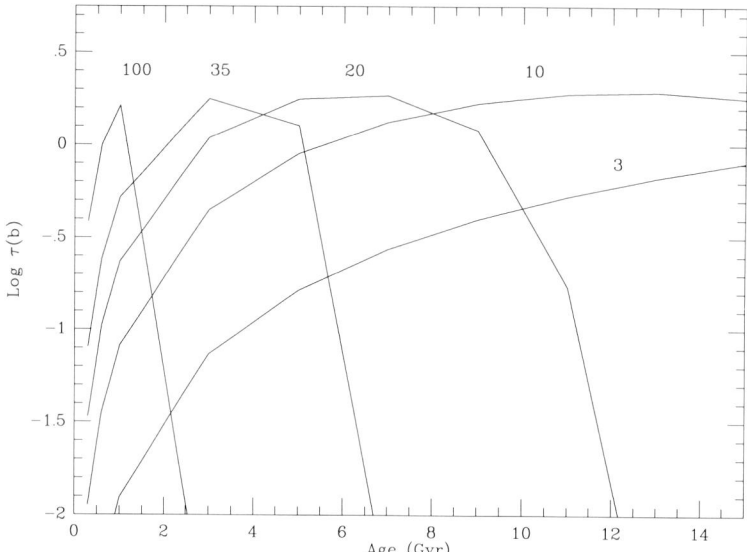

**Figure 2.** *Optical depth in the blue from Mazzei* et al. *(1992). The curves are labelled by the initial star formation rate ($M_\odot/yr$).*

elliptical, and matches observations of local ellipticals well (Mazzei et al. 1993). This figure clearly illustrates the expected very strong evolution in the shape of the spectral energy distribution (SED), with a far higher percentage of the total energy emerging at far-infrared wavelengths at early times than today.

Two models based on the natural chemical and photometric evolution of stellar populations and interstellar medium are summarised in the lower panel of Table 1. Franceschini et al. (1991) have used the models of Mazzei et al. to predict the cosmological background due to evolving galaxies in the near and mid-infrared, however in the far-infrared region they revert to a power law parametric approach similar to those discussed above. This approach is improved in Franceschini et al. (1993), who extend the synthesis modeling to the far-infrared. They obtain good fits to the local 60$\mu$m counts with their model, due for the most part to the strong dust evolution of early-type systems. Chokshi et al. (1993) have used the models of Mazzei et al. (1992) to simulate deep blue and near-infrared galaxy images, and Chokshi et al. (1994, in preparation) will extend their approach to a full spectral synthesis modelling at far-infrared wavelengths.

Wang (1991a,b) has also considered the evolution of the dust and the far-infrared emission from disk galaxies with time, and their contribution to the far-infrared background. Taking an analytical rather than a population synthesis approach, Wang derives the chemical evolution for a prompt initial enrichment (PIE) model and also for an accretion model, and the subsequent evolution of the interstellar dust, which he argues forms principally in molecular clouds. Wang finds that the dust content of young disk galaxies can be up to 4 times larger than today, and the far-infrared luminosity can be two orders of magnitude greater. The PIE model predicts much stronger backgrounds than the accretion model because it shows strong evolution of the dust mass.

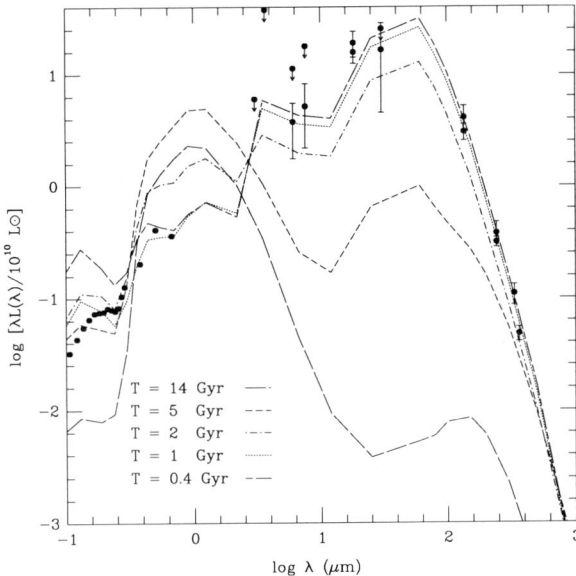

**Figure 3.** *Evolution of the spectral energy distribution of an elliptical galaxy (Mazzei et al. 1993) compared to the observed spectrum of IRAS F10214+4724 at z=2.286 (reproduced from Mazzei and De Zotti 1993). This model has a Salpeter (1955) IMF with a lower mass limit of 0.5 $M_\odot$, a star formation rate proportional to the gas fraction to the 0.5 power, and dust-to-gas ratio proportional to the metallicity. The data are from Rowan-Robinson et al. (1993), Downes et al. (1992) and Telesco (1993).*

## 2.1 Model Results

The results of the various parametric LF translational evolution models for the infrared background emission are summarised in Figures 4 and 5. The Beichman and Helou (1991) models selected for this figure bracket all their models, and incorporate an improved treatment of the energy distributions of local galaxies compared to the models in the published paper (G. Helou, private communication).

Most of the parametric models summarised in Table 1 do not conflict with the currently available DIRBE limits, however the higher evolutionary rates and the higher values for $z_{max}$ are constrained by the longest wavelength DIRBE limits, in particular the higher $z_{max}$ density evolution models of Oliver *et al.* are inconsistent with the DIRBE limits.

The FIRAS distortion limits provide a much stronger constraint, particularly the lower line with the galactic foreground removed by Mather *et al.* As noted above, however, this limit may be too stringent for comparison to the evolving galaxy models, since a portion of the "galactic foreground" that has been removed by Mather *et al.* (1993) may be part of an isotropic extragalactic background due to the integrated light of galaxies (Wright *et al.* 1993). Even our approximate estimate of the FIRAS limit using Wright *et al.*'s more conservative csc$|b|$ galactic foreground subtraction constrains the evolving models quite strongly, *ruling out* the parametric power luminosity evolution models with values of $\alpha$ high enough to fit the 60$\mu$m number counts ($\alpha \sim 3-4$). Thus if pure power law luminosity evolution is the explanation for the steepness of the local

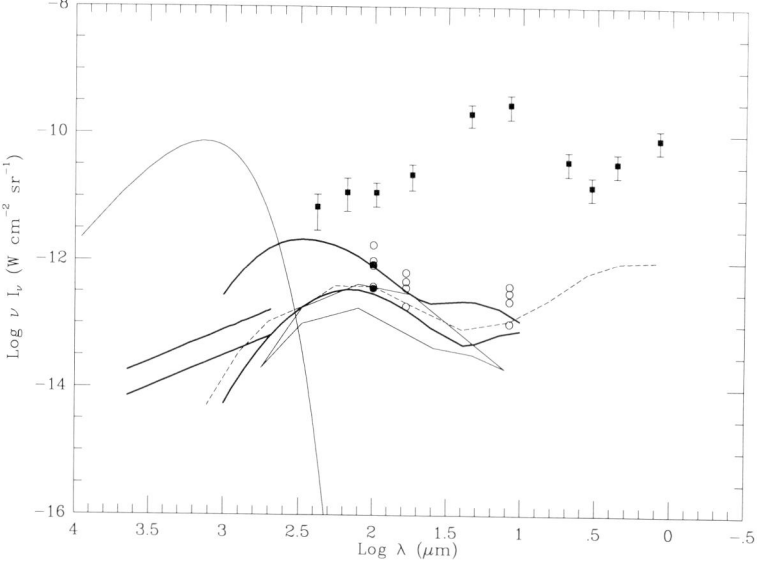

**Figure 4.** *Comparison of various translational evolution models for the integrated light of galaxies to the COBE data. Heavy solid lines – Beichman and Helou (1991) no-evolution model, $H_0=40$, $q=0.25$ (lower), and density evolution model, $\beta=4$, $H_0=100$, $q=0$ (upper); Short dashed line – Franceschini et al. (1991) evolving model, as summarised in Table 1; Light solid lines – Treyer and Silk (1993) no-evolution and evolving model (equation 4); Open circles – Hacking and Soifer (1991); Solid circles – Weedman (1990).*

universe 60$\mu$m number counts, then this evolution cannot continue to cosmological distances without severely violating the FIRAS distortion limits. There is some evidence that power law luminosity evolution with $\alpha > 2$ is also ruled out by the redshift distributions of IRAS 60 and 100$\mu$m-selected galaxy samples (Hacking and Soifer 1991; Fisher et al. 1992).

The power law density evolution models which can fit the 60$\mu$m number counts, $\beta \sim 4-7$, are only consistent with FIRAS for very low values of $z_{max}$; $z_{max} \ll 2.2$ for the Oliver et al. models. This is because higher evolutionary rates are needed to fit the $z < 0.5$ number counts using power law density evolution than power law luminosity evolution. Since this form of translational density evolution involves a scenario in which the fraction of galaxies having starbursts and/or AGN increases with lookback time, a low value for $z_{max}$ is necessary, in any case, to prevent this starbursting (and/or AGN) population from becoming larger than the total galaxy population (Weedman 1990; Hacking and Soifer 1991).

The exponential luminosity evolution models fare somewhat better: both the Franceschini et al. (1991) model with $z_{max} = 5$ and the Oliver et al. model with $z_{max} = 6.9$, which can fit the local number counts with $\kappa \sim 2-3$, are consistent with the upper FIRAS line, though the Oliver et al. model is only barely consistent with the lower of the two FIRAS lines. This difference from the power law luminosity evolution models

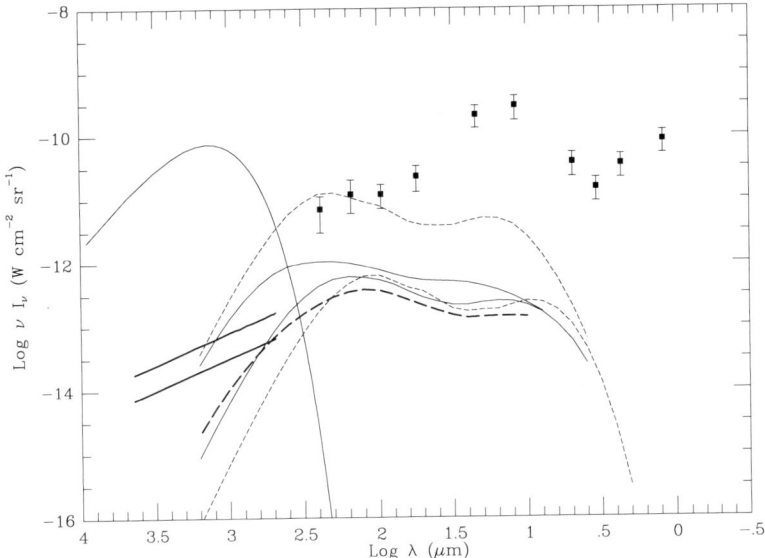

**Figure 5.** *Comparison of Oliver et al.'s (1992) translational evolution models for the integrated light of galaxies to the COBE data. Dashed lines – density evolution, $z_{max} = 1.0$ (lower) and $z_{max} = 2.16$ (upper); Heavy dashed line – exponential luminosity evolution, $Q=3.2$, $z_{max} = 6.94$; Light solid lines – power law luminosity evolution, $\alpha = 3.15$, $z_{max} = 2.16$ (lower) and $z_{max} = 5.31$ (upper).*

is not surprising since exponential evolution was introduced to avoid the high-redshift divergence of power-law evolution (Rowan-Robinson 1968).

Wright et al. (1993) fitted the distortion limits with a $\kappa < 2.3$ ($2\sigma$) exponential model based on the Beichman and Helou models, after subtracting the conservative $csc|b|$ contribution due to the Galaxy that they do not believe likely to include a significant extragalactic component (approximated by the upper FIRAS deviation line in figures 4 and 5). They estimate that this translates to an upper limit on the fraction of the baryon density converted from He to H by such a population of evolving infrared galaxies of $< 0.8\%$.

The fading dwarf model of Treyer and Silk is also comfortably consistent with the FIRAS limits, which is not surprising given that this model assigns most of evolutionary effect to dwarf galaxies at relatively low redshifts.

As noted above, all of the pure translational evolution models are limited in that they allow only translational evolution of the entire LF; they do not allow evolution of any other physical galaxy property, including the dust temperature. Most of them do not treat different galaxy types separately. Given that these models are now coming into direct conflict with the COBE results, it is clear that that the more sophisticated chemical evolution-based models are necessary to get a more realistic understanding of the integrated emission of galaxies in the far-infrared. In particular, it is important to recognise the likely importance of strong evolution of the dust optical depth in early type systems. Such models of course have their own shortcomings, most notably the fact that they have many more physical variables than can be realistically constrained by observables at present. Non-the-less they represent the best approach in the long run since the universe contains these parameters, whether we can measure then now or not!

Figure 6 illustrates the results of the two models based on the natural chemical and photometric evolution of stellar populations and interstellar medium described above. The two models of Wang (1991b) shown in Figure 6 bracket all those displayed in his Figure 2 for $z_f = 5$. Some of these models are also strongly limited by the FIRAS distortion results, in particular the PIE models of Wang with exponentially declining star formation rate. The conflict with the FIRAS data would be lower for the lower $z_f$ PIE model of Wang ($z_f = 2$). The accretion models of Wang are in more acceptable agreement, since they show little evolution of the dust mass with lookback time. Note that the models of Wang have not been constrained to fit the observed local universe far-infrared number counts, unlike the parametric translational models discussed above, and the models of Franceschini et al. (1993). The opaque model of Franceschini et al. (1993) is also marginally in conflict with the more stringent of the two FIRAS limits.

A constraint on the dust content and temperature of young disk galaxies has been derived by Fall and Pei (1993), who have estimated the dust density of the damped Ly$\alpha$ systems, which may be the progenitors of present-day galactic disks, from the observed reddening of quasars seen on the line-of-sight through the damped Ly$\alpha$ systems. From this dust density they can then estimate the contribution of these disks to the far-infrared and submillimeter background in a simple way, with the (unknown) dust temperature as the only important variable. They find limits on the co-moving dust density in the damped Ly$\alpha$ systems of $10^{-6} < h\Omega_{dust} < 10^{-4}$. Using the then available FIRAS and DIRBE limits, these dust densities translated to limits on the dust temperature in the disks of $< 60K$ and $< 25K$ for $h\Omega_{dust} = 10^{-6}$ and $10^{-4}$, respectively, based on an integration of the damped Ly$\alpha$ systems between redshifts of 2 and 3, where they are observed. The newer FIRAS limits of Mather et al. (1993) would lower these temperature limits somewhat. This is an interesting technique, though it is limited by definition to low optical depth lines-of-sight through the foreground systems, thus can tell us little of any denser star forming regions in young disk galaxies. It can also tell us nothing of the dust masses and temperatures in ellipticals and S0s at high redshift, and these are the systems that are most likely to be the most important far-infrared emitters at early epochs.

Some additional constraints on evolving galaxy models can be sought at other wavelengths. In the radio a strong correlation in evolutionary behavior may be expected because of the well-known strong correlation between far-infrared and radio fluxes for most types of radio-quiet galaxies (e.g., Helou, Soifer and Rowan-Robinson 1985). A historical anecdote is of interest here. Far-infrared evolution studies are presently based on IRAS data, and when IRAS was launched few people dreamed it would be sensitive enough to tackle cosmological questions realistically! In fact what motivated Perry Hacking to address this question for his thesis (Hacking 1987; Hacking, Condon and Houck 1987) was the realization that if, because of the radio-infrared correlation, galaxies evolved as much in the far-infrared as they apparently do at cm radio wavelengths ($L \propto (1+z)^{\sim 4}$, Condon 1984) then such strong evolution should be detectable by IRAS, even though it only probed to modest redshifts ($z \lesssim 0.1$). The result of Hacking's study was the conclusion that the faint IRAS 60$\mu$m counts could be fitted reasonably well with the same model that was found to best fit the counts for the sub-millijansky radio source population (i.e., the non-active galaxy radio source population) by Condon (1984).

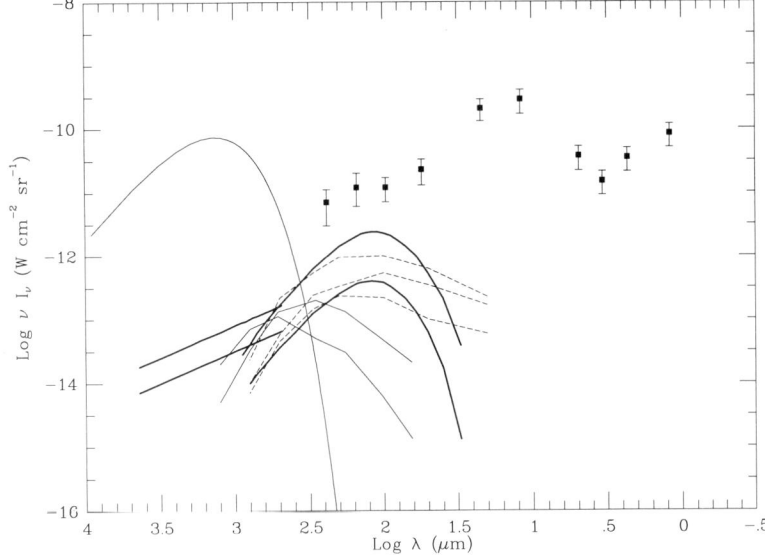

**Figure 6.** *Comparison of various models for the integrated light of galaxies and protogalaxies to the COBE data. Heavy solid lines – Wang (1991b) $z_f = 5$ PIE model with exponentially decreasing star formation rate (upper), $z_f = 5$ accretion model with constant star formation rate (lower). Dashed lines – Franceschini et al. (1993) population synthesis no-evolution (lower), moderately opaque (middle) and opaque (upper) models. Light solid lines – Franceschini et al. (1991) protogalaxy models with $z_f = 4.3$, $\Delta z = 0.3$ (lower), $z_f = 2$ $\Delta z = 0.1$ (upper).*

Similar conclusions have been reached by other authors using different models (e.g., Danese et al. 1987; Treyer and Silk 1993; Franceschini et al. 1993), therefore it seems clear that in the relatively local universe at least, the radio and far-infrared number counts and redshift distributions of non-active galaxies can be used together to constrain evolutionary scenarios. Unfortunately, for cosmological background studies the radio wavelength region is considerably less interesting that the far-infrared, as Malcolm Longair describes eloquently elsewhere in this volume.

There has been some work addressing the implications of evolving far-infrared bright galaxies to the X-ray background (Weedman 1990; Griffiths and Padovani 1990; Lonsdale and Harmon 1991; Treyer and Silk 1993). These studies conclude that the simplest parametric models which fit the far-infrared number counts can produce 50 to 100% of the soft X-ray background. Since much, if not all, of the soft X-ray background seems to be due to AGNs, the actual contribution from starbursting infrared-bright galaxies must be no more than 50% and probably lower. This constraint is likely to prove interesting as evolving models become more sophisticated.

## 3. PRIMEVAL GALAXIES AND THE FAR-INFRARED BACKGROUND

Searches at optical wavelengths for primeval galaxies using the Ly$\alpha$ line have not been successful, although systematic efforts over large ranges in volume and redshift have been going on for many years. An elusive and still hypothetical creature, a primeval galaxy is usually defined by its hunters as an early type galaxy going though a dramatic initial star formation event, perhaps much like the luminous starburst galaxies we see in the local universe. A good recent review of the subject is given by Djorgovsky and Thompson (1992). There are several possible explanations for this lack of success. One is that primeval galaxies must lie beyond a redshift of about 10, which is the approximate limit of large scale searches to date, a result that would be consistent with a baryonic dark matter model with primeval isocurvature fluctuations (Peebles 1987). Alternatively, it may be that the dark matter models in which galaxies grow much more recently by gravitational instability out of a scale-invariant spectrum of primeval adiabatic density fluctuations are correct. However, the most sensitive searches are now coming into conflict even with the predictions of cold dark matter galaxy formation simulations such as those of Baron and White (1987). Finally it may be that if primeval galaxies do exist, they are sufficiently dusty that the Ly$\alpha$ photons are extinguished and much of the energy of the object appears at far-infrared wavelengths (*e.g.*, Kaufman and Thuan 1977, van den Bergh 1990).

Two recent far-infrared/submillimeter PG models are those of Djorgovsky and Weir (1990) and Franceschini *et al.* (1991). The model of Djorgovsky and Weir was actually designed to fit the 700$\mu$m excess emission over the CMB claimed to be detected by Matsumoto *et al.* (1988). That excess has now been shown to be non-existent by COBE (Mather *et al.* 1990), however the model is still of interest for far-infrared bright PGs in general. Based on the observed spectral energy distributions of the nearby far-infrared bright galaxies M82 and Arp 220, the model had a range of possible initial mass functions and burst timescales of 10 to 200 Myr. The models were constrained not to exceed the formation of a solar metal abundance during the burst phase. The numerical results are not directly applicable to the situation I am discussing here, however Djorgovsky and Thompson (1992) used the model to conclude that the then available COBE limits ruled out more than a few percent of the stars in ellipticals and the bulges of spirals having been formed in dusty PGs, unless the redshift corresponding to the epoch of galaxy formation is less than $z_f = 3$, and/or the dust is unusually warm.

The model of Franceschini *et al.* (1991) is also based on the spectral energy distributions of local star forming galaxies, and they constrained the energy output to be that required to produce a solar metal abundance in $2 \times 10^8$ years. Their resulting models for formation epochs of $z_f = 2$ and 4.3 are shown in Figure 6, where it may be seen that, as concluded by Djorgovsky and Thompson, FIRAS strongly limits models with even moderate formation redshifts and dust temperatures like those of local universe starburst galaxies.

### 3.1. IRAS F10214+4724 - A Possible Protogalaxy

The conclusion one may draw from the models described above is that a scenario in which most galaxies went through a dusty early phase similar to local starburst galaxies may be in conflict with the FIRAS limits. However, on the other side of the coin there is some recent evidence in favor of the existence of large amounts of dust in galaxies at early epochs, and possibly even one example of a dusty PG: the extremely luminous galaxy IRAS F10214+4724 at z=2.286 discovered by Rowan-Robinson *et al.* (1991). This galaxy is arguably the most luminous object in the universe with a luminosity of $10^{14}h^{-2}$ L$_\odot$ ($h$=H$_0$/100 km/s/Mpc; q$_0$=0.5), and a dust mass estimated from submillimeter observations of $2.5 \times 10^8 h^{-2}$M$_\odot$ (Rowan-Robinson *et al.* 1993; Downes *et al.*

1992). Evidence for large masses of dust has also been found in several high redshift quasars (Andreani et al. 1993).

The controversy over the interpretation of F10214+4724, as for lower redshift ultraluminous infrared galaxies, concerns the dominant source of the extremely high far-infrared luminosity detected by IRAS. There is little doubt that **both** a luminous starburst and a non-stellar active nucleus are present in the source. There is abundant evidence that F10214+4724 is a primeval galaxy undergoing rigorous star formation, including $\sim 10^{11} h^{-2}$ $M_\odot$ of molecular gas (Solomon, Downes and Radford 1992), a UV-to-radio continuum energy distribution which is most simply interpreted as a powerful starburst (Rowan-Robinson et al. 1993; Mazzei and De Zotti 1993), and a radio source which is extended on a scale of about 2.5 $h^{-1}$ kpc (Lawrence et al. 1993). Likewise, there is substantial evidence for an embedded AGN: high excitation emission lines (Rowan-Robinson et al. 1991), and strong polarization (Lawrence et al. 1993). In addition, new results from near-infrared (rest-frame optical) spectroscopy show [NII]/H$\alpha$ and [OIII]/H$\beta$ emission-line ratios to be typical of those found in type 2 Seyfert galaxies (Eisenhardt et al. 1993).

The near-infrared (rest-frame optical) continuum morphology observed using a $256^2$ InSb array on the Keck telescope shows at least 3 continuum components that appear to be physically associated over a physical scale of 25 $h^{-1}$ kpc (Matthews et al. 1993), suggesting a small cluster since each object is more luminous in the rest frame r band than a local $L^*$ galaxy, and the main, southern object has almost 100 $L^*$ in rest frame r. A number of faint sources (K>21 mag) are also seen within 20″ of the central source that may be galaxies in an associated cluster. This image is reproduced in Figure 7. The Keck results also show that the brightest H$\alpha$ source is now resolved on a scale consistent with that of the radio source, ($0''.5 \sim 2.5$ $h^{-1}$ kpc), supporting the star formation origin for the H$\alpha$ emission.

Can any protogalaxy model plausibly explain the tremendous luminosity of this object? It is clearly enriched in heavy elements already, as evidenced not only by the emission line spectrum which includes lines of C, N, Ne and Mg, but also by the presence of the dust itself, and this enrichment must also be explained by any plausible model. In particular, must the dust have been created in an earlier generation of stars? If the dust was created in the envelopes of evolved stars, as may be the case for much of the dust formed in our galaxy at the present epoch, then we must be seeing F10214+4724 at an age of at least 1 Gyr.

Elbaz et al. (1992) have developed starburst models for F10214+4724. They found that a model with a bimodal initial mass function (IMF) can achieve both the very high observed $L/M_{gas}$ ratio of 750 $L_\odot/M_\odot$ and the strong enrichment, reaching $Z_{Fe,C,O,Si} > Z_\odot$ and $M_{dust} < M_{metals}$, in less than $10^8$ years. Hamann and Ferland (1993) have developed detailed chemical evolution models for QSOs, also concluding that high metallicities can be reached rapidly: $> 10 Z_\odot$ in $< 1$ Gyr. The model of Elbaz et al. has a bursting component with a lower mass limit to the IMF of 3 $M_\odot$ and a star formation rate of 6200 $M_\odot$/yr. They were unable to achieve a fit with a single IMF. The source of the dust in the Elbaz et al. model is not evolved stars but supernova remnants. While it is not known whether supernova remnants can be responsible for significant amounts of dust formation, there is evidence that SN 1987A has produced 0.1 $M_\odot$ of dust (Dwek et al. 1992). At a supernova rate of $1.25 \times 10^{-12}$ $(L/L_\odot)$/yr (Solomon, Radford and Downes 1992) for $10^8$ years, remnants like 1987A could easily produce the the estimated dust mass of $2.5 \times 10^8 h^{-2}$ $M_\odot$.

Mazzei and De Zotti (1993) have successfully modeled the spectral energy distribu-

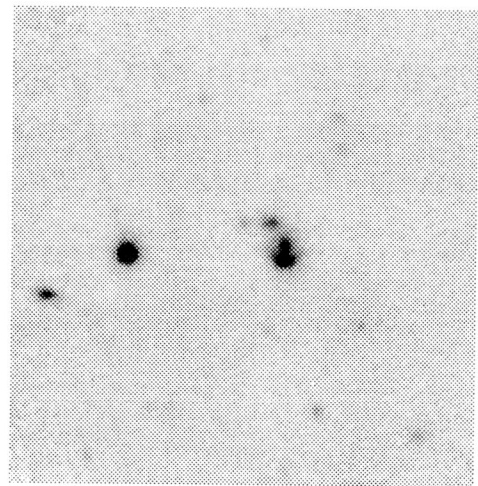

**Figure 7.** *Image of F10214+4724 at 2.2μm from Matthews et al. (1993). The object is just right of center in this 40" × 40" image.*

tion of F10214+4724 using the population synthesis models of Mazzei et al. (1993) (see Figure 3). They find a good fit at an age of 1 Gyr (for H = 50 km/s/Mpc; $q_0 = 0.5$) with a star formation rate of $3 \times 10^4$ $M_\odot$/yr; a fit with a much younger age is also possible. Mazzei and De Zotti show that F10214+4724 could plausibly fade to a z = 0 elliptical with bolometric luminosity less than $10^{13}$ $L_\odot$.

## 3.2. A Protogalaxy Model Based on IRAS F10214+4724

The models described above for the contribution to the far-infrared background by infrared-bright protogalaxies, those of Djorgovsky and Weir (1990) and Franceschini et al. (1991), are based on the spectral energy distributions of *local universe* starburst galaxies. If it is truly powered by star formation, then F10214+4724 provides us with the opportunity of using a high redshift object with known luminosity and spectral energy distribution as a template for protogalaxies, thus eliminating the uncertainties introduced by assuming that local universe objects are good analogs of protogalaxies, or by adopting a model spectral energy distribution with an assumed dust temperature and luminosity. It also allows us to avoid the large K-corrections involved in redshifting local templates to cosmological distances.

I have therefore developed a simple model to determine the contribution to the far-infrared background of a population of galaxies, forming with a protogalactic burst like that observed in F10214+4724. If a QSO contributes significantly to the luminosity of F10214+4724 then the model predictions can be treated as upper limits to the background emission unless the coeval existence of a QSO with the starburst is a common feature in the formative stages of all ellipticals (*cf.* Hamann and Ferland 1993 and references therein).

I assume all galaxies have an SED of similar shape to F10214+4724. Galaxies are assigned luminosities according to a luminosity function. The variable parameters of the

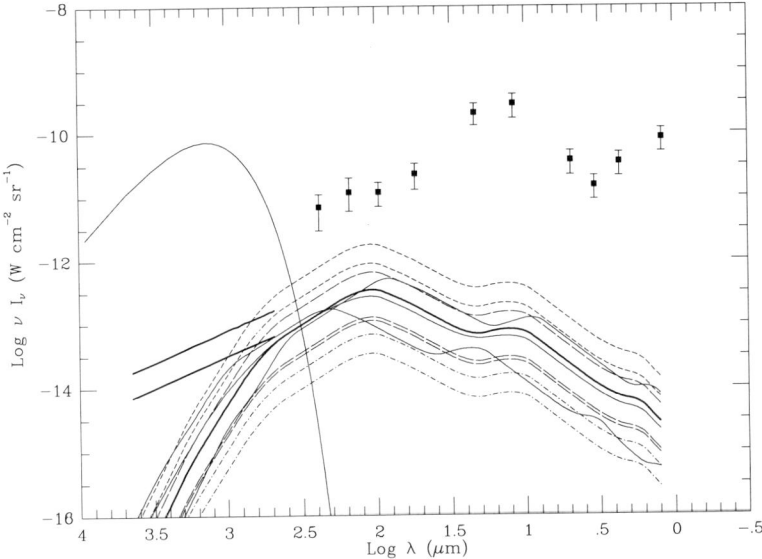

**Figure 8.** *Protogalaxy model predictions compared to the COBE data. Baseline model of Table 2 (heavy solid line); other lines show effects of changing other parameters: redshift range (light solid lines): $z_f = 5 - 10, 2 - 10, 1.5 - 2.5$; cosmology (short dashed lines): $\Omega = 0$ and $H_0 = 100$ (upper line) and 50 (lower line); LF (long dashed lines): Tammann et al. (lower line), Franceschini et al. (middle line), Efstathiou et al. (upper line – includes disk galaxies); f (dot-dash lines): f = 50 (upper line), f = 100 (lower line).*

model are the cosmology, the formation redshift $z_f$, the burst duration $\delta t$, the luminosity function, and the factor, $f$, by which F10214+4724 is assumed to be brighter or fainter than the characteristic luminosity at the knee of the luminosity function, $L_{F10214} = f L^*$, where the luminosity function is given by $\phi(L)dL = \phi^*(L/L^*)^\alpha e^{-L/L^*} d(L/L^*)$ (Schechter 1976). Here $\phi^*$ is the characteristic space density. Four different local luminosity functions were considered. Note that using a local luminosity function to define the distribution of galaxy luminosities at high redshift is equivalent to adopting a mass function, as long as the evolutionary behavior with lookback time of the L/M ratio does not vary greatly with galaxy mass.

The range of parameters considered is given in Table 2. For most models the luminosity function was restricted to elliptical galaxies only, since spirals are not expected to have formed with a dramatic initial burst. The factor $f$ was restricted to 10 or higher because F10214+4724 is undoubtedly a very rare and unusually luminous object. Estimates for the expected surface density of protogalaxies are in the range $10^3$ to $10^5$ per square degree, depending on the cosmology, the epoch of formation and the duration of the bright phase (e.g., Djorgovsky and Thompson 1992). From the detection statistics we can estimate a surface density of objects like F10214+4724 of $1.5 \times 10^{-3}$ per square degree; allowing a factor of $\pm 10$ on this estimate since the statistics are very crude (only one object has been detected, and that very close to the detection limit of the IRAS

## Table 2. Protogalaxy Model Based on IRAS F10214+4724

| Parameter | Range Considered | Baseline Model |
|---|---|---|
| $H_0$ | 50, 100 | 50 |
| $\Omega$ | 0, 1 | 1 |
| $z_f$ | 1.5 – 10 | 2 – 5 |
| $\Delta t$ | 1.0 to $2.0 \times 10^8$ yrs | $10^8$ yrs |
| $f$ | 10 – 100 | 10 |
| Luminosity Function | Shanks et al. 1990, Es only<br>Efstathiou et al. 1988, all galaxies<br>Franceschini et al. 1988a, Es only<br>Tammann et al. 1979, Es only | Shanks et al. 1990 |

| Luminosity Function ($H_0$=100) | $\phi^*$ (# Mpc$^{-3}$) | M* (mag.) | $\alpha$ |
|---|---|---|---|
| Shanks et al. 1990 | 0.0096 | −19.00 | −0.07 |
| Efstathiou et al. 1988 | 0.0156 | −19.68 | −1.7 |
| Franceschini et al. 1988a | 0.0032 | −19.60 | −1.0 |
| Tammann et al. 1979 | 0.0031 | −19.45 | −0.77 |

survey) it follows that objects like F10214+4724 are at least $10^5$ times less numerous that "typical" primeval galaxies. For a Schechter LF, this translates roughly to $f > 10$.

Full details of the model are given in Lonsdale (1994, in preparation). Figure 8 summarises the results of the models compared to the data of Figure 1. Figure 9 illustrates the blue and K-band number counts for the model population of F10214-like protogalaxies, compared to observational data.

The low $\Omega$ protogalaxy models shown in Figure 8 are in conflict with the FIRAS limits. The $\Omega = 1$ models are mostly consistent with FIRAS except the model using the Franceschini et al. (1988a) LF including all galaxies (not only ellipticals), which is in conflict with the more stringent FIRAS limit, and the high redshift range model which is only marginally consistent with the stringent limit. None of the models are in conflict with the current DIRBE observations. Therefore, basically the entire range of parameter space that has been explored is allowed for a high $\Omega$ universe. An acceptable fit for a low $\Omega$ universe would require $z_f = 5$ or lower, and/or $f < 10$, and/or a burst duration shorter than $2 \times 10^8$ yrs.

The number count predictions are small compared to the observed counts, therefore it is not surprising that only one object like F10214+4724 has so far been discovered by serendipitous spectroscopic follow-up studies of faint field galaxies. Systematic surveys of 2.2$\mu$m-selected objects in the 15 to 18th magnitude range, where such objects could account for 10% of the sample, might be the most fruitful.

To summarise, the main result of the model presented here is that it is quite possible that a large fraction of the light of forming galaxies is hidden in the far-infrared wavelength region. Thompson and Djorgovsky (1992) and Franceschini et al. (1991) concluded from their models that to hide galaxy formation in the far-infrared would require quite low values of $z_f$ ($z_f < 5$) and/or warm dust temperatures. Both of these requirements are the result of the FIRAS limits. The F10214+4724 model is consistent with these results because this object does indeed contain relatively warm dust: Downes et al. (1993) derive a dust temperature of 80K for the far-infrared/submillimeter emission. Thus this model demonstrates the plausibility of a significant background from protogalaxies in the far-infrared most convincingly since it based on the real SED of a dust-rich, star-forming galaxy with known luminosity at a known (cosmological) redshift, rather than on a local universe analog or a theoretical thermal spectrum. In particular, the $\lambda > 100\mu m$ spectral shape, which is a critical constraint compared to the FIRAS observations, has been directly measured for this object.

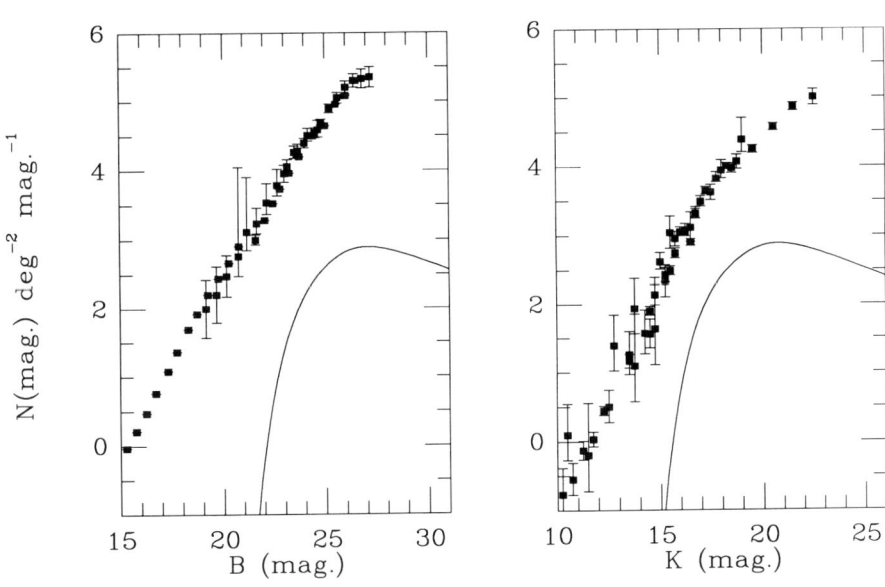

**Figure 9.** *Blue and K-band galaxy number counts compared to the prediction from the baseline protogalaxy model (see Table 2). The references to the data can be found in Chokshi et al. (1993).*

## 4. THE FUTURE IN SPACE: DIRBE, ISO AND SIRTF

Where do we go next to test these ideas? Obviously we anxiously await the revised, foreground-subtracted, DIRBE limits: if the team is able to reliably subtract the foregrounds to levels approaching the design sensitivity of $\nu I_\nu = 10^{-13}$ W cm$^{-2}$ sr$^{-1}$ (Boggess et al. 1992), which is about 1% of the foreground emissions, then they can expect to detect or rule out most of the predicted contributions to the background by evolving galaxies and infrared-bright protogalaxies shown in Figures 4, 5, 6 and 8.

Although the nineties was heralded as the decade of the infrared by the Bahcall committee, the poor funding situation at present has delayed our hopes for the next generation US infrared mission, SIRTF, into the next century. Meanwhile the European satellite ISO is due to launch in late 1995. Both ISO and SIRTF are observatory class instruments, differing from the survey instruments IRAS and COBE in having small beams and fields-of-view. Thus they are optimised for point source work and

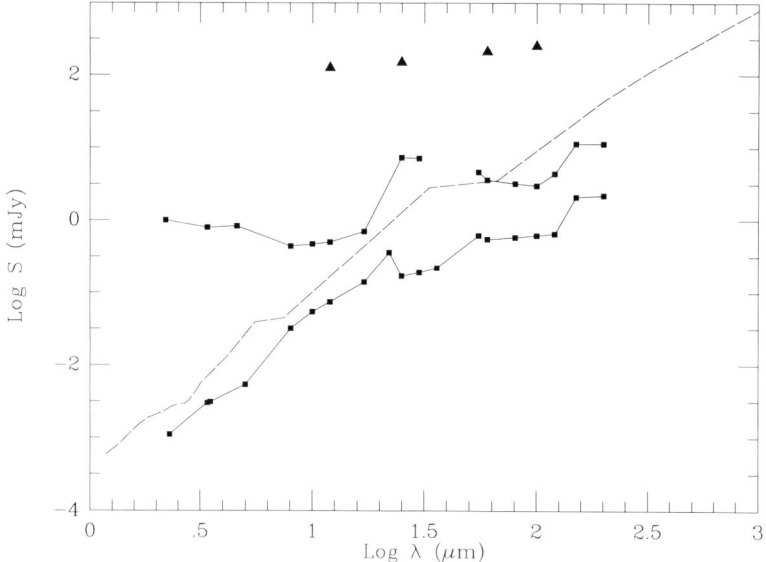

**Figure 10.** *ISO and SIRTF background-limited point source sensitivities, $5\sigma$ in 500 seconds (upper and lower solid lines, respectively) (E. Young, private communication), compared to the IRAS Faint Source Catalog sensitivity limits (90% completeness limits for the most sensitive 10% of the sky; filled triangles), and to IRAS F10214+4724 moved to a redshift of 10 ($H_0 = 50$, $\Omega = 1$).*

fluctuation analyses much more than direct background measurements, although SIRTF may be able to match the surface brightness sensitivity of DIRBE with considerable work. SIRTF will be especially important for cosmological studies because its detector technology will be frozen at a much later date than that of ISO, and because it has larger arrays than ISO. Figure 10 compares the expected point source sensitivities of ISO and SIRTF compared to IRAS F10214+4724 removed to z = 10.

SIRTF may hope to resolve all of any background that DIRBE may detect in the near-infrared window near $3\mu$m. A calculation by E. Wright (M. Werner, private communication) shows that the minimum expected integrated intensity at $3.5\mu$m due to galaxies is about $\nu I_\nu = 3.4 \times 10^{-13}$ W/cm$^2$/sr. This is derived from the measured extragalactic number counts at $2.2\mu$m (Gardner et al. 1993), assuming a temperature of 2000K to estimate the $2.2\mu$m-$3.5\mu$m color. Comparing this to the hoped-for ultimate DIRBE sensitivity of about $10^{-13}$ W/cm$^2$/sr it is clear that the background due to this known population will be detected by DIRBE. At an intensity of $3.4 \times 10^{-13}$ W/cm$^2$/sr, the number of faint galaxies per 0.7 degree DIRBE beam is about $1.2 \times 10^5$, giving an average separation of 10 arcsec. This is within reach of the 1 arcsec beam of SIRTF to resolve. To reach the required limit of 21.6 mag. at $3.5\mu$m with SIRTF will require an integration time of about 10,000 seconds.

A similar calculation at $60\mu$m based on a local $60\mu$m luminosity function and exponential evolution with $\kappa = 1.5$, $H_0 = 75$, indicates that about 60% of the predicted background above an anticipated DIRBE limit of about $2\times10^{-13}$ W cm$^{-2}$ sr$^{-1}$ will be resolvable by SIRTF, assuming a SIRTF confusion limit of about 0.1 mJy at this wavelength (Wright 1993).

Franceschini et al. (1991) have calculated the fluctuations expected in the 0.7 degree DIRBE beam due to their model evolving galaxy populations. Fluctuations $\Delta I_\nu/I_\nu$ range from 2% at 280$\mu$m, through 4–7% at 25 to 100$\mu$m and rise to 50% at 2.2$\mu$m. These high values at short wavelengths are dominated by bright stars in the galaxy. The main limitation at the longer wavelengths will be confusion noise due to galactic cirrus emission (Gautier et al. 1992).

## 5. CONCLUSIONS

The very sensitive new FIRAS CMB-deviation limits severly constrain parametric translational models for galaxy evolution at wavelengths longer than 500$\mu$m. It follows that the evolutionary rates for infrared-bright galaxies implied by such models in the local universe cannot continue to cosmological redshifts for the class as a whole, unless exponential luminosity evolution is adopted. Most likely the current generation of parametric models is too simple in approach to adequately model this complex situation; in particular the likely strong evolution of the dust content of early type galaxies is not taken into account by these models.

The analytical chemical evolution models of Wang (1991a,b) address the evolution of the dust content. His accretion models are not in conflict with the current COBE limits but the prompt initial enrichment model is closely constrained. Similarly, the models of Franceschini et al. (1993), which are based on the closed box population synthesis models of Mazzei et al. (1992, 1993) incorporating chemical and dust evolution, are presently consistent with, but close to being constrained by, the COBE data. The simulations of deep galaxy fields of Chokshi et al. (1994, in preparation), which are also based on the population synthesis models of Mazzei et al. (1993), are also likely to provide interesting constraints (see Chokshi et al. 1993).

The FIRAS limits also constrain models for infrared-bright protogalaxies. If a significant fraction of the light created in primeval galaxies emerges in the far-infrared due to large dust optical depths then the FIRAS limits restrict the epoch or formation to $z < 5$ for low $\Omega$, and/or require relatively warm dust temperatures. A model based on the luminous, warm, $z = 2.286$ IRAS galaxy F10214+4724 can satisfy these requirements.

DIRBE will be able to detect the backgrounds expected from evolving galaxies and from infrared-bright protogalaxies, unless these objects are much less dusty than the models assume.

ISO and SIRTF will be able to detect objects like F10214+4724 to redshifts approaching 10. SIRTF will be able to resolve all of any background that DIRBE can expect to detect at 3 microns, and most of a DIRBE background at 60$\mu$m.

I thank Paola Mazzei, K. Matthews and B. T. Soifer for allowing me to reproduce figures from their papers, and Seb Oliver for making his model data files available. I am grateful to Mike Hauser for discussions regarding the DIRBE observations, Mike Werner and Peter Eisenhardt for discussions regarding SIRTF's potential, and Erick Young for permission to use his calculations of ISO and SIRTF point source sensitivities. I thank Ned Wright for allowing me to quote his unpublished derivations of SIRTF's ability to resolve infrared backgrounds. Rick Shafer kindly estimated the correction to the FIRAS limit for galactic foreground emission at the meeting. I thank George Helou and Chas

Beichman for allowing me to show their revised evolutionary models in Figure 4, and the foreground solar system and galactic emission models in Figure 1. I thank Joe Mazzarella for discussions on the nature of F10214+4724, Boqi Wang and Mike Fall for bringing my attention to some important work, and Perry Hacking for comments on the manuscript. The research described in this review was carried out by the Jet Propulsion Laboratory, California Institute of Technology, under a contract with the National Aeronautics and Space Administration.

## REFERENCES

Andreani, P., La Franca, F., and Cristiani, S. 1993, *Mon. Not. R. Astr. Soc.*, in press
Baron, E., and White, S. M. 1987, *Ap.J.*, **322**, 585
Beichman, C. A., and Helou, G. 1991, *Ap.J.(Letters)*, **370**, L1
Boggess, N., *et al.* 1992, *Ap.J.*, **397**, 420
Bond, J. R., Carr, B. J., and Hogan, C. J. 1986, *Ap.J.*, **306**, 428
Bond, J. R., Carr, B. J., and Hogan, C. J. 1991, *Ap.J.*, **367**, 420
Carr, B. J. 1982, in *The Infrared and Submillimeter Sky After COBE*, ed. M. Signore and C. Dupraz (Kluwer: Dordrecht), p. 213
Chokshi, A., Lonsdale, C. J., Mazzei, P., and de Zotti, G. 1993, *Ap.J.*, in press
Condon, J. J. 1984, *Ap.J.*, **287**, 461
Danese, L., De Zotti, G., Franceschini, A., and Toffolatti, L. 1987, *Ap.J.(Letters)*, **318**, L15
Djorgovsky, S., and Thompson, D. J. 1992, in *The Stellar Populations of Galaxies, IAU Symp. 149*, ed. B. Barbuy and A. Renzini, (Kluwer: Dordrecht), p. 337
Djorgovsky, S., and Weir, N. 1990, *Ap.J.*, **351**, 343
Downes, D. Radford, S. J. E., Greve, A., and Thum, C. 1992, *Ap.J.(Letters)*, **398**, L25
Dwek, E., Moseley, S. H., Glaccum, W., Graham, J. R., Loewenstein, R. F., Silverberg, R. F., and Smith, R. K. 1992, *Ap.J.*, **389**, L21
Efstathiou, G., Ellis, R. S., Peterson, B. A. 1988, *Mon. Not. R. Astr. Soc.*, **232**, 431
Elbaz, D., Arnaud, M., Casse, M., Mirabel, I. F., Prantzos, N., and Vangioni-Flam, E. 1992, *Astr. Astrophys.*, **265**, L29
Eisenhardt, P., Elston, R., McCarthy, P., Dickinson, M., Spinrad, H., Jannuzi, B., and Maloney, P. 1993, *Bull. Am. Astr. Soc.*, **25**, (2)789
Fall, S. M., and Pei Y. C. 1993, *Ap.J.*, **402**, 492
Fisher, K. B., Strauss, M. A., Davis, M., Yahil, A., and Huchra, J. 1992, *Ap.J.*, **389**, 188
Franceschini, A., Danese, L., De Zotti, G., and Toffolatti, L. 1988a, *Mon. Not. R. Astr. Soc.*, **233**, 157
Franceschini, A., Danese, L., De Zotti, G., and Xu, C. 1988b, *Mon. Not. R. Astr. Soc.*, **233**, 175
Franceschini, A., Toffolatti, L., Mazzei, P., Danese, L., and De Zotti, G. 1991, *Astr. Astrophys.*, **89**, 285
Franceschini, A., Mazzei, P., De Zotti, G., and Danese, L., 1993, *Astr. Astrophys.*, in press
Gardner, J. P., Cowie, L. L., and Wainscoat, R. J. 1993, *Ap.J.(Letters)*, in press
Gautier, T. N., Boulanger, F., Perault, M., and Puget, J. L. 1992, *A.J.*, **103**, 1313
Griffiths, R. E., and Padovani, P. 1990, *Ap.J.*, **360**, 483

Hacking, P. B. 1987, *PhD. Thesis*, Cornell University
Hacking, P. B., and Soifer, B. T. 1991, *Ap.J.(Letters)*, **367**, L53
Hacking, P. B., Condon, J. J., and Houck, J. 1987, *Ap.J.*, **316**, L15
Hamann, F., and Ferland, G. 1993, *Ap.J.*, in press
Helou, G., Soifer, B. T., and Rowan-Robinson, M. 1985, *Ap.J.*, **298**, L7
Kaufman, M., and Thuan, T. X. 1977, *Ap.J.*, **215**, 11
Lawrence, A., et al. 1993, *Mon. Not. R. Astr. Soc.*, **260**, 28
Lonsdale, C. J. 1992, in *First Light in the Universe, Stars or QSOS?*, ed. B. Rocca-Volmerange, B. Guiderdoni, M. Dennefeld, and J. Tran Thanh Van, (Frontieres: Gif-sur-Yvette), p. 3
Lonsdale, C. J., and Hacking, P. B. 1989, *Ap.J.*, **339**, 712
Lonsdale, C. J., Hacking, P. B., Conrow, T. P., and Rowan-Robinson, M. 1990, *Ap.J.*, **358**, 60
Lonsdale, C. J. and Harmon, R. 1991, *Adv. Space Res.*, **11**, (2)333
Mather, J. C. et al. 1990, *Ap.J.(Letters)*, **354**, L37
Mather, J. C. et al. 1993, *Ap.J.*, in press
Matthews, K., et al. 1993, *Ap.J.*, in press
Matsumoto, T., Hayakawa, S., Matso, H., Murakami, H., Sato, S., Lange, A., and Richards, P. L. 1988, *Ap.J.*, **329**, 257
Mazzei, P., and De Zotti 1993, in press
Mazzei, P., De Zotti, G., and Xu, C. 1993, *Ap.J.*, in press
Mazzei, P., Xu, C., and De Zotti, G. 1992, *Astr. Astrophys.*, **256**, 45
Oliver, S. J., Rowan-Robinson, M., and Saunders, W. 1992, *Mon. Not. R. Astr. Soc.*, **256**, 15P
Ostriker, J. P., and Heisler, J. 1984, *Ap.J.*, **278**, 1
Partridge, R. B., and Peebles, P. J. E. 1967, *Ap.J.*, **148**, 713
Peebles, P. J. E. 1987, *Nature*, **327**, 210
Rowan-Robinson, M. 1968, *Mon. Not. R. Astr. Soc.*, **138**, 445
Rowan-Robinson, M., et al. 1991, *Nature*, **351**, 719
Rowan-Robinson, M., et al. 1993, *Mon. Not. R. Astr. Soc.*, **261**, 513
Salpeter, E. E. 1955, *Ap.J.*, **121**, 161
Saunders, W., et al. 1990, *Mon. Not. R. Astr. Soc.*, **242**, 318
Schechter, P. 1976, *Ap.J.*, **230**, 297
Shanks, T., et al., as reported in Metcalfe, N., Shanks, T., and Fong, R. 1991, *Mon. Not. R. Astr. Soc.*, **249**, 498
Solomon, P. R., Downes, D., and Radford, S. J. E. 1992, *Ap.J.*, **398**, L29
Tammann, G. A., Yahil, A., and Sandage, A. 1979, *Ap.J.*, **234**, 775
Telesco, C. M. 1993, *Mon. Not. R. Astr. Soc.*, **187**, 73P
Treyer, M., and Silk, J. 1993, *Ap.J.(Letters)*, **408**, L1
van den Bergh, S. 1990, *Publ. Astr. Soc. Pacif.*, **102**, 503
Wang, B. 1991a, *Ap.J.*, **374**, 456
Wang, B. 1991b, *Ap.J.*, **374**, 465
Weedman, D. W. 1990, in *Massive Stars in Starbursts*, ed. C. Leitherer, N. R. Walborn, T. M. Hackman, and C. A. Norman, (Cambridge University Press: Cambridge), p. 317
Wright, E. L. 1990, *Ap.J.*, **353**, 411
Wright, E. L., et al. 1993, *Ap.J.*, in press

## DISCUSSION

**J. Mould**: The parameterized models with $\alpha$'s and $\beta$'s presumably do not allow you to compare the predicted counts at other wavelengths. But the evolutionary synthesis models I guess would allow you to make comparisons with optical counts in the V-band and so on. Is there any conflict between models that just manage to fit the FIRAS limits and optical galaxy counts?

**Lonsdale**: No, not really because the population synthesis models of Franceschini *et al.* have been matched to fit the near infrared and blue counts.

**T. Tyson**: Could you describe your blue galaxy counts as a result of this model?

**Lonsdale**: The standard model I referred to, which has the formation redshift between 2 and 5 and assumes that F10214+4724 is 10 $L^*$ in luminosity, is shown compared to the observed counts in Figure 9. The range of parameters in Table 2 allows the model to move around in this diagram a fair amount. Obviously if you push things to low redshift, you are starting to get some fairly bright galaxies, but they are at most about 1% of the total number of counts.

**S. White**: This seems slightly schizophrenic because we have two completely different models for galaxy formation which have absolutely no overlap, as in this model you are forming all the elliptical galaxies at a redshift of 2.5. Yet they are absolutely invisible in the counts. And yet in the counts we are also seeing things at 2 microns which apparently are a significant fraction of all the stars in the universe and do not have much dust, because we can see them into the blue. So are there two completely separate populations of galaxies?

**Lonsdale**: It is also schizophrenic in that we are calling for this epoch of massive galaxy formation at a pretty low redshift and, if things are happening at low redshift, you might expect them to be not occurring with such a huge burst. So it is a good question. Of course, I am only counting these galaxies going through the huge burst for $10^8$ years or so, but I do have to count up their contributions to producing metals. Rough estimates indicate they could be responsible for half the metals, which may not be inconsistent with the blue galaxies. I do not like to constrain myself too much with theory, but prefer to see how far the observations will take me. Maybe if we take your comment to its logical conclusion, I have to come back and say that means that that F10214+4724 is not a protogalaxy but a quasar, and that would be interesting. Also it may not follow that because galaxies are seen in the blue that they do not have much dust. It could be that one is just seeing the surface layer in the blue: down to one optical depth. Many IRAS galaxies are dusty, yet blue in the blue region of the spectrum.

**S. White**: I did not mean that it was necessarily implausible that there are two populations. You could easily imagine there are things that are shrouded and things that are not. The difference in properties could be enormous. So it could well be.

**E. Feigelson**: You require dust in your primeval galaxies, which implies that there was at least one epoch of massive star formation previously. Does this have any testable consequences?

**Lonsdale**: I do not think it necessarily implies that. I thought so when I started looking at this, but the enrichment model for F10214+4724 of Elbaz *et al.* forms all the dust in the same star formation burst that is responsible for the observed luminosity. There is evidence from SN1987A that enough dust might be created in supernova remnants.

**J. Bahcall**: It is clear from your talk, Carol, that when there is a higher ratio of measured numbers to calculated numbers, it will be even more interesting.

# MICROWAVE BACKGROUND RADIATION (OBSERVATIONS)[1]

John C. Mather
NASA Goddard Space Flight Center
Greenbelt, MD 20771

**Abstract.** The COBE satellite was designed, built, and launched by NASA Goddard Space Flight Center to examine the Big Bang and the first objects to form after that cosmic cataclysm. There are two main cosmological results to date. First, the spectrum of the cosmic microwave background radiation (CMBR) agrees with the pure blackbody form expected for a hot Big Bang within 0.03%, a thousand times better than was known before COBE. Energy release after the first year of expansion is strictly limited. Processes that might have released energy include decay of turbulence, decay or growth of black holes, decay of elementary particles, decay of cosmic strings, the formation of discrete objects such as galaxies, and the emissions from early generations of quasars, and infrared galaxies. Second, the COBE mapped the primordial fluctuations of the CMBR for the first time, with an angular resolution of 7°. These spatial fluctuations of the CMBR agree with a simple scale-invariant form and the amplitude agrees roughly with extrapolations from galactic clustering. These anisotropies map the gravitational potential field of the universe at the decoupling, about 300,000 years after the Big Bang, but must have been produced in the initial conditions or the inflationary period.

## 1. INTRODUCTION

Four years after the COBE satellite (Boggess *et al.* 1992) was launched on 18 November 1989, great advances in both instrumentation and theory are still propelling an exponential growth in cosmology. The Hot Big Bang theory has passed stringent tests, and it seems possible to explain the growth of large scale structures from primordial seeds under the influence of gravity alone (if the right sort of dark matter exists). Inflation explains the uniformity of the cosmos, and high energy physics permits a broken symmetry between matter and antimatter. The initial COBE data sets have been delivered to the National Science Data Center and are available on the Internet and Mosaic.

---

[1] The National Aeronautics and Space Administration/Goddard Space Flight Center (NASA/GSFC) is responsible for the design, development, and operation of the Cosmic Background Explorer (COBE). Scientific guidance is provided by the COBE Science Working Group. GSFC is also responsible for the development of the analysis software and for the production of the mission data sets.

Nevertheless, there are many mysteries left. What are the laws of physics in the early universe? What is the dark matter that seems to provide 90 to 99% of the gravitating mass? What is the history of the intergalactic gas? How and when did galaxies and quasars form? What is the law of gravitation on cosmic scales? How can it be quantized? What are its sources? Are there satisfactory alternatives to the hot Big Bang? What can we measure about inflation?

There are few observables available. The Hubble expansion is the fundamental one, the cosmological discovery of the century. The ages of stars and clusters confirm the age deduced from the Hubble law (with modern values for $H_0$). The properties of elementary particles yield predictions of the abundances of light elements, and even confirm that there are just three flavors of neutrinos. The distribution and velocities of the galaxies test the gravitational field on large scales. The evolution of galaxy luminosities, spectra, clustering, and densities with time tests our understanding of their formation and growth. The cosmic background radiations from microwave to gamma ray record the accumulated energy releases from all kinds of events after the Big Bang. The cosmic backgrounds of weakly interacting particles such as neutrinos are not directly observable.

The cosmic microwave background radiation stands alone in all these observables as the sole radiative remnant of the initial conditions. Since the annihilation of positrons at a redshift of about $3 \times 10^9$, the CMBR photons have outnumbered protons, neutron, and electrons by a factor of the order of $10^9$. From that moment until the decoupling at a redshift of 1000 (or even later for material that remains ionized), the photons have governed the motion of the ordinary matter particles. This prevents their participation in the growth of gravitational instabilities and should have recognizable consequences. Since the gravitational instabilities grow only linearly in the expanding universe, the existence of large scale structure implies primordial seeds of substantial amplitude.

There are two main observable characteristics of the CMBR. First, the spectrum of the radiation establishes its origin in conditions of local thermodynamic equilibrium, followed by adiabatic expansion. If the spectrum is not of blackbody form, then one of these conditions must be violated. The number of photons in the CMBR is essentially fixed by a redshift of $3 \times 10^6$, about a year after the Big Bang, when the double quantum and free-free processes became slow relative to the cosmic expansion. After that time the CMBR spectrum could deviate from a blackbody form if energy is added (Zeldovich & Sunyaev 1969, 1970; Peebles 1971; Sunyaev & Zeldovich 1980). At first, multiple Compton scattering is sufficient to establish a pseudo-equilibrium form, a Bose-Einstein spectrum with a chemical potential. After a redshift of about $10^5$, this process also becomes slow relative to the cosmic time scale, and the CMBR spectrum can be distorted in other ways. One likely form is called "Comptonized," which is equivalent to mixing blackbodies at a range of temperatures. This would be expected for any kind of energy release that is mediated by the electrons, but need not apply for processes like the decay of elementary particles that go directly to photons.

The second observable characteristic of the CMBR is its anisotropy, or difference in brightness between one direction and another. Large angular scale features, greater than a few degrees in size, are presumably primordial and thus of great cosmological significance. This is true if the universe is transparent after the nominal decoupling, because objects separated by larger angles have not been in causal contact since the Big Bang. Two points separated by $ct$ at the decoupling, where $t = 300,000$ years is the age of the universe then, now appear a few degrees apart. A domain this size at decoupling would grow by a factor of 1000 and would now be 100 Mpc across, about the size of the largest observed clustering structures. It is remarkable that the main cause of large scale anisotropy is the gravitational redshift (Sachs & Wolfe 1967). Even though denser regions are intrinsically hotter, the greater gravitational redshift makes them appear cooler to us.

Smaller size structures are affected by movements of dark matter, if it exists, even before the decoupling. Unlike ordinary matter, the dark matter is not coupled to the radiation and can move as soon as it is feels the gravitational forces coming over the horizon. On these smaller scales, the ordinary matter can be moving during the process of decoupling, producing Doppler shifts in the CMBR temperature. The temperature fluctuations of the radiation itself can also exceed the gravitational redshift effects.

A complete understanding of the CMBR fluctuations requires detailed modeling of all these processes. The approach is to assume that the primordial fluctuations are random, and then to calculate their growth using linear perturbation theory. This should be successful and accurate for large scales where the relative amplitudes of the fluctuations are low. The major difficulty in theory is knowing how to treat galaxy formation and luminosity growth, since these are nonlinear phenomena. It is also not known whether or how closely the luminous matter traces the total matter field.

## 2. COBE MISSION

In 1974, NASA released an announcement of opportunity for satellite missions. Little was known about large scale galaxy clustering, although the Shane-Wirtanen survey had clearly shown its importance. The CMBR had been known for 10 years, and its intensity had been measured from 70 cm to 0.5 mm wavelength. It was known to be approximately isotropic except for the dipole, and to have a blackbody spectrum with a precision of tens of percent. Over a hundred experiments had been done at great expense, using radiometers and spectrometers on the ground, on aircraft, on balloons, and on sounding rockets. The majority of these experiments showed tantalizing hints of imperfections at the several sigma level, only to be shown wrong by subsequent events.

Thus it was that three separate groups responded to NASA's announcement with proposals for measurements of the CMBR. I organized a Goddard-MIT-Princeton team that proposed (Mather et al. 1974) four instruments: a far infrared absolute spectrophotometer to test the blackbody nature of the CMBR, a set of differential microwave radiometers to map the CMBR, a far infrared differential radiometer for the same purpose, and a small photometer to measure the cosmic infrared background from the first galaxies. Simultaneously, teams at the Jet Propulsion Laboratory (Gulkis et al. 1974) and the Lawrence Berkeley Laboratory (Alvarez et al. 1974) prepared proposals for microwave radiometers, hoping that their simplicity and lower costs would enable early flights. Initially, NASA considered putting the spectrophotometer in the same cryostat as the IRAS telescope, but that was soon seen as impractical. In 1976, NASA appointed a Mission Definition Study Team, including Samuel Gulkis (JPL), Michael Hauser and myself (GSFC), George Smoot (LBL), Rainer Weiss (MIT), and David Wilkinson (Princeton). We worked with Goddard Space Flight Center to develop a detailed plan that included all but one of the original group of instruments. We removed the far infrared differential radiometer because it seemed too difficult to combine with the others. This experiment has since been done with a balloon-borne radiometer (Ganga et al. 1993), beautifully confirming the COBE DMR anisotropy results and extending them to much higher frequency.

Over the next 13 years, our team grew and the instrument and spacecraft concepts changed. Weiss was elected science team chairman, and Mather was appointed as the NASA Project Scientist for COBE. Hauser is Principal Investigator and Tom Kelsall is Deputy for the Diffuse Infrared Background Experiment (DIRBE), Mather is PI and Rick Shafer is Deputy for the Far Infrared Absolute Spectrophotometer (FIRAS). Smoot is PI and Charles Bennett is Deputy for the Differential Microwave Radiometers (DMR). Edward Wright is the Data Team Leader, and Nancy Boggess and Edward Cheng served as Deputy Project Scientists. Coinvestigators include Eli Dwek, Mike Janssen, Phil Lubin, Stephan Meyer, Harvey Moseley, Thomas Murdock, Robert Silverberg, and Jan Vrtilek. Roger Mattson was Project Manager, and Dennis McCarthy was his Deputy. Approximately 1500 people, mostly at Goddard and its contractors, contributed to the success of the COBE project.

The COBE concept evolved as well. The initial idea was for a small satellite launched by a Delta rocket, but NASA, Congress, and the OMB eventually required almost all space missions to use the Shuttle. In 1982, shortly before the IRAS launch, the COBE was approved for construction. The Challenger explosion in 1986 required a redesign and we found a way to cut the spacecraft weight by half and squeeze it back into the Delta. Despite the spacecraft redesign, we preserved the scientific capabilities of all three instruments with few changes. The launch was at dawn from Vandenberg Air Force Base in Lompoc, California on 18 November 1989. Four years of scientific operations were completed on 23 December 1993. The COBE is now in use for satellite communications tests, and has greatly exceeded its design life of one year.

The COBE orbit plane is nearly perpendicular to the Sun. It is inclined at 99° to the equator and has a 6 PM ascending node. With an altitude of 900 km, the orbit plane precesses at the rate of one revolution per year, dragged by the gravitational quadrupole of the Earth. This favorable orbit is similar to the IRAS orbit and is chosen for the same reasons. Its altitude is a compromise between cosmic ray bombardment from the radiation belts and atmospheric condensation on cryogenic surfaces at lower altitudes.

The spacecraft provides a very protected environment for all three instruments. It spins at 0.8 rpm about its symmetry axis, which is oriented away from the Earth and about 94° to the Sun. A radiation shield prevents the Sun from shining on the instruments. The Earth limb remains below the plane of the shield except for brief intervals near the June solstice, when the spacecraft passes between Earth and Sun. The FIRAS and DMR data taken during those periods were not as high quality, but the DIRBE data were almost unaffected. A liquid helium cryostat, very similar to the IRAS cryostat, carried 600 liters of superfluid at 1.4 K to cool the FIRAS and DIRBE. A porous plug kept the liquid inside the tank using the osmotic pressure of the superfluid, and allowed evaporation and cooling at its outer surface. An insulated cover was ejected after 3 days, to allow for the outgasing of the spacecraft to diminish. The helium lasted for 10 months, when the FIRAS and the long wavelength detectors for the DIRBE were turned off. The long life was achieved with good insulation, careful design, and a cold outer cryostat shell (about 135 K). The sunshade and the aperture of the cryostat are approximately coplanar, so the shade emits almost no radiation toward the cryostat aperture.

The spin is generated by a pair of large momentum wheels that spin in the opposite direction from the spacecraft, so that the net angular momentum about the spin axis is zero. There is no gyroscopic effect and no tendency for the spacecraft to flip over and spin about another axis. It is not known how the liquid helium spins, since it is a superfluid with nearly zero viscosity and therefore not well coupled to the cryostat.

The attitude is controlled by three servoloops, using Sun and Earth sensors and gyroscopes to generate error signals. Three reaction wheels with their axes 120° apart and perpendicular to the spacecraft spin axis provide short term control. They absorb angular momentum as needed, and are capable of spinning up and down once each spin. Short term sensing is provided by rate integrating gyros, one for each servo loop and three along the spin axis. Magnetic torquer bars act on the Earth's field to remove angular momentum from the spacecraft over long time periods. These three servo loops are highly redundant, which is fortunate since four of the 6 gyros eventually failed. No data were lost because of the failures, and the spacecraft never rolled over toward the Sun. That would have led to a rapid loss of liquid helium. The control system is capable of orienting the spin axis at angles from 90 to 98° from the Sun. The spin axis is kept nearly vertical but can tilted back up to 30° from the orbital direction to minimize the impingement of the residual atmosphere. No effects of the atmosphere were seen, but the thrust of the helium vent along the spin axis caused a gradual reduction of the orbit altitude.

The three instruments look out in different directions. The FIRAS views along the spin axis, and its field of view traces out a circle on the sky about 94° from the Sun. As the orbit precesses and the Earth moves around the Sun, the FIRAS maps the entire sky in 6 months. The DIRBE beam is oriented 30° from the spin axis and traces a cycloidal pattern in each orbit, covering half the sky in each day. Each DMR receiver uses two antennas, each pointed 30° from the spin and covering half the sky on each orbit.

The orientation of the COBE is determined from the attitude sensors and from star sightings by the DIRBE. The gyro readings have a resolution of the order of one arcsecond and are used to smooth and interpolate the readings from the other sensors. The solutions are good to 1.5 to 2 arcmin rms in the first part of the mission but are degraded by a factor of a few after the failure of the fourth gyro on 10 July 1993. This accuracy is quite sufficient for the DMR and FIRAS instruments, which have large beamwidths, but is just adequate for the DIRBE.

## 3. FIRAS INSTRUMENT AND MEASUREMENTS

The COBE FIRAS (Far Infrared Absolute Spectrophotometer) is designed to compare the CMBR spectrum to a blackbody with great precision. The most recent results show that the CMBR deviates from blackbody form by less than 0.03% of the peak intensity over the wavelength range from 0.5 to 5 mm (Mather et al. 1994). The interpretation of these results is given by Wright et al. (1994a). The FIRAS also showed that the cosmic dipole has the expected shape (Fixsen et al. 1994a). The primary basis of the comparison is a full beam external blackbody calibrator that can be adjusted to match the temperature of the sky. This calibration is described by Fixsen et al. (1994b). The FIRAS is the first instrument to have this opportunity in the protected space environment. The instrument is designed with multiple modes of operation and multiple detectors to test for possible systematic errors.

The instrument is a polarizing Michelson interferometer used as a spectrometer (Mather, Shafer, & Fixsen 1993). It has a beamwidth of 7°, defined by a compound parabolic concentrator. It is symmetrical, with differential inputs and differential outputs. It is a two-beam interferometer, whose output signal (called an interferogram) depends on the path difference between the two beams. A monochromatic input produces an output proportional to the cosine of the phase difference between the two beams. A general input produces an interferogram which is the Fourier transform of the input spectrum. These interferograms are detected, amplified, filtered, digitized, filtered and averaged again, and transmitted to the computers on the ground. They are then Fourier transformed numerically and calibrated to determine the input spectra. The mirror mechanism can scan at two different scan speeds and over two different stroke lengths, to obtain low and high spectral resolution.

Since the instrument is differential, the effective input is the difference in spectra between the two input ports. The two output ports are complementary, in the sense that light which does not reach one must reach the other, so the two output ports receive interferograms of opposite sign. The differential operation of the instrument is necessary to reduce the dynamic range and improve the accuracy, since it would be very difficult to maintain absolute gain stability to a part in $10^4$. The reference input observes a second calibrator through an antenna that is optically similar to that used for the sky.

Each output beam is further divided by a dichroic filter and sent to two detectors, so there are four altogether. The frequency ranges are from 2 to 21 cm$^{-1}$ and from 23 to 95 cm$^{-1}$. The detectors are composite bolometers, using diamonds 7.8 mm across to collect the heat that is absorbed in a chrome-gold blackening film. The temperature fluctuations are measured by doped silicon resistance thermometers, and amplified by JFET transistors operated around 70 K. Each detector receives light from a compound elliptic concentrator that illuminates it over a whole hemisphere. The resulting light grasp (étendue) is $A\Omega = 1.5$ cm$^2$ sr. This large value enables excellent sensitivity to the CMBR even with operation at 1.5 K. It is required for operation at long wavelengths, since the effective number of modes of the photon field is approximately $n = A\Omega/\lambda^2$. If $n$ is smaller than unity, then diffraction is very important and little light can be coupled through the optics to the detector.

The photometric calibration of the instrument (Fixsen 1994b) starts with the assumption that the optical input to each detector is a sum of terms, one from each of the radiating objects in the system. Each radiates an intensity proportional to the Planck function for its temperature, times an effective emissivity that depends on frequency. These objects include the external calibrator (XCAL) or the sky, the internal reference body (ICAL), the main sky horn antenna, the reference horn antenna, the bolometer housing, the physical support structure, and the moving mirrors. These are the only objects expected to be important, and thermometers were provided for each. For each object, an effective emissivity is determined relative to the external calibrator, which is defined to be unity. The ICAL has an effective emissivity of approximately $-0.95$ to $-1.0$, where the negative sign indicates that the instrument is differential. The sky horn and reference horn have emissivities of a few percent, increasing with frequency, and are also of opposite sign. The sum of all the effective emissivities is required to be zero by the Kirchhoff condition. This must be so because an isothermal cavity would yield a perfectly balanced and therefore null interferogram. There was an apparent drift of the calibration offset during the mission that had a time constant of about two months. The effects of this drift were minimized by calibration at monthly intervals at the beginning of the mission, and weekly intervals in the last six weeks.

The accuracy of the external calibrator is sufficiently good that it is difficult to measure directly. The measured temperature gradients are a few millidegrees, so there can be a similar uncertainty in the absolute temperature scale. The temperature gradient produces only a second order effect in the blackbody nature of the spectrum (of the same shape as the Comptonization $y$ distortion but very small), and is therefore not important for the determination of the cosmic spectrum distortion. The measured reflectivity of the calibrator is approximately $10^{-4.5}$ for the two frequencies at which it was measured with a microwave interference apparatus. Calculations show that it should be even better at higher frequencies where diffraction is less important. Leakage around the edge was reduced by flexible leaves of aluminized plastic. It was tested and seems unimportant.

The calibration model also includes a description of the detector gain as a function of temperature, bias, background power, and modulation frequency. The model was derived from ground-based measurements. It was confirmed and new values for the parameters were determined from flight data. The calibration process also allows determination of some errors of the thermometric temperature scale, based on the photometric signals. Other errors that are modeled include multiple passes of the radiation through the interferometer, and the effects of coherent vibrations.

The calibration constants are applied to the sky data to produce a data set composed of one or more spectra for every pixel, every detector, and every mirror scan mode. The sky data for every pixel are then analyzed as a sum of three terms: an isotropic cosmic background, a dipole due to the Earth's motion, and a galactic signal. The simplest galactic model seems almost adequate and was used for our most recent published results. This simple model assumes that the galactic signal is separable, as a product of a function $g(\nu)$ of frequency and a function $G(l, b)$ of direction. It is clearly inadequate in special regions of the galaxy, such as the molecular clouds and the galactic center, so this portion of the sky was not used for measuring the cosmic spectrum. To determine the intensity of the Galactic signal in every direction, we used the FIRAS measurements in the high frequency band, typically around 40 cm$^{-1}$. We also used measurements from the 240 $\mu$m channel of the DIRBE instrument on COBE, smoothed to a wider beamwidth. These two methods give results that agree very well.

After this modeling has been done for every pixel, the cosmic spectrum is derived from a mean of those parts of the sky not too close to the Galactic plane. To make sure that all calibration drifts have been removed as well as possible, we choose the data taken in the last 6 weeks of operation, when the calibrations were taken weekly and had a duty cycle of about 50%. We determine a final calibration adjustment by applying the calibration algorithm to the calibration data that were taken in conditions like those of the sky observations. The residual correction is small but not negligible. There is one remaining problem, which is a coherent vibration that is not entirely suppressed by this correction process. It increases the error bar at 11 cm$^{-1}$.

To quantify the distortion of the mean spectrum, it is fitted to a sum of terms. The first two terms are a Planck function for the mean temperature, and a temperature adjustment times the derivative of the Planck function with respect to temperature. The second term just allows for the possibility that the first may not be calculated for the correct mean temperature, since we do not know it a priori. The third term is a small number times the function $g(\nu)$ that represents the mean galactic spectrum. This allows for the possibility that the initial subtraction done on each pixel might have been incomplete or biased. The fourth term represents the cosmic distortion. There are two interesting forms for this: a $\mu$ form and a $y$ form (Zeldovich & Sunyaev 1969, 1970). The $\mu$ form is a chemical potential distortion, that might arise between the redshifts of about $3 \times 10^6$ and $10^5$. The second form is a Comptonization form that could arise after $z = 10^5$.

Figure 1 (Mather et al. 1994) shows the results of these least squares fits. The residuals are for the case where $y = \mu = 0$ and show that the CMBR spectrum is the

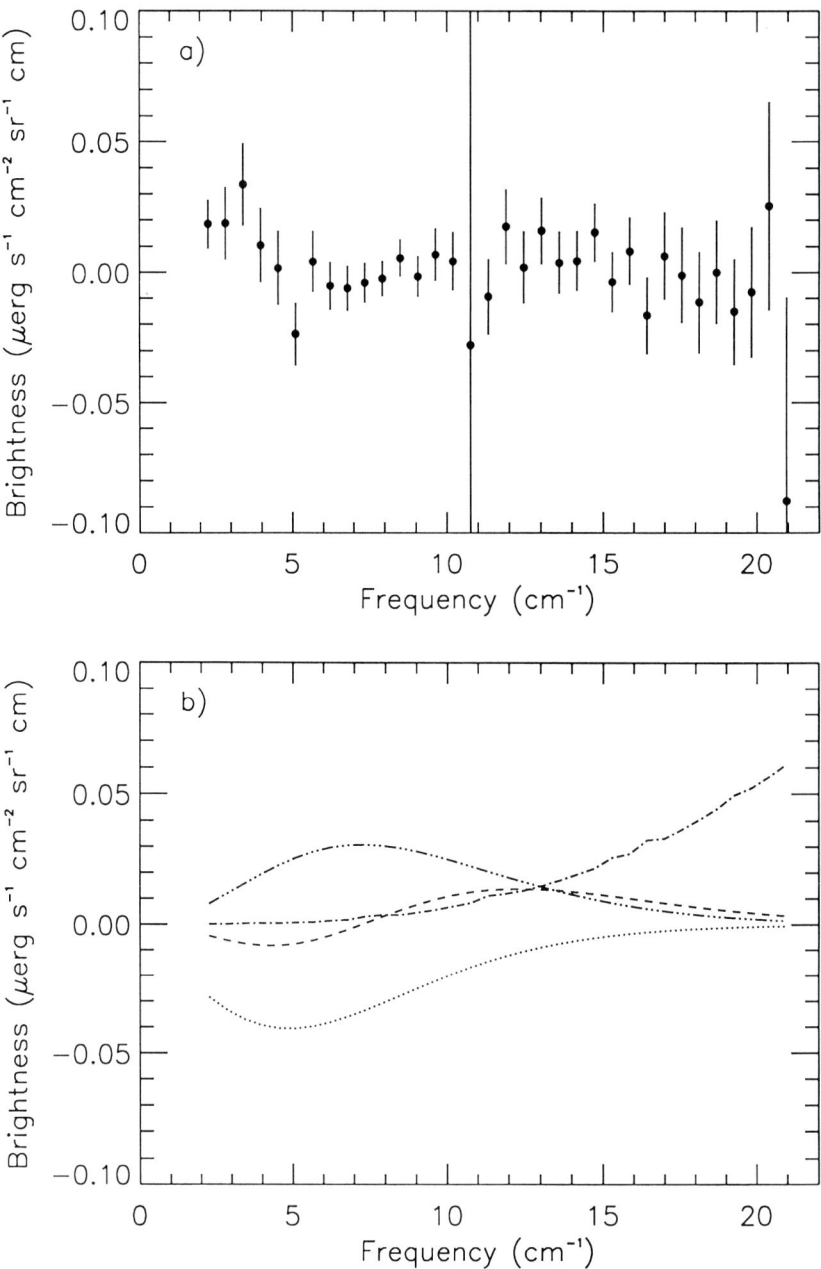

**Figure 1.** a) FIRAS measured CMBR residuals, $I_0 - B_\nu(T_0) - \Delta T(\partial B_\nu/\partial T) - G_0 g(\nu)$. b) Spectrum model components: the maximum allowed distortions (95% CL) $y = 2.5 \times 10^{-5}$ (— — —) and $|\mu| < 3.3 \times 10^{-4}$ (· · ·); the Galaxy spectrum $g(\nu)$ scaled to one fourth the flux at the Galactic pole (· -), and the effects of a 200 $\mu K$ temperature shift in $T_0$, $0.0002(\partial B_\nu(T)/\partial T)$, (· · · —). (Mather et al. 1994)

same as a blackbody within 0.03% of the peak intensity. The weighted rms residual is only 0.01% of the peak brightness. Nevertheless, these residuals are about twice as large as those expected from detector noise alone. We do not presently know the cause of this but have arbitrarily increased the size of the error bars throughout. The $y$ and $\mu$ curves show the shapes of distortion that would be produced if these coefficients have the 95% confidence limit values of $|y| = 2.5 \times 10^{-5}$ and $|\mu| = 3.3 \times 10^{-4}$. These limits allow for the increased uncertainty noted above, and for variations of the derived $y$ and $\mu$ as the boundary of the selected data set moves farther from the galactic plane.

Other possible distortions can also be explored. As an example, if the CMBR spectrum is a graybody, then we can limit its emissivity to $1 \pm 0.00041$ (95% CL). If the true cosmic signal contains a distortion with the same shape as the galactic signal, then we cannot determine it with this method alone. Similarly, a cosmic energy release that simply changes the CMBR temperature could not be determined.

The temperature of the CMBR is also important, although few cosmological calculations require a more precise number. The measured temperature is primarily needed for comparisons with different experiments, such as ground-based measurements and interstellar CN measurements. The FIRAS has two ways to determine this number. First, the thermometers in the external calibrator are the primary scale. When the calibrator is set to match the sky brightness, its thermometers measure the sky temperature. This method gives 2.730 K. The second method uses the wavelength scale of the FIRAS to determine color temperatures. In effect, we measure the frequency of the peak brightness of the CMBR and use fundamental constants to compute a temperature. The wavelength scale was determined from interstellar [C II] observations and confirmed with other lines of [N II], [C I], and CO. The color temperature method yields a CMBR temperature of 2.722 K. These numbers differ by much more than the random errors, which are only a few $\mu$K. We have no obvious explanation for the difference and simply take the mean, obtaining T = $2.726 \pm 0.010$ K (95% confidence). Clearly this is not a statistical determination of the error bar, which is entirely systematic error.

Fortunately, there are other confirmations that this number is approximately correct. A sounding rocket experiment launched only weeks after the COBE by Gush, Halpern, and Wishnow (1990) with a similar instrument and greatly superior detectors obtained the result $2.736 \pm 0.017$ K. Interstellar CN results are also in agreement with these numbers (Roth, Meyer, & Hawkins 1993, and Kaiser & Wright 1990). We can also determine the temperature from the spectrum of the dipole anisotropy. This can be done from both DMR and FIRAS data. The FIRAS result is determined only from the color temperature, since the absolute value of the Earth's velocity is not known. The FIRAS yields $2.714 \pm 0.022$ K (95% CL). The DMR result (see below) is $2.76 \pm 0.18$ K, and is derived from the known variation of the Earth's velocity around the Sun. The spectrum of the FIRAS dipole is given in Fig. 2 along with residuals from the fit to an amplitude of 3.343 mK. The dipole residuals are very small and are close to the value expected from detector noise alone.

## 4. FIRAS INTERPRETATION

Our interpretation of these spectrum distortions is given by Wright *et al.* (1994a) and summarized here. Large CMBR spectrum distortions are very difficult to produce in plausible versions of the hot Big Bang universe. After the annihilation of positrons, the CMBR energy density far exceeded even the rest mass energy density of the baryonic matter until quite recently. Consequently, there are few processes involving the baryonic

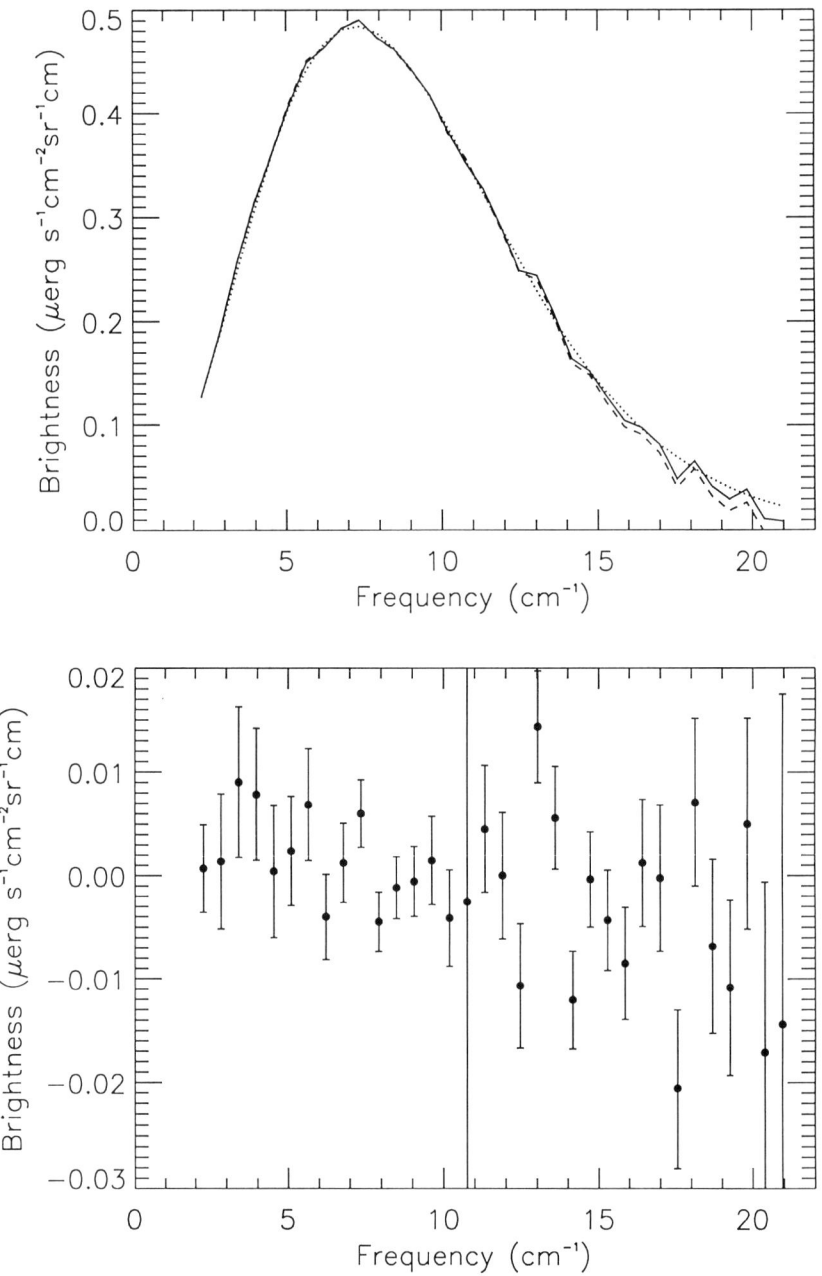

**Figure 2.** a) *Dipole spectrum. Solid line, using FIRAS high frequency Galactic spatial model; dashed line, using csc|b| Galactic spatial model; dotted line, best fit derivative of Planck function to solid line. b) Residual of dipole spectrum from best fit derivative of Planck function. (Fixsen et al. 1994a)*

matter that can liberate much energy and change the CMBR spectrum significantly. It is even more difficult to produce enough energy to create the whole CMBR radiation from anything except a hot Big Bang.

Therefore, the most immediate conclusion of the FIRAS spectrum distortion measurements is that the hot Big Bang is the only natural explanation for a nearly perfect blackbody. Carefully tailored models are required if one desires to produce the whole energy content of a blackbody spectrum by adding up non-blackbody contributions at different redshifts, as required by alternatives to the hot Big Bang. If one imagines that dust in intergalactic space can thermalize the radiation, then that dust must have substantial optical depth over an interval of cosmic history. That moment cannot be recent, or we would not be able to see distant galaxies at far infrared wavelengths. The IRAS galaxy at $z = 2.286$ demonstrates that one can see very far, and if the millimeter wave optical depth were large we would not see even such a spectacular object.

The next conclusion is that rather little of the energy of the CMBR was added to it after the first year of the expansion. The fraction of the CMBR energy added is approximately $0.71\mu$ in the redshift range $3 \times 10^6 > z > 10^5$. For later redshifts, the fraction is $4y$. A more precise calculation gives the results shown in Fig. 3.

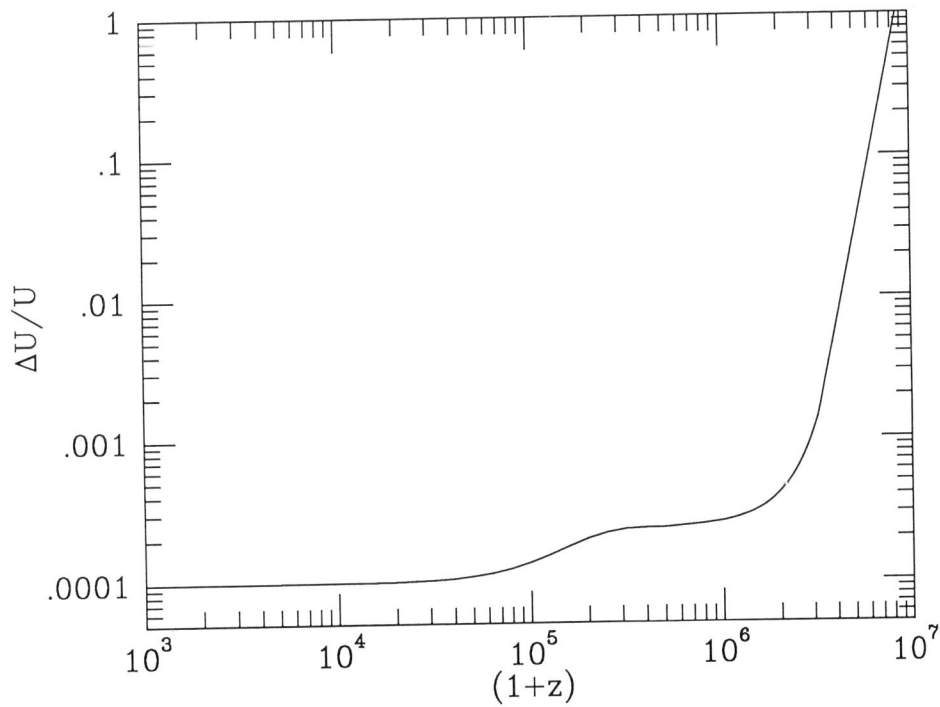

**Figure 3.** *Limits on Energy Release Prior to Recombination (Wright et al. 1994a)*

There are many possible sources of such energy augmentations, including decay of primeval turbulence, elementary particles, cosmic strings, or black holes. The growth of black holes, quasars, galaxies, clusters, and superclusters might also convert energy from other forms. The FIRAS data, together with the DMR anisotropy measurements, provide a limit on the spectral index of primordial density fluctuations. Wright *et al.*

(1994a) found an upper limit of 1.9, while a preprint by Hu, Scott, and Silk (1994) claims an upper limit of about 1.55. It is interesting that these calculations give tighter limits than existing direct measurements, even though the FIRAS numbers are a non-detection. These results are dependent on assuming that a power law is the correct form for the fluctuations over 7 orders of magnitude of scale. There is little possibility of observational evidence to confirm this assumption over such a wide range, since small scale fluctuations have long since been replaced by nonlinear phenomena.

Wright et al. (1994a) also give limits on hydrogen burning following the decoupling. These results depend on using geometrical arguments (a csc $|b|$ fit) to estimate the maximum amount of extragalactic energy that could have a spectrum similar to that of our own Galactic dust. We found a limit that is a factor of about 3 smaller than the polar brightness of the Milky Way. A better understanding of the galactic dust would help produce a tighter limit on these extragalactic signals.

Consider first population III stars liberating energy that is converted by dust into far infrared light (using an optical depth of 0.02 per Hubble radius), and assume that $\Omega_b h^2 = 0.015$. In that case less than 0.6% of the hydrogen could have been burned after $z = 80$. As a second example, consider evolving infrared galaxies as observed by the IRAS. For reasonable assumptions, we found that less than 0.8% of the hydrogen could have been burned in evolving IR galaxies.

We also obtained limits on the heating and reionization of the intergalactic medium. It does not take very much energy to reionize the medium, relative to the CMBR energy, because there are so few baryons relative to CMBR photons. Even the strict FIRAS limits permit a single reionization event to occur as recently as $z = 5$. More detailed calculations by Durrer (1993) show that the energy required to keep the intergalactic medium ionized over long periods of time is much more substantial and quite strict limits can be obtained. If the FIRAS limits were about a factor of 5 more strict, then it would be possible to test the ionization state of the IGM all the way back to the decoupling.

If the IGM were hot and dense enough to emit the diffuse X-ray background light, it should distort the spectrum of the CMBR by inverse Compton scattering. This is a special case of the Comptonization process, with small optical depth and possibly relativistic particles. Calculations show that a smooth hot IGM could have produced less than $10^{-4}$ of the X-ray background, and that the electrons that do produce the X-ray background can have a filling factor of less than $10^{-4}$.

## 5. DMR INSTRUMENT AND OBSERVATIONS

The initial results from the DMR were reported by Smoot et al. (1992), Bennett et al. (1992b), and Wright et al. (1992), based on the error analysis of Kogut et al. (1992) and the calibration reported by Bennett et al. (1992a). More recent results have been given by Bennett et al. (1994) and Wright et al. (1994b). The Differential Microwave Radiometers (DMR) instrument is designed to search for primeval fluctuations in the CMBR brightness, small temperature differences between different parts of the sky. The DMR observes at three frequencies, 31.5, 53, and 90 GHz, and has two redundant radiometers at each frequency. These frequencies lie within the range covered by the FIRAS, and are chosen to minimize the contamination by emission from galactic electrons and dust. Like the FIRAS, the DMR has a beamwidth of 7°.

The DMR is also differential, but in this case both inputs are received from the sky. The instrument is always observing the sky, so there is only a modest penalty in signal-to-noise ratio due to the insertion loss of the Dicke switch chopper. The two beams for each channel are both 30° from the spacecraft spin axis and 60° apart. The receivers use mixer–preamplifier front ends, and Gunn diode local oscillators. The Dicke switches are ferrite and operate at 100 Hz. The output of the front end is amplified, detected by a square law diode, and then synchronously demodulated. The sampling rate is 2 Hz as the beams scan the sky at the rate of 2.5° per second.

The antennas are corrugated circular conical horns that lead to polarization splitters. For the 31.5 GHz radiometers, the redundant receivers share the horns by using opposite circular polarizations, but for the other two bands each radiometer has its own pair of horns and uses linear polarization. No polarization of the CMBR anisotropy is expected at this sensitivity level, but software has been written to test for it. Some polarization would be expected by the electron scattering of slightly anisotropic radiation at the time of decoupling. Synchrotron emission from our own galaxy might also contribute some polarization at the longer wavelengths.

The receivers are mounted on massive aluminum blocks to make the temperatures steady and symmetrical. The 31.5 GHz receivers run at room temperature, but the other two are radiatively cooled to 140 K for improved sensitivity. This sensitivity improvement has proven very desirable, since the signal to noise ratio of the observed anisotropy is small. Since the radiometers do not require liquid helium cooling, they continue to operate.

The analysis of the data begins with careful review of the raw data to ensure that no bad data are analyzed. Then corrections are made for baseline drift (instrument asymmetry) and for known instrument errors, including primarily the sensitivity of the ferrite switches to the Earth's magnetic field. Then a least squares fit is made to all the measured brightness differences. These fits determine the relative brightness of 6144 pixels from about 60 million observations per year per receiver. The least squares solution can be done directly from inverting a sparse matrix, or iteratively. The leading term in the iterative solution is quite simple: the brightness of each pixel is the mean difference observed between it and all the pixels 60° away to which it is connected by differential observations, plus the mean temperature of all those pixels. This produces a small and unimportant correlation of the recovered intensities.

The gain and stability of the instrument were carefully monitored throughout the mission using noise diode excitations every two hours. In addition, the CMBR dipole was determined every day to search for changes in the calibration. The Moon was observed frequently during the two weeks of each month that it was possible to see it on every orbit. It was used to determine that the spacecraft pointing information was correct, and to map the response function of each antenna in flight. There was good agreement with the ground measurements of the antennas.

A very extensive review and simulation of possible systematic errors of the DMR was given by Kogut et al. (1992). We found that the leading error is due to magnetic sensitivity of the ferrite Dicke switches. Much smaller errors are added by diffraction of the Earth over the edge of the shield, and by a memory effect in the lockin amplifier. Simulation showed that after correction there could be no residual error larger than a few microkelvin on the map. We have performed extensive searches for other kinds of errors but none have proven significant.

## 6. DMR RESULTS

Recent papers by Bennett et al. (1994) and Wright et al. (1994b) have reported the results from analysis of two years of flight data. Figure 4 shows maps of the sky after subtracting the dipole, and after subtracting a model of the galaxy. The map with the galaxy removed is a suitable linear combination of the maps obtained at the three DMR frequencies, using the assumption that the spectral indices of the electrons and the dust are constant enough. The results from the two year data are quite consistent with those from the first year alone. They are also consistent with a far IR balloon survey of the CMBR anisotropy, as reported by Ganga et al. (1993). They showed that there is a good statistical correlation between their maps and the one-year DMR maps. As we reported previously, the angular correlation function of the maps is also consistent with the scale-invariant Peebles-Harrison-Zeldovich power law spectrum of primordial density fluctuations.

The statistical properties of the map can be presented in several forms. The angular correlation function shown in Fig. 5 shows the results when the portion of the sky at galactic latitude greater than 20° are analyzed. The shaded area gives the range of simulated skies drawn from a distribution of primordial fluctuations with a spectral index of 1.6 and an amplitude (normalized to the quadrupole) of 12.6 $\mu$K. There is very good agreement between the model and the observations.

It is also possible to analyze the maps into spherical harmonics. There is considerable subtlety and difficulty in this since the removal of the galactic plane causes the harmonics to be non-orthogonal, and produces strong correlations among the fitted amplitudes. Nevertheless, it is informative to show Fig. 5, taken from Wright et al. (1994b). They use the Hauser-Peebles (1973) method to obtain the harmonic amplitudes. The figure also shows the results of other instruments. A scale-invariant power spectrum would produce a horizontal line on this plot.

Analyses of both methods of presenting the data lead to similar results for the underlying Gaussian random noise process assumed to generate the primordial fluctuations. The direct analysis of the angular correlations produces $n = 1.59^{+.49}_{-.55}$ with a quadrupole normalization of $Q_{rms} = 12.4^{+5.2}_{-3.3}$ $\mu$K. Analysis of the spherical harmonics yields $n = 1.46^{+0.41}_{-0.44}$. In both cases these are quite consistent with the theoretically preferred $n = 1$, although the error bars are wide. These results can not be caused by galactic emissions, or by systematic errors in the instrument, or by populations of point sources and nearby astrophysical objects. The error bars are dominated by cosmic variance rather than instrument noise.

There are many theoretical papers showing how variations of the inflation potential could affect the spectral index of the fluctuations. Similarly, some papers show that primordial large scale gravitational waves would affect the fluctuations as well. It is clear from the variety of models that there can be no unique interpretation of the spectral index or the amplitude of the fluctuations. Additional data and theoretical constraints will be needed to establish a single picture of the early universe. There are several parameters to be determined, including $\Omega$, $\Lambda$, $\Omega_b$, and $H_0$. There are also choices about the relative quantities of hot and cold dark matter, and the inflation potential. The observable quantities do not yet and probably could never determine them individually even if the observational data were known perfectly.

The DMR measurements of the spectral index $n$ are already limited primarily by cosmic variance, since we do not measure the statistical properties of the entire universe but only our own local region. Some improvement can be obtained by observing a

## DMR 53 GHz Two-Year Sky Map

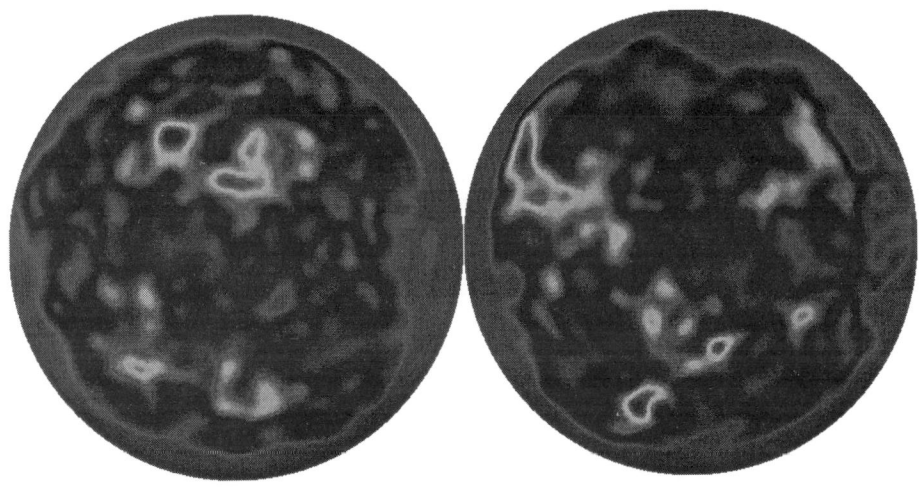

## DMR *Combination* Two-Year Sky Map

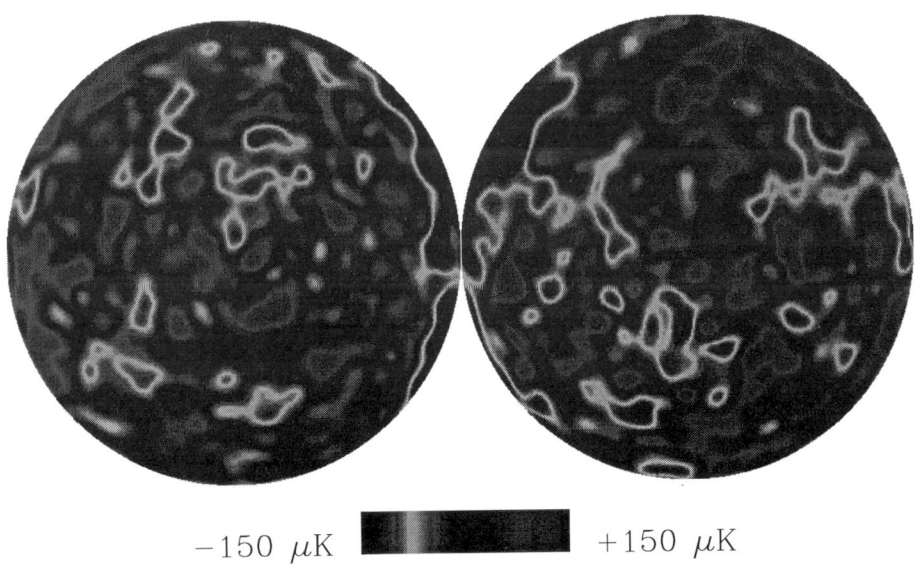

$-150~\mu K$      $+150~\mu K$

**Figure 4.** *Azimuthal equal area projections of the dipole removed 53 GHz $(A+B)/2$ (sum) data and the combination technique Galaxy-reduced data, smoothed to $10°$. The North Galactic polar cap is on the left, the South Galactic polar cap is on the right, and the point where they meet is the Galactic center. $\ell = 90°$ is at the bottom of each projected polar cap. (Bennett et al. 1994)*

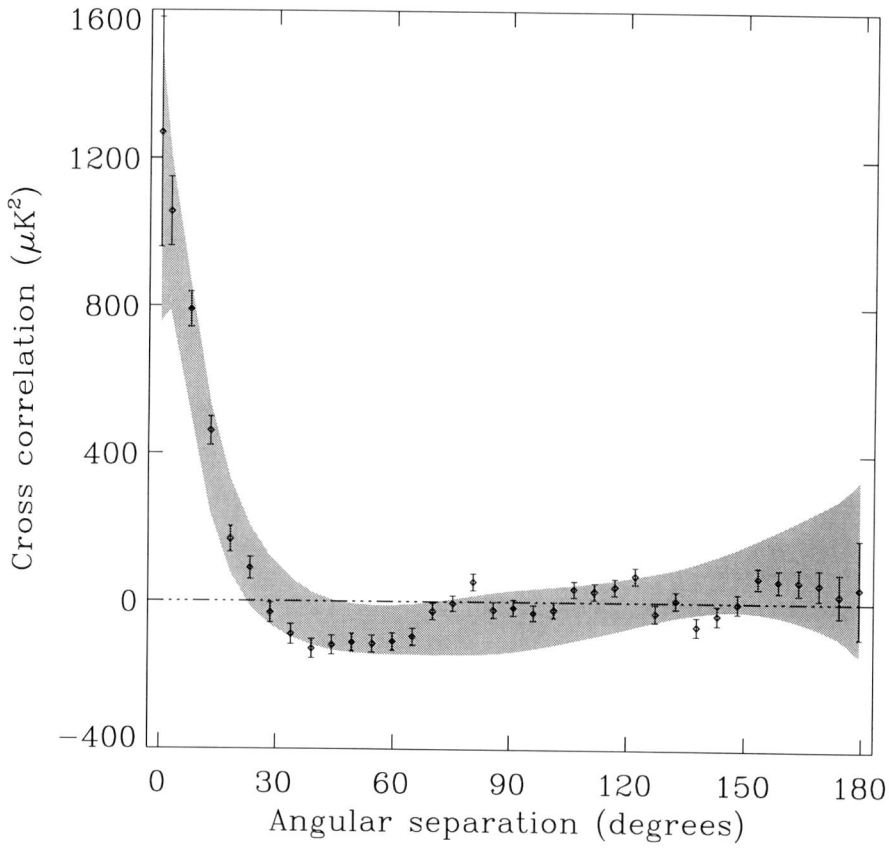

**Figure 5.** *The two-point 53 GHz (A + B)/2 × 90 GHz (A + B)/2 cross correlation function with the dipole (but not the quadrupole) removed, for |b| > 20° in thermodynamic temperature units. The error bars on the individual points include only instrument noise. The shaded region is the 68% confidence region expected from a 12.4 µK, n = 1.6 spectrum of fluctuations, including cosmic variance and instrument noise. (Bennett et al. 1994)*

wider range of angular scales. A large number of experiments are being done from the ground and using balloons with a beam size of the order of 0.5°. Most of these groups are reporting detections of anisotropy of the order of a few parts in $10^5$. This angular scale does not probe unmodified primeval fluctuations, so calculation and interpretation are necessary to connect the data with the COBE data and galaxy clustering data. The ground based projects include observations at Tenerife, Saskatoon (Princeton), the South Pole (UCSB and CARA), and Owens Valley (OVRO). Balloon instruments include the MAX (Berkeley and UCSB), MSAM (GSFC, MIT, Princeton), and the ULISSE (Rome). There is some sign of contamination by point sources at both radio and far IR wavelengths, so these measurements could reveal a new population of objects or a new understanding of the process of galaxy clustering and formation. Eventually an all-sky survey will be required to elucidate the statistical properties of the primordial

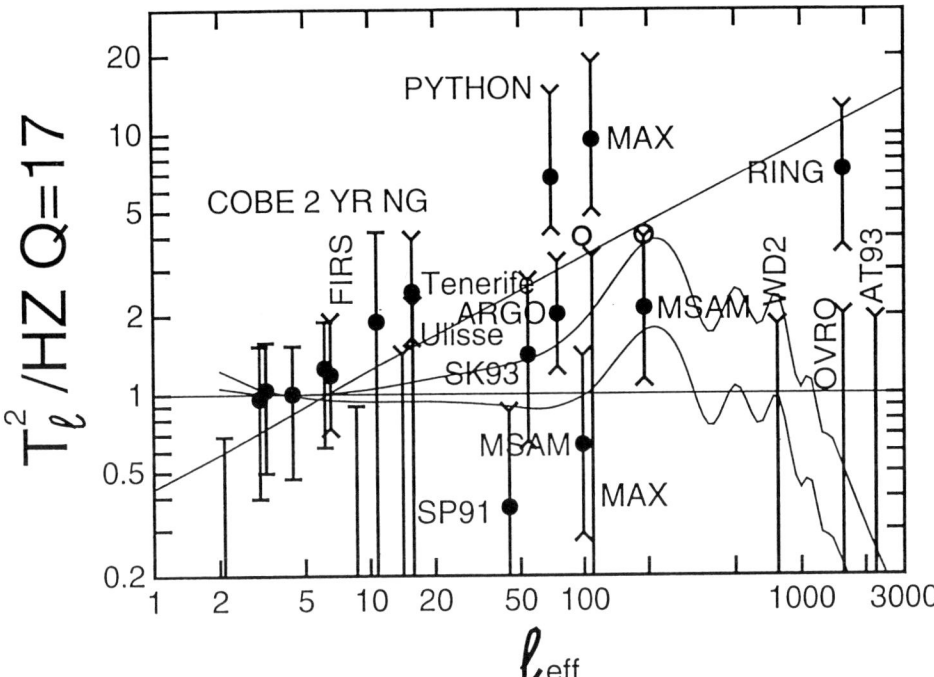

**Figure 6.** Power spectra for spherical harmonic number $\ell$, taken from Wright et al. 1994b, q.v. for references. Spectra are normalized to the mean of 17 $\mu K$ Harrison-Zeldovich Monte Carlo skies. COBE data point from the 2 year no galaxy DMR maps. Models shown as thin curves: $n = 1$, $Q = 17$ $\mu K$ is the horizontal line, the best fit $n = 1.5$ power law is the slanted line, & tilted CDM including the effects of gravitational waves with the long dashed curve showing $n = 0.96$ (predicted by $\phi^4$ chaotic inflation), and the short dashed curve showing $n = 0.85$ where the tensor and scalar quadrupoles are equal (Crittenden et al. 1993). Points with bent ends on their error bars are from other experiments: FIRS (Ganga et al. 1993), (from left to right) ULISSE (de Bernardis et al. 1992), Tenerife (Watson et al. 1992), the South Pole (Schuster et al. 1992), Saskatoon (Wollack et al. 1993), and the Python experiment (Dragovan et al. 1994), MSAM single subtracted (Cheng et al. 1993), MAX (Gunderson et al. (1993) and Meinhold et al. (1993)), MSAM double subtracted, White dish second harmonic (Tucker et al. 1993), OVRO (Readhead et al. 1989), OVRO RING (Myers et al. 1993), and the Australia Telescope (Subrahmanyan et al. 1993). The open circles above the MSAM points show the effects of not removing sources.

fluctuations, and improved angular resolution will be needed to rule out or understand the point sources.

It is also important to determine whether the primordial fluctuations are really Gaussian. This is still difficult because even with two years of DMR data the signal to noise ratio is only about 1.4 on an average pixel. In other words, only 2/3 of the variance is due to the sky, and the other 1/3 is from the instrument. There are a number of tests. The raw distribution of the temperature residuals should be close to Gaussian if the sky variance is Gaussian and the receiver noise is Gaussian. The receiver noise varies somewhat from pixel to pixel because the observation times are not all the same, but when this is taken into account there is extremely good agreement between these histograms (Smoot et al. 1994). There is no evidence that there is an excess of large deviations, as would be expected if there is an unknown population of point sources. On the other hand, considering the large beam of the instrument and the variance of both cosmic signals and receiver noise, it would be quite possible for interesting signals to be hidden in the data. Cosmic strings and so forth could well produce maps that are similar to those we observed. Strings might be much more observable with higher spatial resolution. Kogut et al. (1994) report a search for point sources in the two-year maps, and the result is negative.

It is also interesting to test the genus statistic of the maps. These results are also consistent with Gaussian statistics. The genus statistic was explored by Gott et al. (1990) and the COBE DMR calculations in Smoot et al. (1994) were based on their work. Smoot et al. found no evidence for deviations from Gaussian statistics, but most popular alternative theories of cosmic structure now produce similar statistics. A three point correlation function also tests the statistical properties of the maps. These results are also consistent with Gaussian statistics.

A more important question is whether the fluctuations are sufficient to produce the present large scale structure of the galaxy correlations. This is a difficult question. It seems desirable to introduce dark matter of both cold and hot types in the same model. Our initial interpretations were given by Wright et al. (1992), who used the tabulations of many cases by Holtzman (1989). However, there are many computations between concept and realization. I expect much further progress in this area before conclusions are firm.

## 7. COBE DATA PRODUCTS

Initial COBE data products were released in June 1993, and a full set of products are planned for release in June 1994. They are available online with documentation by anonymous FTP from nssdca.gsfc.nasa.gov with the username "anonymous" and your e-mail address as password. Change to directory [000000.cobe] and get the file aareadme.doc. Data and documentation may also be obtained on tape by request to the Coordinated Request and User Support Office (CRUSO), NASA/GSFC, Code 633.4, Greenbelt, MD 20771, phone: 301-286-6695, e-mail: request@nssdca.gsfc.nasa.gov. The data will also be made available under the MOSAIC interface, and can be located using the URL http://www.gsfc.nasa.gov/astro/cobe_home.html to give access.

## 8. SUMMARY

The COBE has demonstrated that the protected environment and long observing time of a space mission can have wonderful benefits for cosmology. The long observing time permits building up signal to noise ratio for the faint whispers of the great explosion 15 Gyr ago. The protected environment eliminates most of the systematic errors that limit lower altitude experiments. The very tight limits on deviation from a blackbody spectrum effectively eliminate alternatives to a simple hot Big Bang. They also rule out large energy conversion from interesting sources such as primeval black holes, cosmic strings, elementary particle decay, and primeval turbulence. The amount of hydrogen burned in Population III stars and in evolving infrared galaxies is limited to less than about 1%.

Anisotropy measurements from the COBE are consistent with the scale-invariant form suggested by Peebles, Harrison, and Zeldovich, and by inflation scenarios. The amplitude is roughly consistent with extrapolation from galaxy clustering catalogs, but discrepancies are still interesting. There is no evidence for non-Gaussian statistics in the underlying random process, but the tests are not strong. One reason is that there are not enough samples of the cosmic distribution to test it well, and another is that the receiver noise still contributes 1/3 of the measured variance.

I thank the members of the COBE Science Working Group and the Goddard Space Flight Center civil servants and contractors for making the COBE a reality. It took commitment, creative thought, and hard work in a team context to turn the thin proposal of a handful of scientists into a mission that could probe the early universe. It also took the continued support of NASA Headquarters through many trials, including the rebuilding of the COBE after the Challenger explosion. In particular, I thank Nancy Boggess, the NASA Program Scientist and later Deputy Project Scientist for COBE, who served as mentor, advocate, and active participant for the entire COBE Science Team from its inception.

## REFERENCES

Alvarez, L. W., Buffington, A., Gorenstein, M. V., Mast, T. S., Muller, R. A., Orth, C. D., Smoot, G. S., Thornton, D. D., & Welch, W. J., *Observational Cosmology: The Isotropy of the Primordial Black Body Radiation*, UCBSSL 556/75 Proposal to NASA (1974)

Bennett, C. L. et al. 1992a, ApJ, 391, 466.

Bennett, C. L. et al. 1992b, ApJ, 396, L7.

Bennett, C. L. et al. 1994, COBE preprint 94-01, submitted to ApJ.

Boggess, N. et al. 1992, ApJ, 397, 420.

Durrer, R. 1993 preprint ZU-TH 28/93, "Early Reionization in Cosmology"

Fixsen, D. J. et al. 1994a, ApJ, 420, 445–449.

Fixsen, D. J. et al. 1994b, ApJ, 420, 457–473.

Ganga, K. et al. 1993, ApJ, 410, L57

Gott, R. et al. 1990, ApJ, 352, 1.

Gulkis, S., Carpenter, R. L., Estabrook, F. B., Janssen, M. A., Johnston, E. J., Reid, M. S., Stelzried, C. T., & Wahlquist, H. D., *Cosmic Microwave Background Radiation Proposal*, NASA/JPL Proposal (1974)

Gush, H. P., Halpern, M., & Wishnow, E. H., 1990, PRL, 65, 537.

Hauser, M. G., & Peebles, P. J. E., 1973, ApJ 185, 757–785.

Holtzman, J. A., 1989 ApJS, 71, 1.

Hu, W., Scott, D., & Silk, J. 1994, "Power Spectrum Constraints from Spectral Distortions in the Cosmic Microwave Background," preprint CfPA-THJ-94-12, submitted to ApJL.

Kaiser, M. E. & Wright, E. L. 1990 ApJL, 356, L1–L4

Kogut, A., et al. 1992, ApJ, 401, 1.

Kogut, A, et al. 1994, COBE Preprint 94-06, submitted to ApJ.

Mather, J., Thaddeus, P., Weiss, R., Muehlner, D., Wilkinson, D. T., Hauser, M. G., & Silverberg, R. F., *Cosmological Background Radiation Satellite*, NASA/Goddard Proposal (1974)

Mather, J. C., et al. , 1994 ApJ, 420, 439–444.

Mather, J. C., Shafer, R. A., & Fixsen, D. J., COBE Preprint 93-10, Proc. SPIE, vol 2019, pp. 168–179, conf. on Infrared Spaceborne Remote Sensing, in San Diego, CA, 11–16 July 1993, (SPIE: Bellingham, WA)

Peebles, P. J. E. 1971, *Physical Cosmology*, (Princeton: Princeton University Press)

Roth, K. C., Meyer, D. M. & Hawkins, I. 1993, ApJL, 413, L67.

Sachs, R. K. & Wolfe, A. M., 1967, ApJ, 147, 73.

Smoot, G. F., et al. 1992, ApJ, 396, L1

Smoot, G.F., et al. 1994, COBE Preprint 94-03, submitted to ApJ.

Sunyaev, R. A. & Zel'dovich, Ya. B. 1980, Ann Rev Astron Astrophys, 18, 537

Wright, E. L., et al. 1992, ApJ, 396, L13

Wright, E. L., et al. 1994a, ApJ, 420, 450–456.

Wright, E. L., et al. 1994b, COBE Preprint 94-02, submitted to ApJ.

Zeldovich, Ya. B. & Sunyaev, R. A. 1969, Astrophys Sp. Sci. 4, 301.

Zeldovich, Ya. B. & Sunyaev, R. A. 1970, Astrophys Sp. Sci. 7, 20.

## DISCUSSION

**M. Rees**: I wonder if you could clarify a little bit the significance of the quadrupole. The question is: if you do not make the $n = 1$ assumption, can you really say there is a quadrupole in the data?

**Mather**: Yes, we can do a direct spherical harmonic transform out of the map that we have and the quadrupole is quite significant all by itself. The value here is correct. There is a small error bar which is not dependent on the assumptions of the power-law spectrum. If you do assume a power-law spectrum then the quadrupole, as a normalization for that, gives this number. But even without making that assumption, one can make a direct calculation.

**R. Sunyaev**: About the new measurements and the new data on the $\mu$ potential, I remember the words of Zel'dovich in 1969, the first time he wrote about this. Zel'dovich told me that the $\mu$ potential is much more important than the $y$ parameter, because the $\mu$ parameter gives you the possibility to see as deep as possible in the universe, providing information of the cosmological nucleosynthesis. There is an additional effect that we considered at the time. It is a distortion of the spectrum due to the recombination of hydrogen at redshift 1,300. These distortions are very weak and at rather high frequencies. Is it possible today to give an upper limit from your data or is this frequency out of your range?

**Mather**: The question is: Can we see the effects of the recombination on the spectrum? If the Lyman $\alpha$ itself at recombination were to be directly observed it would be approximately at $\lambda =150$ $\mu$m; actually, 100–150 depending on details, such as the redshift of the recombination epoch. Unfortunately, at these wavelengths the Galaxy is very bright. There is a strong carbon line at 158 $\mu$m. About a few tenths of a percent of the total luminosity of our galaxy appears in this one line. So it is very difficult for us to say that there is anything there. We have looked, but the sensitivity of the experiment is one or two orders of magnitude less than at longer wavelengths and we cannot make a meaningful statement about that.

**R. Daly**: What is the uncertainty in $N$? Can you put some error bars on any close one?

**Mather**: Yes, the uncertainty is about $\pm 0.5$ based on our data. I think Phil Lubin will probably be able to show you that we have better ability to constrain $N$ when we use smaller angular scale data. Our problem here is in part that the universe has a random number generator, so even if our data were perfect we would not be able to reduce that error bar a lot because we do not cover the smaller angular scales.

**M. Rees**: This is just a comment about early reionization. It is important, as you say, that although you have these very severe constraints on heat input at early times, you cannot rule out early reionization, if that does not produce a high temperature. I think it is important to keep this in mind because uncertainty about early reionization is very important in interpreting the angular fluctuation data. For instance, if there was enough early reionization from, say, decaying particles to maintain a few percent of the material ionized back to recombination, that would in itself be enough to affect the interpretation of small angular scale microwave background fluctuations. Of course, full ionization has a even bigger effect, so I think it is important to keep in mind that although this limit on the $y$ parameter constrains early heating to high temperatures (see Sunyaev), it does not severely contrain the idea of the medium being ionized at early epochs, providing it is only photoionized at low temperature. We still have this uncertainty in interpreting the small angular scale fluctuations until we can sort that out.

**Mather**: Yes, that is my understanding of it also. I should also mention that if there were reionization, then whatever made reionization would have to be isotropic or we would see *it* also. It would be interesting if it were there. And smaller angular scale information is of great importance to help solve that question.

**A. Kogut**: It is also worth pointing out, in relation to that question, that if you had a recent reionization at a fairly low temperature, you can avoid the Compton $y$ constraint from FIRAS, but you would expect to see the thermal bremsstrahlung from that same plasma as a distortion at long wavelengths. This is one of the reasons behind a proposed Small Explorer which would be able to detect this effect. If IRAS does not see it at high frequencies and you do not see a distortion at low frequencies, you can pretty much rule out a reionization at a redshift of more than $z \sim 5$. One thing that would be very useful for people to calculate would be, in fact, the thermal history of universe, the thermal bremsstrahlung expected from models in which you have a fairly recent reionization. This is not well established in the literature.

**Q.**: This is not a very polite question. If you had the possibility to build COBE tomorrow, what would be the sensitivity which is possible to reach? You gave enormously good upper limits for the $y$ and $\mu$ parameters. But today, at the present level of technique, is it our Galaxy which does not permit us to go farther, or is it possible to do better with a better device? Which are the perspectives?

**Mather**: That is a good question and I have been thinking about it, too. It is clear that it is possible to build a better instrument. On the other hand, my impression of our data is that we are limited in the $y$ and $\mu$ parameters approximately equally by instrumental uncertainties and by the galactic spectrum. We tried to examine that question by changing the amount of galactic data that were included. We have a cut off angle where we exclude the galactic plane, and the thickness of the plane that we exclude makes a difference to the $y$ and $\mu$ limits; this uncertainty is comparable to the range that we see from the instrument noise. So, a factor of 2 better, I think, could be obtained; a factor of 4 or more would be hard.

**Q.**: You have mentioned ROSAT. The main result there has been obtained in the Lockman Hole, the region where we have a very low density or column density of neutral hydrogen. I believe that in the same region the dust content is minimum, and we can check this and also the intensity of the galactic synchrotron emission. My question is: if you take only such regions as Lockman Holes, special regions of the sky, is it possible to go farther with your already existing data or do you prefer to integrate over bigger areas?

**Mather**: That is certainly a good question. The result that we announced previous to this result, about a year before, was based on the data in the direction of the Lockman Hole; we used to call it the Baade Hole. Maybe we have to go back and check our literature to see if Baade said it first. But, we do not have enough data or enough sensitivity in that one small area by itself. A more ideal instrument that had a higher sensitivity could certainly do better by concentrating on a special spot like that.

**M. Livio**: May I ask you a somewhat naive question? Presumably you have now in your hands what may be the best black body curve ever obtained. Is it possible in fact to use this, for example, to set upper limits on changes in time of fundamental constant?

**Mather**: Good question. We have not really thought much about that. It seems to me that we would have a hard time telling about a change in the value of Planck's constant as long as the radiation still follows Planck's law. So I am not sure that we could ever see that. I should, on the other hand, say that we did think about whether we could test where the plot was right about Planck's law. We show you something that looks like a perfect Planck function. I should emphasize that we did not measure that Planck was right. We measured the differences between Planck's curve and the sky.

**Q.**: Is this present-day blackbody? How would it change at redshift, say, 7?

**Mather**: We did approach that question by measuring the effective emissivity of the external blackbody sky. You may suppose that it has really a different temperature with an emissivity a little different from unity, and the result is that it is a blackbody within a hundredth of a percent of unity. This is a very good test that it is a real blackbody. At another time we will be examining our calibration data very carefully to see whether we can test Planck's law as well. There are some theories I have been hearing that suggest that quantum mechanics might have very tiny corrections that would modify it a little bit.

# DETECTION OF DEGREE SCALE ANISOTROPY

P. M. Lubin
Physics Department
University of California
Santa Barbara, CA 93106-9530, USA
and
Center for Particle Astrophysics
University of California
Berkeley, CA 94720, USA

**Abstract.** Degree Scale Anisotropy Measurements are a crucial testing point for cosmological models. Giving us one of the few probes into density perturbations in the early universe, they promise to be a watershed for future progress in understanding structure formation. Because of the extreme sensitivities needed (1–10 ppm) and the difficulties of foreground sources, these measurements require not only technological advances in detector and measurement techniques, but multi spectral measurements and careful attention to low level systematic errors. This field is advancing rapidly and in a true discovery mode. Our own group has been involved in a series of ten experiments over the last five years using the ACME (Advanced Cosmic Microwave Explorer) payload which has made measurements at angular scales from 0.3 to 3 degrees and over a wavelength range from 1 to 10 mm. I will review some of the challenges and potential involved in these measurements, both present and future.

## 1. INTRODUCTION AND CURRENT STATUS

The Cosmic Background Radiation (CBR) provides a unique opportunity to test cosmological theories. It is one of the few fossil remnants of the early universe to which we have access at the present. Spatial anisotropy measurements of the CBR in particular can provide a probe of density fluctuations in the early universe. If the density fluctuation spectrum can be mapped at high redshift, the results can be combined with other measurements of large scale structure in the universe to provide a coherent cosmological model.

Recent measurements of CBR anisotropy have provided some exciting results. At the largest angular scales, NASA's Cosmic Background Explorer (COBE) satellite has provided the first measurements of large scale CBR anisotropy at a level $\Delta T/T = 10^{-5}$ at $10°$. This result may have been corroborated by a balloon survey, but much more remains to be done. While the large scale measurements are useful as a normalization for the fluctuation spectrum, they do not define the spectrum. For this, measurements must be made at smaller angular scales.

At 4°, recent ground-based measurements from Tenerife have set an upper limit to CBR fluctuations of $\Delta T/T \leq 1.6 \times 10^{-5}$. However, new data may have resulted in a possible detection of anisotropy with an amplitude $\Delta T/T \approx 2 \times 10^{-5}$.

At scales near 1 degree, close to the horizon size, results from the South Pole using the ACME (Advanced Cosmic Microwave Experiment) with a High Electron Mobility Transistor (HEMT) based detector place an upper limit to CBR fluctuations of $\Delta T/T \leq 1.4 \times 10^{-5}$ at 1.2° (Gaier et al. 1992). This data set has significant structure in excess of noise but was unlikely to be CBR given the spectrum. A conservative upper limit for a Gaussian autocorrelation function sky was computed from the highest frequency channel. A four channel average of the bands yields a detection at the level of $\Delta T/T = 1 \times 10^{-5}$.

Additional analysis of the 1991 ACME South Pole data using another region of the sky and with somewhat higher sensitivity shows a significant detection at a level of $\Delta T/T = 1 \times 10^{-5}$ (Schuster et al. 1993). The structure observed in the data has a relatively flat spectrum which is consistent with CBR but could also be Bremsstrahlung or synchrotron in origin. This data sets an upper limit comparable to the Gaier et al. upper limit, but can also be used to place a lower limit to CBR fluctuations of $\Delta T/T \geq 8 \times 10^{-6}$, if all of the structure is attributed to the CBR. The $1\sigma$ error measured per point in this scan is 14 $\mu$K or $\Delta T/T = 5 \times 10^{-6}$. Per pixel, this is the most sensitive CBR measurement to date at any angular scale and will be used later in a discussion of systematics for possible future experiments from sub-orbital platforms. Recent measurements by the Princeton Big Plate experiment using a detector and beam size very similar to Gaier et al. (1992) and Schuster et al. (1993) but with a different chopping scheme, has found detection at levels consistent with the Schuster et al. 1993 results but in a completely different region of the sky and at lower galactic latitude (Wollack et al. 1994).

At scales near 0.5°, balloon-borne and South Pole based experiments have made very sensitive measurements. In 1988-89 the ACME experiment oufitted with a sensitive SIS (Superconductor–Insulator–Superconductor) detector set an upper limit of $\Delta T/T \leq 3.5 \times 10^{-5}$ at 0.5° for a Gaussian sky. Further refinements using sensitive bolometers on ACME resulted in the joint UCSB-UCB ACME-MAX balloon-borne experiment which has now had four successful flights. One scan (the $\mu$ Peg scan) resulted in an upper limit of $\Delta T/T \leq 2.5 \times 10^{-5}$ (Meinhold et al. 1993). This particular scan is notable in that it contained a strong detection of dust as well. After subtraction of the dust component a residual detection consistent with a CBR thermal spectrum remains at a level of $\Delta T/T = 1.5 \times 10^{-5}$. A scan from another region of the sky (the GUM scan) during the same flight resulted in a detection consistent with a CBR spectrum with an amplitude $\Delta T/T = 4.2 \times 10^{-5}$ (Gundersen et al. 1993). An ADR cooled bolometer based receiver has also been recently flown on MAX (June 1993) resulting in three deep CBR scans. One of the scans, the GUM scan, resulted in a detection consistent with that we previously saw with MAX in Alsop et al. (1992) and Gundersen et al. (1993). This data is presented in Devlin et al. 1994. The other two scans are reported in Clapp et al. 1994 and give $\Delta T/T = 3.1 \times 10^{-5}$ for the Sigma Hercules scan and $\Delta T/T = 3.3 \times 10^{-5}$ for the Iota Draconis region for a Gaussian autocorrelation function with a coherence angle of 0.5 degrees. Recent results from the Goddard-Chicago-Princeton MSAM (Medium Scale Anisotropy Measurement) balloon experiment using a beam size also near 0.5 degree also shows evidence of a possible

CBR structure at the few $\times 10^{-5}$ level using a multi-wavelength He-3 cooled bolometric detector (Cheng et al. 1993).

At scales smaller than 0.1°, the results have come from ground-based radio telescopes. No significant CBR detections have been reported and current upper limits to fluctuations are $\Delta T/T \leq 1.8 \times 10^{-5}$ at 5 arcminutes and at 1 arcminute.

Theoretical arguments predict CBR on intermediate and large angular scales at a level $\Delta T/T \approx 1 \times 10^{-5}$. Different models predict a variety of power spectra. Recent arguments about foreground emission suggest that a per pixel sensitivity of $\Delta T/T \leq 1 \times 10^{-6}$ (3 $\mu$K) may be required to separate foreground contaminants from true CBR signals at a level where the power spectrum can be determined. (An order of magnitude more sensitivity may be required to do this separation well.) This number will be important for later discussions of future experiments.

It is clear from the existing results that in order to fully map out the primordial fluctuation spectrum, more data are needed. By taking advantage of rapidly evolving technology, and building low noise receivers at several frequencies using single detector elements at first, and then in focal plane arrays it should be possible to reach the required sensitivity in the next five years.

## 2. CBR ANISOTROPY MEASUREMENTS

The spectrum of the cosmic background radiation peaks in the millimeter-wave region. Figure 1 shows a plot of antenna temperature vs. frequency, demonstrating the useful range of CBR observation frequencies and the various backgrounds involved. The obvious regime for CBR measurements is in the microwave and millimeter-wave regions.

In the microwave region, the primary extra-terrestrial foreground contaminants are galactic synchrotron and thermal bremsstrahlung emission. Below 50 GHz, both of these contaminants have significantly different spectra than CBR fluctuations. Because of this, multi-frequency measurements can distinguish between foreground and CBR fluctuations (provided there is large enough signal to noise).

Above 50 GHz, the primary contaminant is interstellar dust emission. At frequencies above 100 GHz, dust emission can be distinguished from CBR fluctuations spectrally, also using multi-frequency instruments.

At all observation frequencies, extra-galactic radio sources are a concern. For an experiment with a collecting area of 1m$^2$ (approximately a 0.5° beam at 30 GHz for sufficiently under-illuminated optics), a 10 mJy source will have an antenna temperature of 7.3 $\mu$K, which will produce a significant signal in a measurement with a sensitivity of $\Delta T/T \approx 1 \times 10^{-6}$. Extra-galactic radio sources have the disadvantage that there is no well known spectrum which describes the whole class. For this reason, measurements over a very large range of frequencies and angular scales are required for CBR anisotropy measurements in order to achieve a sensitivity of $\Delta T/T \approx 1 \times 10^{-6}$.

## 3. INSTRUMENTAL CONSIDERATIONS

Sub-orbital measurements differ from orbital experiments in at least one important area, namely our terrestrial atmosphere is a potential contaminant. A good ground-based site like the South Pole has an atmospheric antenna temperature of 5 K at 40 GHz, for example. For a measurement to reach an error of $\Delta T/T \approx 1 \times 10^{-6}$,

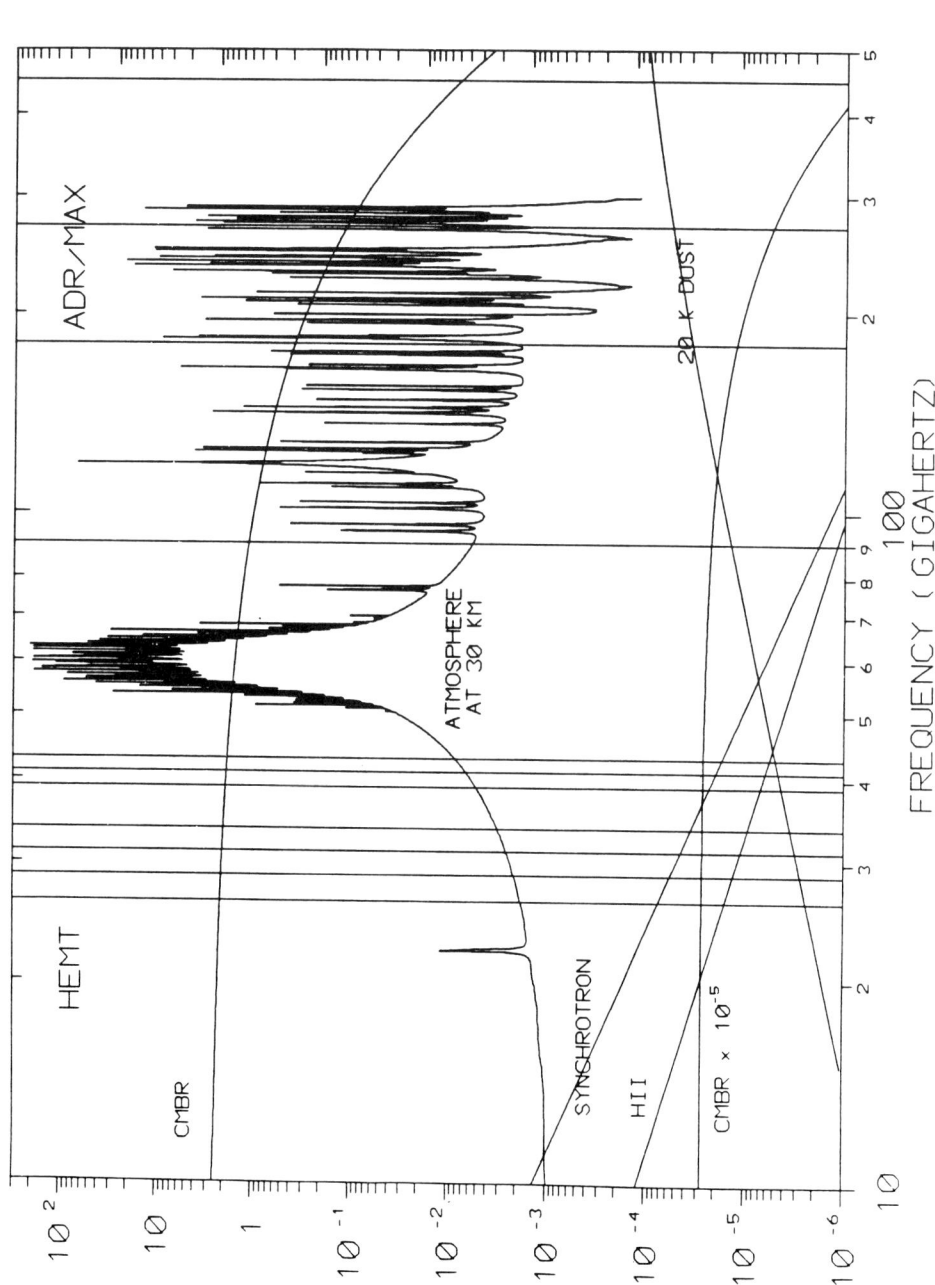

**Figure 1.** *Atmospheric and galactic emission as a function of frequency.*

the atmosphere must remain stable over 6 orders of magnitude. In addition to this, the atmosphere will contribute thermal shot noise. At balloon altitudes, atmospheric emission is 3–4 orders of magnitude lower and much less of a concern. In addition, the water vapor fraction is extremely low at balloon altitude. Satellite measurements avoid this problem altogether. Another consideration for CBR anisotropy measurements is the sidelobe antenna response of the instrument. Astronomical and terrestrial sources away from foresight can contribute significant signals if the antenna response is not well behaved. Under-illuminated optical elements and off-axis low blockage designs are typically employed for the task. The sidelobe pattern can be predicted and well controlled with single-mode receivers, but appears to be viable for multi-mode optics as well. Even with precautions, sidelobe response will remain an area of concern for all experiments.

Most of the measurements discussed in the previous section were limited by receiver noise when atmospheric seeing was not a problem. It is possible to build receivers today with sensitivities of 200–400 $\mu K - \sqrt{s}$ using HEMTs or bolometers. A balloon flight obtaining 10 hours of data on 10 patches of sky, for example, could achieve a 1 $\sigma$ sensitivity of 6.7 $\mu K$ or $\Delta T/T = 2.5 \times 10^{-6}$ per pixel using one such detector.

To map CBR anisotropy with a sensitivity of $\Delta T/T = 1 \times 10^{-6}$ requires more integration time, lower noise receivers or multiple receivers. A 14-day, long duration balloon flight launched from Antarctica could result in a per pixel sensitivity of $\Delta T/T = 5 \times 10^7$, if 10 patches could be observed with a single detector element or $\Delta T/T = 5 \times 10^{-6}$ on 1000 patches as another example.

Measurements from the South Pole are also very promising. The large atmospheric emission (compared to the desired signal level - few million times larger!) is of great concern and based upon actual experience, even in the best weather, there is significant atmospheric noise. Estimated single difference atmospheric noise with a 1.5 degree beam is about 1 $mK\sqrt{s}$ at 30 GHz during the best weather. This added noise, as well as the overall systematic atmospheric fluctuations, make ground-based observations challenging but so far possible, and, in fact, yielding the most sensitive results.

Another approach to the problem is to use very low noise receivers and obtain the necessary integration time by flying long duration balloons. These receivers can be tested from ground-based observing sites like the South Pole. Should the long duration balloon effort prove inadequate, the only means toward the goal of mapping CBR anisotropy at this level may be a dedicated satellite. Again, the receivers on such a satellite would have to be low noise. The minimal cryogenic requirements for HEMT amplifiers make them an obvious choice for satellite receivers, but bolometric receivers using ADR coolers or dilution refrigerators offer significant advantages at submillimeter wavelengths.

## 4. HISTORY OF THE ACME EXPERIMENTS

In 1983, with the destruction of the 3 mm mapping experiment (Lubin et al. 1985), we decided to concentrate on the relatively unexplored degree scale region. Motivated by the possibility of discovering anisotropy in the horizon scale region where gravitation collapse would be possible and with experience with very low noise coherent detectors at balloon altitudes, we started the ACME program. A novel optical approach, pioneered at Bell Laboratories for communications, was chosen to obtain the extreme sidelobe rejection needed. In collaboration with Robert Wilson's group at Bell Labs, a 1 meter off-axis primary was machined. A lightweight, fully-automated, stabilized, balloon platform capable of directing the 1 meter off-axis telescope was constructed. As the initial detector we chose a 3 mm SIS receiver. Starting with lead alloy SIS junctions and GaAs FET pre-amplifiers we progressed to Niobium junctions and a first generation of HEMTs to achieve chopped sensitivities of about 3 mK $\sqrt{\sec}$ in 1986 with a beam size of 0.5 degrees FWHM at 3 mm.

The first flight was in August 1988 from Palestine, Texas. Immediately afterwards, ACME was shipped to the South Pole for ground-based observations. The results were the most sensitive measurements to date (at that time) with 60 $\mu$K errors per point at 3 mm. The primary advantage of the narrow band coherent approach is illustrated in Figure 1 where we plot atmospheric emission versus frequency for sea level, South Pole (or 4 km mountain top) and 30 km balloon altitudes. With a proper choice of wavelength and bandpass, extremely low residual atmospheric emission is possible. (Total < 10 mK. The differential emission, over the beam throw, is much smaller.) Another factor of 10 reduction is possible in the "troughs" in going to 40 km altitude. The net effect is that atmospheric emission does not appear to be a problem in achieving $\mu$K level measurements, if done appropriately.

Subsequently, ACME has been outfitted with a variety of detector including direct amplification detectors using HEMT (High Electron Mobility Transistor) technology. These remarkable devices developed largely for communications purposes are superb at cryogenic temperatures as millimeter wavelength detectors. Combining relatively broad bandwidth (typically 10–40%) with low noise characteristics and moderate cooling requirements (including operation at room temperature) they are a good complement to shorter wavelength bolometers allowing for sensitive coverage from 10 GHz to 200 GHz when both technologies are utilized. The excellent cryogenic performance is due in large part to the efforts of the NRAO efforts in amplifier design (Pospieszalski 1990). We have used both 8–12 mm and 6–8 mm HEMT detectors on ACME, these observations being carried out from the South Pole in the 1990 and 1993 seasons. The beam sizes are 1.5 degrees and 1 degree FWHM for the 8–12 and 6–8 mm HEMTs respectively. Units using both GaAs and InP technology have been used. The lowest noise we have achieved to date is 10 K at 40 GHz, this being only 3.5 times the quantum limit at this frequency. These devices offer truly remarkable possibilities. Figure 2 shows the basic experiment configuration.

## 5. THE MAX EXPERIMENT

During the construction of ACME, a collaboration was formed between our group and the Berkeley group (Richards/Lange) to fly bolometric detectors on ACME. This fusion is called the MAX experiment and subsequently blossomed into the extremely successful Center for Particle Astrophysics' CBR effort. Utilizing the same basic experimental configuration as other ACME experiments, MAX uses very sensitive bolometers from about 1–3 mm wavelength in 3 or 4 bands. Flown from an altitude of 35 km, MAX has had four very successful flights. The first flight occurred in June 1989 using $^3$He cooled (0.3 K) bolometers, and the most recent flight occurred in June 1993 using ADR (Adiabatic Demagnetization Refrigeration) cooled bolometers. All the MAX flights have had a beam size of near 0.5 degrees.

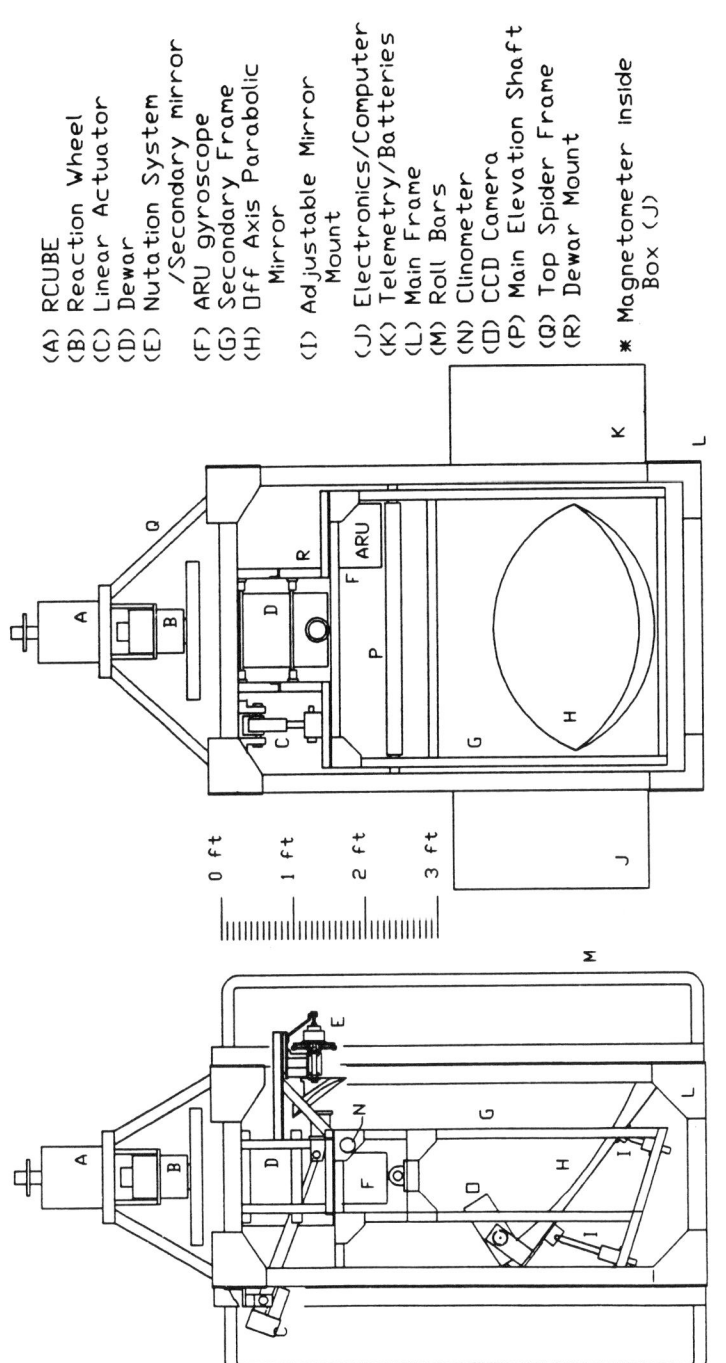

**Figure 2.** The ACME stabilized telescope.

## 6. RESULTS

There have been a total of ten ACME observations/flights from 1988 to 1993. Over twenty articles and proceedings have resulted from these measurements as well as seven Ph.D. theses. A summary of the various observations is given in Table 1.

Table 1
CBR Measurements With the UCSB ACME Platform

| Date | Site | Detector System | Beam FWHM (deg) | Sensitivity |
|---|---|---|---|---|
| 1988 Sep | Balloon$^P$ | 90 GHz SIS receiver | 0.5 | 4 mK s$^{1/2}$ |
| 1988 Nov–1989 Jan | South Pole | 90 GHz SIS receiver | 0.5 | 3.2 |
| 1989 Nov | Balloon$^{FS}$ | MAX photometer (3, 6, 9, 12 cm$^{-1}$) $^3$He | 0.5 | 12, 2, 5.7, 7.1 |
| 1990 Jul | Balloon$^P$ | MAX photometer (6, 9, 12 cm$^{-1}$) $^3$He | 0.5 | 0.7, 0.7, 5.4 |
| 1990 Nov-1990 Dec | South Pole | 90 GHz SIS receiver | 0.5 | 3.2 |
| 1990 Dec-1991 Jan | South Pole | 4 Channel HEMT amp (25-35 GHz) | 1.5 | 0.8 |
| 1991 Jun | Balloon$^P$ | MAX photometer (6, 9, 12 cm$^{-1}$) $^3$He | 0.5 | 0.6, 0.6, 4.6 |
| 1993 Jun | Balloon | MAX photometer (3, 6, 9, 12 cm$^{-1}$) ADR | 0.55-0.75 | 0.6, 0.5, 0.8, 3.0 |
| 1993 Nov-1994 Jan | South Pole | HEMT 25-35 GHz | 1.5 | 0.8 |
| 1993 Nov-1994 Jan | South Pole | HEMT 38-45 GHz | 1.0 | 0.5 |
| 1994 Jun | Balloon | MAX photometer (3, 6, 9, 14 cm$^{-1}$) ADR | 0.55-0.75 | 0.4, 0.4, 0.8, 3.0 |

Sensitivity does not include atmosphere which, for ground-based experiments, can be substantial.

P - Palestine, TX

FS - Fort Sumner, NM

ACME articles by Meinhold & Lubin (1991), Meinhold et al. (1992), Gaier et al. (1992), Schuster et al. (1993), and ACME-MAX articles by Fischer et al. (1992), Alsop et al. (1992), Meinhold et al. (1993), Gundersen et al. (1993), Devlin et al. 1994 and Clapp et al. 1994 summarize the results to date.

Significant detection by ACME at 1.5 degrees is reported by Schuster et al. (1993) at the $1 \times 10^{-5}$ level and by Gundersen et al. (1993) at 0.5 degrees at the $4 \times 10^{-5}$ level in adjacent issues of ApJ Letters. The lowest error bar per point of any data set to date is in the Schuster et al. 1.5° data with 14 $\mu$K while the largest signal to noise signal is in Gundersen et al. with about a 6 $\sigma$ detection (at the peak). Recently Wollack et al. (1993) report a detection at an angular scale of 1.2 degrees of about $1.4 \times 10^{-5}$ consistent with Schuster et al. and using a detector nearly identical to ours. At 0.5 degrees, the MSAM group reports detection of a "CBR component" at a level of about $2 \times 10^{-5}$ but with "point like" sources that are being reanalyzed and which may contribute additional power. Our most recent results from the June 1993 ACME-MAX flight give significant detections at the $3-4 \times 10^{-5}$ level at angular scales near 0.5 degrees.

It is remarkable that over a broad range of wavelengths that most degree scale measurements report detection at the one to a few $\times 10^{-5}$ level. Even more remarkable is the fact that both degree scale and COBE scale detections were published within six months of each other (Smoot et al. 1992; Schuster et al. 1993; Gundersen et al. 1993).

In historical retrospective, the degree scale detection in the Gamma Ursa Minoris region ("GUM data") was first published in Alsop et al. (1992) **prior** to the COBE detections. In any case, 1992 and 1993 were historical years in cosmology and CBR studies in particular. The ACME degree scale results are summarized in Table 2.

Table 2
Recent ACME Degree Scale Results

| Publication | Configuration | Beam FWHM (deg) | $\Delta T/T \times 10^{-6}$ |
|---|---|---|---|
| Meinhold & Lubin 91 | ACME-SIS | 0.5 | $<35$ |
| Alsop et al. 92 | ACME-MAX (GUM) | 0.5 | $45^{+57}_{-26}$ |
| Gaier et al. 92 | ACME-HEMT | 1.5 | $<14$ |
| Meinhold et al. 93 | ACME-MAX ($\mu$ Peg - upper limit) | 0.5 | $<25$ |
| Meinhold et al. 93 | ACME-MAX ($\mu$ Peg - detection) | 0.5 | $15^{+11}_{-7}$ |
| Schuster et al. 93 | ACME-HEMT | 1.5 | $9^{+4}_{-2}$ |
| Gundersen et al. 93 | ACME-MAX (GUM) | 0.5 | $42^{+17}_{-11}$ |
| Devlin et al. 94 | ACME-MAX (GUM) | 0.55–0.75 | $37^{+19}_{-11}$ |
| Clapp et al. 94 | ACME-MAX (Iota Draconis) | 0.55–0.75 | $33^{+11}_{-11}$ |
| Clapp et al. 94 | ACME-MAX (Sigma Hercules) | 0.55–0.75 | $31^{+17}_{-13}$ |

## 7. GOALS FOR THE FUTURE

We adopt as a long term goal the measurement of CBR anisotropy to a level of 1 micro-Kelvin per pixel. A near term goal of mapping a large number of pixels to 10 micro-Kelvin per pixel would also be highly desirable. This is somewhat but not completely arbitrary as recent analysis indicates that such (1 micro-Kelvin per pixel) sensitivity may be needed to allow good multi-parameter galactic subtraction. A word of warning is appropriate here. The galactic and extra-galactic backgrounds are not well understood at the levels and wavelengths needed. The same is true of our understanding of the actual signals we are attempting to find. Different physics scenarios in the early universe if known *a priori* would yield different search and experimental configurations. We are groping for the light here and any such search will be a modified random walk with frequent turns in direction after hitting the cosmic lamp posts. It will be amusing to review this ten years from now. It is believed that much of the current theoretical ideas will be well tested with such a sensitivity over an angular range of a few tenths of a degree to about ten degrees. The very large and small angular scales should not be totally neglected either, however, as evidenced in that even with the COBE DMR results after 4 years of data, many (most) of the pixels in the sky maps will not be galactic limited or even show a significant galactic signal. Another large scale anisotropy experiment with perhaps a 3 degree beam and a sensitivity 10 times better than COBE should allow a near background limited measurement at the large angular scales.

An important reference point to consider is that 1 micro-Kelvin is about an order of magnitude smaller than the currently most sensitive experiments. This comparison is important in what follows as we will discuss the future experiments in terms of the current experiments and indicate the magnitude of the needed improvements and the viability of these being accomplished. Keep in mind that in the past decade CBR anisotropy experiments have improved their sensitivity by about an order of magnitude as well. In what follows we will take a "devil's advocate" position to assume a worst case analysis that all of the presently measured signals are due to various systematic errors.

## 8. ATMOSPHERE

For coherent detectors, atmospheric emissions of a few Kelvin at the South Pole and milliKelvin or less at balloon altitudes. For incoherent detectors, similar emission at lower altitudes though usually substantially larger at balloon altitude due to the broad bandwidths. Note that the gain is going from the South Pole or mountain top to balloon altitudes is about a factor of $10^3$–$10^4$ reduction in emission.

For ground-based measurements, weather is definitely a problem. Typically, at sea level and lower altitude sites, the number of "good" days can be quite limited. For example, the many year experience at Owens Valley was that perhaps only a few handfuls of days per year were suitable. Experience at the South Pole indicates that in a typical summer season, perhaps 30% of the days are usable. Other ground-based sites, such as Mauna Kea, are also usable. If we assume all of the structure in the recent South Pole experiments is of atmospheric origin, then reducing the atmosphere by two orders of magnitude (going to balloon altitudes) should suffice. There is no evidence that atmospheric emission limit the current experiments except as the amount of "good" days available, however there are viable solutions (balloons) if this is a fundamental limitation for some experiments.

## 9. GALACTIC EMISSION

Understanding the emission from our galaxy both as diffuse and compact sources is of utmost concern for CBR anisotropy experiments as it is expected (and already is) to be problematic. Emission from charged particles in the complex galactic magnetic fields as well as collisions between charged particles as well as interstellar dust is complicated and not well enough understood to assess its full impact. Here again we draw on actual data. A variety of experiments from centimeter wavelengths to the millimeter and submillimeter bolometric experiments show regions where at least at the $\Delta T/T \sim 10^{-5}$ level galactic emission is not overwhelming (but may be present) in the "best" regions. Significant debate and uncertainty remains as to the best wavelength range to make measurements in. Most groups have adopted multiple wavelength measurements to allow the discrimination of galactic from cosmological sources due to the different spectral nature of the signals. Planned and existing experiments covering a factor of 2–3 in wavelength are typical.

## 10. EXTRA-GALACTIC SOURCES

For most of the beam sizes being discussed extragalactic sources are essentially point sources (sub beam size). Many extragalactic sources can be distinguished easily on the basis of their spectrum, but relatively flat spectra sources are known to exist. Most sources, however, do not have spectra that are well characterized at centimeter and millimeter wavelengths. This is going to be a challenging problem for CBR experiments requiring careful broad wavelength design and most likely follow up ground-based measurements. Fortunately, besides the spectral discrimination, the morphological (point source like) characteristic of extra-galactic sources will be very helpful. This is an area where much closer coordination is necessary with ground-based radio and submillimeter telescopes.

## 11. SIDELOBE ISSUES (OFF AXIS RESPONSE)

The response of the beam includes contributions from directions other than the target direction. Problematic sources of pickup include atmospheric emission (especially near the horizon where it can be quite large), solar and lunar emission, galactic (plane) and generally the most important being terrestrial earth emission. A simple calculation shows for a 0.5 degree experiment that rejection of the order of $10^{13} - 10^{14}$ is desired *if* we want the total radiometric emission picked up on the back lobe (earth) to be less than 1 $\mu$K. This is a formidable requirement on any antenna system, indeed one that is very difficult to even measure. It is also an unduly pessimistic requirement. Here again we are guided by actual experience and current data. Again, doing a worst case analysis, if we assume that all the structure seen in the recent South Pole and balloon experiments is due to earth sidelobe pickup, we conclude that another factor of $10^2$ is needed to get the desired goal of 1 $\mu$K. A factor of $10^2$ should be available with modest redesign and additional ground shields. Going to a balloon or low earth orbit does not help here (unless it is atmospheric sidelobe contamination that is the concern). Going to a multi-AU orbit (trajectory) does help greatly here since the earth subtended solid angle becomes much less.

## 12. DETECTOR LIMITATIONS—PRESENT AND FUNDAMENTAL

Detectors can be broadly characterized as either coherent or incoherent being those that preserve phase or not, respectively. Masers, SIS and HEMTs are coherent. Bolometers are incoherent. SIS junctions can also be run in an incoherent video detector mode. Phase preserving detectors inherently must obey an uncertainty relationship that translate into a minimum detector noise that depends on the observation frequency, the so called quantum limit. Incoherent detectors do not have this relationship but are ultimately limited by the CBR background itself. At about 40 GHz, these fundamental limits are comparable. Current detectors are not at these fundamental limits, though they are within an order of magnitude for both HEMTs and bolometers when used over moderate bandwidths. Here we include all effects including coupling efficiencies. Currently both InP HEMTs and ADR and $^3$He cooled bolometers exhibit sensitivities of under 500 $\mu$K sec$^{1/2}$. This assumes no additional atmospheric noise, true at balloon altitudes. For ground-based experiments at the South Pole, atmospheric noise is significant however.

Significant advances have been made in recent years in detector technology with effective noise dropping by over an order of magnitude over the past decade. With moderate bandwidths the fundamental limits for detectors are about a factor of 5 below the current values, so fundamental technology development is to be highly encouraged for both coherent and incoherent detectors.

With current detectors, achieving 1 $\mu$K sensitivity requires roughly one day per pixel for a single detector. This is appropriate for detector limited not atmospheric limited detection. This would be appropriate for balloon altitudes.

Small arrays of detectors are currently planned for several experiments. This should allow $\mu$K per pixel sensitivity over, say, 100 pixels in time scales of a few weeks, suitable for long duration ballooning or polar observations. If the fundamental detector limits could be achieved, the effective time would drop to about a day. Factors of 2–3 reduction in current detector noise are not unreasonable to imagine over the next five years, and if they could be achieved, the above time scale would drop to less than a week. Multiple telescopes are also possible. If we are willing to accept a goal of 3 $\mu$K per pixel (1 part per million of the CBR) instead of 1 $\mu$K then roughly 10 times as many pixels can be observed for the same integration time allowing significant maps to be made from balloon-borne detectors. A 10 $\mu$K error per pixel measurement would allow 100 times as many pixels to be measured in the same time. As we learn more about the structure of the CBR and about the nature of low level foreground emission the choice of sensitivity for a given angular scale will become clearer.

## 13. SPECTRUM MEASUREMENTS

The spectrum of the CBR has been extremely well characterized by the COBE FIRAS experiment in the millimeter wavelength range. However, in the range of about 1–100 GHz, where interesting physical phenomenon may distort the spectrum, much work remains to be done; particularly, at the longest wavelengths. Fortunately, the atmospheric emission is quite low over much of this range from both good ground-based sites and extremely low at balloon altitudes. Galactic emission and sidelobe contamination are of primary concern at the longest wavelengths, but it is expected that a number of ground-based and possibly balloon-borne experiments will be performed and should be encouraged.

A recent balloon-borne experiment, Schuster and Lubin (1994), is an example of what might be done in the future from balloon spectrum experiments. With all cryogenic optics and no windows, this experiment measured $T = 2.71 \pm 0.02$ K at 90 GHz with negligible atmospheric contamination ($\sim$ a few mK) and no systematic corrections. Errors of order 1 mK should be obtainable. The basic configuration could be extended to longer wavelengths where much remains to be done. In particular coherent measurements at 10–50 GHz from a balloon could be done. Such experiments are now being explored.

## 14. POLARIZATION

Very little effort has been directed towards the measurement of the polarization of the CBR compared to the effort in anisotropy detection. In part, this is due to the low level of linear polarization expected. Typically, the polarization is only 1–30% of the anisotropy and depends strongly on the model parameters. This is an area which in theory can give information about the reionization history, scalar and tensor gravity wave modes and large scale geometry effects. In the future, this may be a very fruitful area of inquiry.

## 15. TO SPACE

The question of whether or not a satellite is needed to get the degree scale "answer" is complex. There is no question that the measurements can be done from space and given sufficient funding this is the preferable way. It is unclear at this time what the limitations from sub-orbital systems will be and vigorous work is planned for sub-orbital platforms over the next decade. The galactic and extragalactic background problem remains the same for orbital and sub-orbital experiments. The atmosphere can be dealt with, particularly from balloon-borne experiments, with careful attention to band passes. Per pixel sensitivities in the $\mu K$ region are achievable with current and new technologies, HEMTs, and bolometers over hundreds to thousands of pixels. The major issue will be control of sidelobes and getting a uniform dataset. Ideally full sky coverage would be best and this is one area where a long term space based measurement would be ideal. In the control of sidelobe response a multi AU orbital satellite would be a major advance. This advantage is lost for near Earth orbit missions, however. By the end of the millennium, degree scale maps over a reasonable fraction of the sky at the $10^{-6}$ level should be possible from balloons and the ground. The potential knowledge to be gained is substantial, and I can think of few areas of science where the potential "payoff" to input (financial and otherwise) is so high.

This work was supported by the National Science Foundation Center for Particle Astrophysics, the National Aeronautics and Space Administration, the NASA Graduate Student Research Program, the National Science Foundation Polar Program, the California Space Institute, the University of California, and the U.S. Army. Its success is the result of the work of a number of individuals, particularly the graduate students involved with the experiment and our collaborators. The exceptional HEMT amplifier was provided by NRAO. Robert Wilson, Anthony Stark, and Corrado Dragone, all of AT&T Bell Laboratories, provided critical support and discussion regarding the early design of the telescope and receiver system. We would like to thank Bill Coughran and all of the South Pole support staff for highly successful 1988–1989 and 1990–1991 polar summers. In addition, we want to acknowledge the crucial contributions of the entire team of the National Scientific Balloon Facility in Palestine, Texas for their continued excellent support.

## REFERENCES

Alsop, D. C., et al. 1992, ApJ, 317, 146
Cheng, E. S., et al. 1993, ApJ Lett, submitted
Clapp, A., et al. 1994, ApJ Lett, submitted
Devlin, M., et al. 1994, ApJ Lett, submitted
Fischer, M., et al. 1992, ApJ, 388, 242
Gaier, T., et al. 1992, ApJ, 398, L1
Gundersen, J. O., et al. 1993, ApJ, 413, L1
Lubin, P., et al. 1985, ApJ, 298, L1
Meinhold, P. R., & Lubin, P.M. 1991, ApJ, 370, L11
Meinhold, P., et al. 1992, ApJ, 406, 12
Meinhold, P., et al. 1993, ApJ, 409, L1
Pospieszalski, M. W., et al. 1990, IEEE MTT-S Digest, 1253

Schuster, J., *et al.* 1993, ApJ, 412, L47
Schuster, J., & Lubin, P. 1994, in preparation
Smoot, G. F., *et al.* 1992, ApJ, 396, L1
Wollack, E., *et al.* 1994, ApJ, 419, L49

## DISCUSSION

**J. Peacock**: Did you deliberately not give a number for the detection of the Cheng *et al.*'s experiment? If you can give a number, what is it?

**Lubin**: You can analyze it in two different ways, as a 2-beam or a 3-beam experiment. There are basically lower and upper limits. One could say that the Cheng *et al.*'s experiment is bounded by lower and upper limits for the CBR component of that signal: something of the order of $1 \times 10^{-5}$ to a few $10^{-5}$. What we see on MAX is a $\Delta T/T$ about $4 \times 10^{-5}$ in this one region of the sky, although it is interesting that in the other region of the sky we placed an upper limit which is less than $2.5 \times 10^{-5}$. In the South Pole we have a detection of about $1 \times 10^{-5}$ at $1.5°$. So these three experiments show detections of some sort or another on the order of $10^{-5}$ or a few times that value.

**A. Kogut**: What do you think it will take to actually get a concerted effort by various people to look at the same spot on the sky?

**Lubin**: I do not think it will take much, assuming that the various groups are willing to talk to each other; and they are, in this field. I think you will see it happen on a short timescale. The future experiments I alluded to are really experiments that are going to happen within the next five years in my own opinion, because they are already on the books and many of them are under way.

The south pole is a little more complex because obviously we cannot reach from the south pole the northern hemisphere targets, and conversely.

**Q.**: Your cost estimate to get to 1 $\mu$K is a factor of 300 *better* than we were getting 30 years ago when we first started this game. Experiments to get to a few milliKelvin were costing tens of thousands of dollars.

**Lubin**: Well, to give you an interesting comparison on the way detectors technology has proceeded, you might ask what we could do if we were to launch COBE today or to design COBE today. We achieve, as I have showed you before, about 13 $\mu$K per data point, which is roughly a day worth of data from the south pole. If you take that value and divide it by the squared number of data points, you get something like 3 $\mu$K per data point, which is roughly the same sensitivity achieved with a year worth of COBE data. This is obvious because we can cool our detectors to the temperature of liquid helium, and they are much more exotic than what we were able to fly on COBE. So, if we were to launch a satellite today, we could achieve enormous increases in sensitivity. Whether we could utilize that in terms of rejection of systematics is a more complex question. It is not obvious.

**R. Daly**: You mentioned that in order to be able to believe the fluctuations you detect are from the CBR there should be some smoking guns. Could you be more explicit about that?

**Lubin**: Well, any of you who has followed the history of this field knows that it is a very checkered history in terms of believability of any particular measurement. My own personal philosophy is: I do not want to be the only person to find a detection because I do not trust anyone in the field that well. The systematic errors are very scary to deal with and there is a high potential for making mistakes. This is an historical fact. So what I would really like to see is the various groups to coordinate their efforts, make these multigroup measurements, and work with each other independently as well as concertedly. It would be a disaster if only one group in this field works.

We now have a couple of detections. If you go through the same region of the sky and do not see something, then it is not a fluctuation from the CBR; it is something else and this is what we want to know. On the next flight we will go back a third time and look at this target. We have seen it twice now, but we are still not satisfied. On this next flight we are at sensitivity levels of about 200 $\mu$K in a second, so we can achieve high sensitivity very quickly. But systematic errors are the problem. The reason I call it smoking guns is that I would like to see a bump in the COBE map when we will get the Fourier maps. I am quite sure that we will see a bump. And then we can look at that bump and see what it is.

**C. Bennett**: What was the dust index that you used on the plot you showed from IRAS 100 $\mu$ down to the millimeter, for the Meinhold et al. dataset?

**Lubin**: The index $n$ is about 1.4–1.6.

**C. Bennett**: If one ignored the significant possibility that there are systematics in some of the experiments and just took it for granted that these were all CMB fluctuations, could you comment on the statistics of those fluctuations?

**Lubin**: Yes. We have an interesting situation. In one dataset from the South Pole we set an upper limit of 14 parts per million. If you wish to interpret that as a detection, you obtain structure of the order of about $1 \times 10^{-5}$, which is less than the upper limit, so there is no conflict between the upper limit and interpreting this as a detection. If you go to the other strip, which is the most sensitive strip, we have the so-called 14 point-scan structure, again of the order of $1 \times 10^{-5}$. Fairly high significance. If it is CBR, then it is obviously an interesting detection; if it is not CBR, it is perhaps even more interesting, because we are led to some more exotic physics in the early universe.

From the MAX balloon flights we have what one could see as a dilemma. For one of the scans, where we see a strong correlation with IRAS dust, we set an upper limit of about 24 parts per million at about 0.5 degrees. However, in the same flight on a different part of the sky we see a structure which does not appear to be warm dust and has a rms of an order of $4-5 \times 10^{-5}$. If it is CBR, then it has a rms which is larger than the upper limit we set from another part of the sky.

To reiterate your question for others, if the sky is a Gaussian distribution of blobs, as one might expect from the standard CDM models with Gaussian random phases, one might be a little stressed by the fact that the detection was larger than the upper limit; but it is a statistics game and you have to be careful. Naively it appears to be a conflict, but if you want to do it right you have to do a Monte Carlo simulation and look at multiple skies and say to what level is it in conflict. It has not been done yet.

The other thing people are looking at is the question of the Gaussianity of the South Pole 13 point-scan or 15 point-scan depending on how you want to call it. Well, if you were a fan of something like cosmic strings or domain walls you might like to see non-Gaussian statistics on the sky. The South Pole is another case where the structure appears to be non-Gaussian at first order. Again, one has to play the game correctly by invoking a model sky and then saying to what degree is it non-Gaussian. Several people are looking at this problem theoretically right now. I would say there is not a clear cut answer, but it is interesting.

**M. Longair**: I would like to mention the interesting results from the Tenerife experiment. The group (from the Astronomical Observatory in Cambridge) working at the experiment has been rescanning for very very long periods over the same patch of sky; this is the *best* example of a sort of repeatable fluctuation which is appearing in their results. Again, I can simply quote what they are saying and leave it to the experts to decide how much they believe to all of this. This fluctuation does appear to be at the level of $\Delta T/T$ roughly $2.4 \times 10^{-5}$ at that particular point in the sky and they are claiming that the general background would then be consistent on the same scales.

**Lubin**: The scale of this experiment is comparable in angular scale to the COBE DMR experiment. I think it is important to point out here is that it is possible to check the individual bumps on the COBE map. With this and other experiments.

**B. Wang**: Recently the FIRAS map has shown a lot of cold dust at high latitude. I was wondering if this cold dust can mimic some of the signals you detect.

**Lubin**: The question is: could we be misreading cold dust, for which there is some evidence in the FIRAS data? If the dust is of the FIRAS type, you know, Wright *et al.* (1991) dust, we can distinguish that fairly well because it has an emissivity with a spectral index which is not zero. Now, one could imagine that the emissivity is flat, and if this is the case, then you are going to have real problems with it. But we should be able to distinguish the conventional dust with spectral indices different from zero for the emissivity. What is obviously worrisome is if there is some other component out there as well; or maybe one could imagine a conspiracy where you have bremsstrahlung coming down and dust going up. Since we do not have continuous wavelength coverage over a decade, we cannot know. This is why we are observing the same target over and over. But to the first order, I think your question of the Wright *et al.* dust is not a problem. It is the other things that worries me: the things we do not know about.

**M. Rees**: If you thought the fluctuations were non-Gaussian, would you then adopt a different observing strategy and look at more patches for a shorter time? And would it be worth gambling on that?

**Lubin**: That is a good question. Let me show you a couple of slides and then I will come to your question. This is a CDM simulation of galaxies with random phases by Dick Bond; it is the kind of map one might expect to see if provided with high angular resolution, with infinite pencil beam resolution. Even when you smear it out, for instance you can imagine early re-combination, you still see a rather random distribution of bumps. This is clear from the analysis of the Fourier components. However, if I show you a cosmic strings simulation, from its Fourier components you see very strong jumps as you cross a boundary and yet it is relatively flat in between. So the string model has statistics which are fairly highly non-Gaussian and here it would probably be better to take a wide cut in distribution rather than banging away at a small point very deeply. Now, what we have adopted so far is sort of a compromise; we have shot multiple points very deeply but taking a number of them. At the South Pole we actually did a box scan of about 100 square degrees, trying to get enough statistics. So, it depends on the type of model: if it is a cosmic string model or something similar, you can do pretty well just by scanning over reasonable size boxes; if it is monopoles or texture, the problem is even more complicated, because here you really want to scan large areas. You have to place your money somewhere and just go for it.

# COSMIC MICROWAVE BACKGROUND ANISOTROPIES AND STRUCTURE FORMATION IN THE UNIVERSE

Nicola Vittorio
Dipartimento di Fisica
Università di Roma "Tor Vergata"
Via della Ricerca Scientifica, 00133 Roma

**Abstract.** We discuss the implications of the COBE/DMR detection of Cosmic Microwave Background anisotropies on inflationary models for the large scale structure of the universe. We also discuss the constraints set on these models from current upper limits on, and recent detection of Cosmic Microwave Background anisotropies at small and intermediate angular scales, and from the local abundance of X-ray galaxy clusters.

## 1. INTRODUCTION

The Cosmic Microwave Background (CMB) discovery in 1964 (Penzias & Wilson 1965) marks the beginning of modern cosmology. Since then, the Big-Bang model based upon the Friedmann-Robertson-Walker metric became the standard model of the universe. This model has two main predictions: the thermal nature of the CMB radiation and the number of neutrino families. The first one has been clearly shown, at least in the mm region, by the COBE/FIRAS (Mather et al. 1990) and the COBRA (Gush, Halpern & Wisshnow 1990) experiments. The second one found a direct experimental confirmation by the LEP result.

The last year discovery of CMB anisotropies (Smoot et al. 1992) on large angular scales constitutes another milestone in modern cosmology and marks the beginning of a new era in the study of the large scale structure (LSS) of the universe. In fact, the COBE/DMR detection of CMB anisotropy on super-horizon scales at decoupling allows to infer the amplitude and shape of the primordial power spectrum of density fluctuations in the framework of a linear theory, without any uncertainty due to possible biases in the distribution of galaxies with respect to the dark matter. Much as it is a giant step forward, the COBE/DMR measurement does not directly probe those density fluctuations which gave rise to the structures we observe today on scales $\lesssim 100h^{-1}$ Mpc. These fluctuations are expected to induce CMB anisotropy at angular scales $\lesssim 1°$. The importance of these scales has been discussed by several authors, with particular emphasis to the existence of the Doppler peak in the CMB radiation power spectrum. The prominence of this peak is quite model dependent and provides useful constraints on the model parameters. For example, it critically depends upon the baryonic content of the universe (see e.g., Vittorio & Silk 1992), upon the amount of late reheating of the intergalactic medium (see *e.g.*, Sugiyama, Silk & Vittorio 1993), and upon the presence of a primordial background of gravitational waves (GWB; see *e.g.*, Lucchin, Matarrese & Mollerach 1992). Also, the theoretical implications of low resolution ($\approx 7°$ FWHM as in COBE/DMR) anisotropy measurements can be limited by the cosmic variance. On the contrary, at degree scales, the effect of cosmic variance can be strongly suppressed by performing high resolution observations of large regions of the sky. In any case, degree scale observations combined with the COBE/DMR data extend the lever arm in determining the slope of the primordial power spectrum. So, a combined analysis of the anisotropy data at intermediate ($\simeq 1°$) and large ($> 3°$) scales may in principle address fundamental questions, such as the total abundance of baryons in the universe, the existence of a secondary last scattering surface, the value of the primordial spectral index and the existence of a GWB.

In this paper we discuss the significance of the COBE/DMR result and the status of the inflationary LSS models [we will not consider here string (Bennett, Stebbins & Bouchet 1992) or texture (Silk & Juszkiewicz 1991) scenarios] on the light of current ($\leq 1992$) upper limits to and recent detection of CMB anisotropy at intermediate angular scales. We also discuss the constraints on LSS models set by X-ray observations of clusters of galaxies.

## 2. CMB ANISOTROPY ON LARGE ANGULAR SCALES

The CMB temperature pattern can be expanded in spherical harmonics: $\delta T/T(\hat{\gamma}, \vec{x}) = \sum_{\ell,m} a_\ell^m(\vec{x}) Y_{\ell m}(\hat{\gamma})$. Here $\hat{\gamma}$ identifies a given line of sight around $\vec{x}$, the observer position. The coefficients $a_\ell^m(\vec{x})$ of the spherical harmonic expansion are stochastic variables of the observer position, Normally distributed with zero mean and variance $\langle |a_\ell^m(\vec{x})|^2 \rangle \equiv a_\ell^2$. If temperature fluctuations are induced by adiabatic density fluctuations with an initially scale free power spectrum ($P(k) = Ak^n$), then (Bond & Efstathiou 1987; Scaramella & Vittorio 1990)

$$a_\ell^2 = \frac{2^{n-1} A}{4\pi r_0^{n+3}} \frac{\Gamma[3-n]}{\Gamma^2[2-n/2]} \frac{\Gamma[\ell+(n-1)/2]}{\Gamma[\ell+(5-n)/2]} \tag{1}$$

where $r_0 = 2c/H_0$ is the horizon radius. The statistical properties of the CMB pattern on the single observable sky can be written in terms of an angular correlation function (acf):

$$C(\alpha, \sigma_B, \vec{x}) = \sum Q_\ell^2(\vec{x}) P_\ell(\cos\alpha) \exp[(2\ell+1)^2 \sigma_B^2/4] \tag{2}$$

where $Q_\ell^2(\vec{x}) = \sum_m |a_\ell^m|^2/(4\pi)$, $\alpha$ is the angle between two lines of sight and $\sigma_B$ is the dispersion of a gaussian approximating the angular response of the receiver. The observed $Q_\ell^2(\vec{x})$ may differ from their expected values, $\langle Q_\ell^2(\vec{x}) \rangle$. In other words, the specific shape and amplitude of the acf will depend on the observer position, *i.e.*, each cosmic observer deals with one realization of last scattering surface. The mean and variance of the acf over the ensemble of cosmic observers is given by:

$$\langle C(\vec{x}, \alpha, \sigma_B) \rangle = \sum_{\ell=2}^\infty \langle Q_\ell^2(\vec{x}) \rangle P_\ell(\cos\alpha) \exp[(2\ell+1)^2 \sigma_B^2/4] \tag{3}$$

$$Var[C(\vec{x}, \alpha, \sigma_B)] = 2\sum_{\ell=2}^\infty \frac{\langle Q_\ell^2 \rangle^2}{2\ell+1} P_\ell^2(\cos\alpha) \exp[(2\ell+1)^2 \sigma_B^2/2] \tag{4}$$

Note that these quantities are uniquely defined in terms of only two parameters: the amplitude and shape of the primordial power spectrum, which fix through Eq. (1) the multipole expected values $\langle Q_\ell^2(\vec{x})\rangle = (2\ell+1)a_\ell^2/4\pi$.

The cross correlation between the 53 GHz and 90 GHz COBE/DMR maps exhibits a signal which is consistent with the theoretical predictions of gravitational instability theory, where adiabatic density fluctuations generate CMB anisotropies through the Sachs-Wolfe effect (Sachs & Wolfe 1967). For scale free density fluctuations, Smoot et al. (1992) found $n = 1.1 \pm 0.6$, consistent with scale invariant initial conditions (i.e., $n = 1$). Under this assumption, $A = 6\pi^2 r_0^4 Q_{ps-rms}^2/5$, where $Q_{ps-rms} = (17 \pm 5)$ $\mu K/2.7K$ is the ensemble averaged quadrupole anisotropy (Smoot et al. 1992). It is worth to briefly discuss which are the assumptions behind this result. These are: i) the universe has a critical density, i.e., $\Omega_0 = 1$; ii) density fluctuations are adiabatic; iii) the CMB temperature correlation function (acf) is characterized only by its mean and dispersion over the theoretical ensemble; iv) any contribution to the CMB anisotropy from a possible background of gravitational waves (GWB) is neglected.

Assumption i) is not crucial as long as we restrict to flat universes. In fact, the relative scaling of multipoles of order $\ell > 2$ remains basically unchanged passing from a critical universe to a vacuum dominated flat cosmologies, provided we restrict ourselves to values of $n$ not very different from unity (Kofman & Starobinskii 1985; Górski, Silk & Vittorio 1992; Efstathiou, Bond & White 1992). Assumption ii), suggested by the inflationary scenario, is not at all mandatory. Primordial density fluctuations could be of the isocurvature variety, as in the pure baryonic, open models considered by Peebles (1987). In this case however the shape of the acf is expected to be much steeper than observed, for any reasonable choice of the primordial spectral index (Scaramella & Vittorio 1993a).

Assumption iii) is in principle the more critical. Mean and variance of the acf are necessary but not sufficient to fully characterize the distribution over the ensemble of cosmic observers: the distribution of acf values at fixed $\alpha$ is not gaussian (Bennett, Stebbins & Bouchet 1992). So for each choice of amplitude and shape of the power spectrum we have to consider the whole ensemble of possible realizations of $\mu$-wave skies. Here enters the problem of the so-called cosmic variance (Abbott & Wise 1984a), which has two main effects. First, the statistics of the temperature fluctuation field on the single observable sky may be non Gaussian even if the primordial density fluctuations, out of which CMB anisotropies are generated, are indeed gaussian distributed (Scaramella & Vittorio 1991), as generally expected from the inflationary scenario. This implies that observing a non gaussian CMB pattern on the sky does not necessarily imply non gaussian initial condition for the density field. Second, as the pattern of the single observable sky is determined by the relative amplitude of low and high order multipoles it may happen that a single realization of microwave sky can be consistent with the angular correlation function measured by COBE/DMR, even if the ensemble of the initial conditions is described by a spectral index $n$ different from unity. We investigate the relevance of this effect by performing for each value of $n$ Monte Carlo simulations of the CMB anisotropy pattern (Scaramella & Vittorio 1993b). For each simulated sky we evaluate the acf profile and we estimate its amplitude by a direct fit to the COBE/DMR data. In this way for each $n$ we reconstruct the whole distribution of CMB anisotropy amplitudes and the fraction $f$ of acf profiles that are consistent with the COBE/DMR data. The main result of this analysis is that $f(n)$ is quite broad, with a maximum at $n \sim 1$. If $n = 1$ basically all the possible realizations of the microwave sky are consistent with the data. However if $n = 2$ or $n = -2$, $\simeq 50\%$ of the simulated

$\mu$−wave skies are still consistent with the COBE/DMR data (Scaramella & Vittorio 1993b). So the best model that fits the COBE/DMR result is the scale invariant one. However, we may not be able to reject values of the spectral index quite different from unity (e.g., $n = -2$ or $n = +2$) on the basis of a robust statistical argument.

If we assume, because of our theoretical prejudice, that $n = 1$, then a direct fit to the COBE/DMR acf data gives (Scaramella & Vittorio 1993b): $Q_{ps-rms} = (14.5 \mu K)(1 \pm \epsilon_{cv})(1 \pm \epsilon_{exp})$. This value is slightly smaller than that found by Smoot et al. (1992), but yields a variance of $(29 \mu K)^2$ on 10° (FWHM), in excellent agreement with the COBE/DMR value of $(30 \mu k)^2$. Note that $\epsilon_{cv} = 0.12$ is the uncertainty at the 1 sigma level due to the cosmic variance only, while $\epsilon_{exp} = 0.06$ is the uncertainty due the errors in the observed acf. If confirmed, this result will imply that our knowledge of the large scale CMB anisotropy is already limited by the cosmic variance. We will comment on assumption iv) in the next Section, as neglecting a GWB does not change the result of the COBE/DMR analysis, but rather the implications of the COBE/DMR result on the amplitude of density fluctuations.

## 3. LSS MODELS

For a flat cold dark matter (CDM) dominated universe the COBE/DMR result implies a value of $\sigma_\rho$ (the rms value of the density fluctuations on $8h^{-1}$ Mpc scale) of the order of unity, with an uncertainty of roughly 20% at the one sigma level (Efstathiou, Bond & White 1992; de Gasperis, Muciaccia & Vittorio 1993). Such a value of $\sigma_\rho$ is too high to be consistent e.g., with the observed pairwise galaxy peculiar velocities on small scales (Davis et al. 1985).

Clearly, there is the need of reducing the small scale power of the CDM model. This can be done in two ways. First, we can reduce the small scale power simply by "tilting" the primordial power spectrum and by using "quasi" scale invariant initial conditions (i.e., $n \lesssim 1$). The advantage is twofold. On the one hand, reducing $n$ reduces the small scale power of the model, once we normalize to the COBE/DMR result. On the other hand, "quasi" scale invariant initial conditions can be obtained in the framework of power law inflationary models (Abbott & Wise 1984b; Lucchin & Matarrese 1985), which also predict a non vanishing gravitational wave background (Lucchin, Matarrese & Mollerach 1992; Davis et al. 1992). As the latter contributes to the large scale anisotropy, the amplitude of density fluctuations can be further reduced, allowing for values of $\sigma_\rho \simeq 0.5$, as generally required by small scale observations. Second, we can keep scale invariant initial conditions and reducing the small scale power by adding to the cold component a hot component in the form of massive neutrinos, keeping $\Omega_0 = 1$. In this "hybrid" or mixed dark matter (MDM) models, the small scale power is reduced because of the free-streaming of the neutrinos. If the fraction of cold to hot dark matter is roughly 7:3, there is a good agreement (see Sachs & Wolfe 1967) with the COBE/DMR detection, with the IRAS QDOT redshift survey (Rowan-Robinson et al. 1990) and with the APM angular correlation function (Maddox, Efstathiou & Sutherland 1990). Moreover N-body simulations of this hybrid scenario seem to suggest the possibility of reconciling the different values of $\Omega_0$ observed on small and large scales (Davis, Summers & Schlegel 1992).

We will discuss in the next section the more general case of "tilted hybrid" models (Liddle & Lyth 1993; de Gasperis, Muciaccia & Vittorio 1993). We restrict ourselves to a flat universe with a negligible abundance of baryons (we fix the baryonic density parameter to be $\Omega_B = 0.03$). Hence, the only free parameter of the model is the density parameter of the massive neutrino component, $\Omega_{\nu m}$ (the CDM density parameter is uniquely determined: $\Omega_{CDM} = 1 - \Omega_{\nu m} - \Omega_B$), and the spectral index $n$. The question that we ask is: can CMB anisotropy measurements constrain both these two parameters?

## 4. CMB ANISOTROPY ON SMALL AND INTERMEDIATE ANGULAR SCALES

### 4.1 Upper Limits

As stated in the Introduction, here we will discuss mainly the status of the models on the light of current ($\leq$ 1992) upper limits on the CMB anisotropy at small and intermediate angular scales. In particular we will discuss the results of the OVRO (Readhead et al. 1989), the ACME-HEMT (Gaier et al. 1992; Schuster et al. 1993), and the ULISSE (de Bernardis et al. 1992) experiments.

The OVRO experiment operates at the frequency of $20 GHz$ with an antenna beam of dispersion $\sigma_B = 0'.76$. It tests the differential (double subtracted) CMB anisotropy at the angular scale of $7'.15$. The observations were carried out at a constant declination ($\delta = 89°$), and the data set is constituted by seven temperature differences. The reduced $\chi^2$ of the data is 7.2 and the mean error bar is $\bar{\sigma} = 31 \mu K$. The ACME-HEMT experiment is a multifrequency, single subtraction experiment. We consider the data of the high frequency channel at $34 GHz$. The antenna beam has $\sigma_B = 0°.56$ and there is a sinusoidal modulation, with a beam switching angle of $\theta = 3°$. The experiment provides 9 temperature differences, $2°.1$ away on the sky, out of which a linear gradient is subtracted. The reduced $\chi^2$ of the data is 4.8 and the mean error bar is $\bar{\sigma} = 27 \mu K$. Finally the ULISSE experiment is a balloon borne experiment, operating with two broad band bolometers at millimetric and sub-millimetric wavelengths. It is a single subtraction experiment, with a sinusoidal modulation. The beam switching angle and the antenna beam size are $\theta = 6°$ and $\sigma_B = 2°.2$, respectively. The observations were carried out at constant Galactic latitude ($b = 40°$) and generated 25 temperature differences, corrected for both the CMB dipole and Galactic contribution. The reduced $\chi^2$ of the data is 25.5 and the mean error bar is $\bar{\sigma} = 24 \mu K$.

In the following we will normalize the amplitude of density fluctuations of tilted hybrid models in order to reproduce the $rms$ temperature fluctuation of $30 \mu K$ of the COBE/DMR maps (Smoot et al. 1992). This normalization yields for any chosen pair $\Omega_{\nu m}$ and $n$ a well defined value for $\sigma_\rho$ (de Gasperis, Muciaccia & Vittorio 1993):

$$\sigma_\rho = 1.05 \frac{\exp[-2.49(1-n) - n\Omega_{\nu m}(0.55 + 0.36\Omega_{\nu m})]}{G_\rho(n)(1 + 2.02\Omega_{\nu m} - 2.52\Omega_{\nu m}^2)} \quad (5)$$

where $G_\rho(n) = \sqrt{(14 - 12n)/(3 - n)}$. If $n = 1$ and $\Omega_{\nu m} = 0$ (standard CDM model), we have $\sigma_\rho = 1.05$ (c.f. de Bernardis et al. 1994). When $\Omega_{\nu m} = 0$, lowering $n$ reduces $\sigma_\rho$ for two reasons: the tilt in the spectrum [described by the exponential at the numerator of Eq. (5)] and the presence of a GWB (described by the function $G_\rho$). Note that for $n = 1$ $G_\rho = 1$: the GWB vanishes for scale invariant initial conditions. When $n = 1$, increasing $\Omega_{\nu m}$ (hybrid models) also reduces $\sigma_\rho$. For tilted hybrid models this effect is clearly maximized.

Because of similar mechanisms, we expect that in tilted hybrid models the CMB anisotropy at intermediate angular scale is also reduced. If we ignore a late reheating of the intergalactic medium, the anisotropy at the angular scales probed by the OVRO, ACME-HEMT and ULISSE experiments are (de Gasperis 1993):

$$\Delta_{rms}(7'.15, 0'.76) = (30\mu K)\frac{1 - 2.23y + 1.63y^2}{\sqrt{1 + 11 G_\Delta e^{-0.08y}}} \quad (6)$$

$$\Delta_{rms}(3°, 0°.56) = (48\mu K)\frac{1 - 1.23y + 0.70y^2}{\sqrt{1 + 5.39 G_\Delta e^{-2.19y}}} \quad (7)$$

$$\Delta_{rms}(6°, 2°.2) = (28\mu K)\frac{1 - 0.66y + 0.36y^2}{\sqrt{1 + 1.98 G_\Delta e^{-2.92y}}} \quad (8)$$

where $y = 1 - n$ and $G_\Delta = y/(2 + y)$. Lowering $n$ has indeed a similar effect both for $\sigma_\rho$ and for $\Delta_{rms}$. Once we normalize the amplitude of density fluctuation on large scales, lowering $n$ implies reducing the small scale inhomogeneities of the model and, then, the CMB anisotropy. This is monitored by the quadratic form at numerator (which describes the tilting of the spectrum) and by the function $G_\Delta$ (which describes the effect of a GWB). Note also that in the above formulae the dependence on $\Omega_{\nu m}$ is completely lost. This is quite unfortunate, as it implies that with the COBE/DMR normalization the pattern of the $\mu$-wave sky little depends upon the chemistry of the universe: CMB anisotropy experiments by themselves can not discriminate among different hybrid models with the same spectral index.

After the release of the ACME-HEMT data, different results appeared in the literature. From one hand it has been claimed that the CDM standard model with the COBE/DMR normalization was inconsistent with the ACME-HEMT experiment (Górski, Stompor & Juszkiewicz 1993). Because of this, much interest has been focused on possible late reheating of the intergalactic medium. Such reheating smooths out the primordial temperature fluctuations on scales smaller than a few degrees while it creates new fluctuations on larger scales, corresponding to the horizon scale at last scattering, as well as on much smaller scales by the Visnhiac effect (Vishniac 1987). The need for a reheating of the intergalactic medium is less urgent according to other authors (Dodelson & Jubas 1993), which found that a COBE/DMR normalized CDM model is indeed consistent with the ACME-HEMT experiment. Most of the disagreement arises from using different statistics in analyzing the ACME-HEMT data. These data have a reduced Chi-squared quite less than unit, so we must carefully interpret the results of the data analysis. de Gasperis et al. (1993) have compared different method of analysis and found that the $\chi^2$ analysis [used by Gorski et al. (1993)] of the ACME-HEMT data provides a very stringent upper limit, while the Bayesian approach of Dodelson and Jubas (1993) give a less constraining upper bound to the CMB anisotropy.

The OVRO, ACME–HEMT and ULISSE experiments provide independent data sets, as they observe different regions of the sky at different angular scales and with different resolution. This suggests to perform a joint analysis of the three data sets. The observed reduced $\chi^2$ of the combined data set is very near to unity: in other words, the combined data set is the "typical" outcome of a null experiment. Moreover, using the combined data set allows to test three different angular scales at once, possibly setting better constraints on the theoretical models. Because of the reasonable value of the $\chi^2$ of the total data set, we found at 3° an upper limit of 34.3$\mu K$, *independently* of the statistics used: we got essential the same result with a Chi-squared analysis, with a likelihood ratio test, and with a Bayesian analysis. This gives confidence on the reliability of the upper bound we found. This upper limit constrains the tilted hybrid models considered here. As the theoretical predictions are independent on $\Omega_{\nu m}$, an upper limit to the CMB anisotropy implies an upper limit to the spectral index, independently, we stress it again, of $\Omega_{\nu m}$. For the case of a vanishing GWB we find at the 95% confidence level: $n \leq (0.73 \pm 0.15)$ (the uncertainty in the upper bound corresponds to the 1 sigma uncertainty in the COBE/DMR normalization). If we take into account the GWB, we find $n \leq (0.87 \pm 0.07)$.

## 4.2 Detections

Schuster et al. (1993) provide 13 data points at 27.5, 30, 32.5 and 35GHz. They also give a combination of the four channels, the full band data set, and claim detection of CMB anisotropy. In fact, the analysis of the full band data, done assuming tilted hybrid models, provides (de Gasperis, Muciaccia & Vittorio 1993) at the 90% confidence level $\Delta_{rms}(3°, 33.6') = (26^{+18}_{-12})\mu K$. Most of the signal in the full band data set comes from two adjacent data points. Neglecting these two points, assuming that they are fully contaminated, provide only an upper limit: $\Delta_{rms}(3°, 33.6') \lesssim 29\mu K$. Note that the Schuster et al., full band data set is consistent with the upper limit derived in the previous sub-Section by analyzing the combination of the OVRO, ACME and ULISSE data sets.

Comparing the level of anisotropy detected in the full band data set by Schuster et al. with the theoretical predictions allows to estimate the value of the spectral index. An anisotropy of $26\mu K$ would imply $n \simeq 0.6$ and $n \simeq 0.5$, with and without a GWB. Also, at the 95% confidence level, $n < 1$ (with a GWB) and $n < 0.9$ (without a GWB). Taking into account the uncertainty in the COBE normalization [17% at the 1 sigma level (Smoot et al. 1992)] does not allow to reject scale invariant initial conditions, although they are only marginally consistent with the data. Note however that these upper limits would be strongly weakened if we use conservatively only the Schuster et al., high frequency channel data. In this case, independently of the statistical approach, $\Delta_{rms}(3°, 33.6') \lesssim 50\mu K$.

We want to conclude this Section by mentioning the result of the more recent ARGO experiment (de Bernardis et al. 1993; de Bernardis et al. 1994), also reporting detection of CMB anisotropy. ARGO is a balloon borne, 1.2 m Cassegrain telescope with wobbling secondary mirror, featuring large throughput bolometric detectors cooled at 0.3 K. This multiband (2.0, 1.2, 0.8, 0.5 mm) experiment has a beam well fitted by a gaussian of dispersion $\sigma_B = 22'$. The sky chop is sinusoidal with a frequency of 14 Hz, and with a peak to peak amplitude of 1.8°. The signal detected at 0.5 mm is well correlated with the 100 $\mu m$ IRAS map. This allows to estimate the spectral index of dust emissivity. With such an index, dust emission is responsible for less than 3% of the signals detected at 2 mm. In this channel, the $rms$ fluctuation of the 63 temperature differences is $\sim 20\mu K$, while the typical one sigma error bar is $\sim 10\mu K$.

For the tilted hybrid models considered here, the anisotropy expected at the ARGO scale, taking into account the sinusoidal modulation, is well fitted by the following expression, valid to better than few percent for $0.5 \lesssim n \lesssim 1.2$ and $\Omega_b = 0.05$:

$$\Delta_{rms}(1.8°, 22') = (22\ \mu K)\frac{1 - 1.54y + 1.00y^2}{\sqrt{1 + 7.80G_\Delta e^{-1.03y}}}$$

As above, $G_\Delta = y/(2+y)$, $y = 1-n$ and the fit does not depend upon $\Omega_{\nu m}$. For $n \geq 1$ the GWB is supposed to be vanishing (i.e., $G_\Delta = 0$).

Even if the sky region observed by ARGO has also been observed by COBE/DMR, the ARGO and the COBE/DMR data can be considered independent. In fact, the COBE/DMR experiment has a low angular resolution, and it is insensitive to the detailed structure of the CMB pattern at degree scales. The combined analysis (de Bernardis 1994) of the ARGO and COBE/DMR data select a range of values for the primordial spectral index: $n = 0.95^{+0.25}_{-0.15}$ (values of $n$ outside this range can be rejected at more than 95% confidence level). These bounds are basically independent of the cosmological abundance of baryons (at least in the range allowed from primordial nucleosynthesis) and of the ratio of cold to hot dark matter. So, flat, cold or mixed dark matter models, with "quasi" scale invariant initial conditions and a standard recombination history, successfully take into account the CMB anisotropy detected at intermediate and large angular scales.

Other degree-scale experiments compare well with the results of this analysis. To verify this let us consider the MSAM (Cheng et al. 1993), the MAX (Meinhold et al. 1993; Gunderson et al. 1993) and the Python (Dragovan et al. 1993) experiments. These experiments report detection of anisotropy at the level of $(72 \pm 41)\mu K$, $45^{+25}_{-45}\mu K$, $114^{+46}_{-30}\mu K$, and $(60 \pm 20)\mu K$, respectively. These rms values are in good agreement with the predictions of scale-invariant, flat hybrid models, favoured by our ARGO & COBE/DMR analysis: $53\mu K$, $85\mu K$, $85\mu K$ and $51\mu K$, respectively.

## 5. X-RAY CLUSTERS

As already mentioned, the results of the previous Section apply to both CDM and hybrid models, as the CMB pattern of a flat universe does not depend on the relative amount of cold to hot dark matter. For this reason, a cleaner test of hybrid or tilted hybrid models must rely on the study of the LSS of the local Universe. These observations probe density fluctuations at scales below $\simeq 100h^{-1}$ Mpc, where the shape of the matter power spectrum is strongly model dependent (see e.g., Holtzmann 1989).

Among others, clusters of galaxies provide an efficient test for LSS models. Being the largest structures which have reached or are near to reach a virial equilibrium, clusters provide a natural link between the linear and non-linear stages of structure formation, and their distribution can be still described within the linear or mildly non-linear gravitational instability theory. For example, the present abundance of rich clusters of galaxies is basically set by $\sigma_\rho$, the rms density fluctuation at $8h^{-1}$ Mpc. This scale roughly contains the mass corresponding to the cut off in the observed mass distribution of these cosmic structures. In particular, as $\sigma_\rho \equiv b^{-1}$, the rich cluster abundance constitutes a measure of the biasing factor, $b$. Moreover, the evolution in redshift of the cluster population constrains the power spectrum $P(k)$ on scales $1h^{-1}$ Mpc $\lesssim \lambda \lesssim 10h^{-1}$ Mpc, providing informations that, on this range of scales, are complementary to those derived by the correlation analysis of the galaxy distribution. So, to compare the theoretical predictions with the X-ray observations of galaxy clusters we need to specify few assumptions. First, we need to specify a model and to use the detailed form of the density fluctuation power spectrum, properly normalized. Second, we need predictions for the mass distribution of groups and clusters of galaxies. The pioneering work by Press and Schechter (hereafter PS; 1974) provides a simple approach for doing this: for power-law power spectra [i.e., $P(k) = Ak^n$], we have the well known expression for the number density of objects of mass $M$ at redshift $z$.

$$N_{rv}(M,z) = N_0 \frac{\mathcal{I}}{\sqrt{2\pi}} \frac{(n+3)}{6} \frac{\delta_c b}{M_*} \left(\frac{M}{M_*}\right)^{-2+a} \frac{1}{D(z)} \exp\left[-\frac{\delta_c^2 b^2}{2D(z)^2}\left(\frac{M}{M_*}\right)^{2a}\right] \quad (9)$$

In the previous expression $D(z)$ describes the growth of linear density fluctuations, $a = (n+3)/6$, $N_0 = \rho_b/M_* \approx 1.8 \cdot 10^{-4} \, h^3 \, \text{Mpc}^{-3} \, (10^{15} M_\odot)^{-1}$, and $M_* \approx \Omega_0 h^2 (R_*/8h^{-1} \, \text{Mpc})^3 10^{15} M_\odot$ is the mass of those structures which are going non linear today. The power law behaviour of the mass distribution at $M < M_*$ has a slope little sensitive to the shape of the power spectrum. On the contrary, the high mass cut off at $M \gtrsim M_*$ still reflects the linear or mildly non-linear evolution of the underlying density field. So the mass distribution has two parameters: the product $\delta_c b$ and the constant $\mathcal{I}$. Here $\delta_c$ is the assumed linear threshold for identifying non linear ($\delta_c = 1.07$), collapsed ($\delta_c = 1.7$) or virialized ($\delta_c = 2.2$) objects, while $1 < \mathcal{I} < 2$, depending on the amount of mass accreted by secondary infall onto the already formed structure. Finally, we have to parameterize the dependence from mass and redshift of the X-ray luminosity of clusters of galaxies. Then, assuming a mass to light ratio, we obtain the comoving X-ray luminosity function directly from the mass distribution: $N(L, z) \equiv N(M, z) dM/dL$ Note that also $N(L, z)$ depends on both $\delta_c b$ and $\mathcal{I}$, that can then be estimated by best fitting the theoretical predictions (Colafrancesco & Vittorio 1994) to the observed X-ray luminosity function (Kowalski et al. 1984).

For the standard CDM flat model we find $\mathcal{I} = 1.0^{+0.57}_{-0.0}$ and $\delta_c b = 2.6^{+0.14}_{-0.12}$ [here and below the uncertainties in the fit are at the 1 sigma level]. The reduced Chi-squared has the minimum value $\chi^2_{min} = 2.5$. Flat, "tilted" CDM models with primordial spectral index $n = 0.8$ provide a fit comparable with the case obtained for the standard flat CDM model. In fact, for this model we obtain $\delta_c b = 2.8^{+0.12}_{-0.23}$, $\mathcal{I} = 1.8^{+0.2}_{-0.7}$ and $\chi^2_{min} = 2.6$. The predictions for flat hybrid models with $\Omega_{\nu m} = 0.3$ do not provide a good fit to the observed XRLF, especially at the faint end. In this model we obtain $\delta_c b = 2.8^{+0.06}_{-0.15}$, $\mathcal{I} = 2.0^{+0.0}_{-0.3}$ and $\chi^2_{min} = 5.9$. Reducing the fraction of neutrinos to $\Omega_{\nu m} = 0.1$ improves the fit: $\delta_c b = 2.8^{+0.14}_{-0.12}$, $\mathcal{I} = 2.0^{+0.0}_{-0.5}$ and $\chi^2_{min} = 2.9$. So, a hybrid model with $\Omega_\nu = 0.3$ provides an unacceptable large value of $\chi^2_{min}$: the probability $Q$ of having a $\chi^2$ larger than $\chi^2_{min}$ for noise fluctuations only is $Q = 2.8 \cdot 10^{-8}$. A hybrid model with $\Omega_\nu = 0.3$ is clearly ruled out from the present data on the local X-ray luminosity function. For the standard CDM model and for tilted CDM model ($n = 0.8$) we find $Q = 0.014$ and $Q = 0.012$, respectively. The flat hybrid model with $\Omega_\nu = 0.1$ is intermediate between the standard CDM and the $\Omega_\nu = 0.3$ model and provides $Q = 3.7 \cdot 10^{-3}$.

## 6. CONCLUSIONS

The conclusions presented in Section 4 must be considered preliminary. We did not consider the effects of a late reheating of the intergalactic medium, which could relax the estimates of the primordial spectral index. However, the results presented in Sec. 4.2 suggest that a flat non baryonic dark matter universe, with scale invariant initial conditions and with a standard recombination history is consistent with the level of anisotropy detected at intermediate ($1°\!.8$) and large ($> 3°$) angular scales. As the pattern of CMB anisotropy is independent of the relative amount of cold and hot dark matter, the ARGO and COBE/DMR results do not help in specifying the "chemistry" of our universe. However, observations of the large scale structure on small scales can do this quite efficiently, as the matter power spectrum on those scales is strongly model dependent. The analysis shown in Section 5 suggests that the ratio of hot to cold dark matter can not be large: a ratio of 3:7 is already inconsistent with the X-ray luminosity function data. Moreover, preliminary analyses suggest that tilted CDM models have difficulties in accounting at once for the CMB anisotropy upper bounds and for the peculiar velocity measurements (Muciaccia et al. 1993), while tilted hybrid models reproduce quite well the local universe only if the primordial spectral index is near to unity (Liddle & Lyth 1993). All this requires further investigation from both a theoretical and an observational point of view. However, it seems extremely difficult to take into account the bulk of the available observations without considering a flat universe, dynamically dominated today by non-baryonic dark matter, most of which is in the form of cold dark matter.

The results presented here were obtained in collaboration with S. Colafrancesco, G. C. de Gasperis, P. F. Muciaccia and R. Scaramella. Financial support from MURST grant is acknowledged.

## REFERENCES

Abbott, L. F. & Wise, M. 1984a, ApJ, 282, L47
Abbott, L. F. & Wise, M. 1984b, Nucl. Phys., B244, 541
Bennett, D. P. Stebbins, A., & Bouchet, F. R. 1992, ApJ, 399, L5
Bertschinger, E. & Jain, R. 1993, preprint
Bond, J. R. & Efstathiou, G. 1987, MNRAS, 226, 655
Cheng, E. S. et al. 1993, ApJ, in press
Colafrancesco, S. & Antonuccio-Delogu, V. 1994, ApJ, in press
Colafrancesco, S. & Vittorio, N. 1994, ApJ, in press
Davis, M., Efstathiou, G., Frenk, C., & White, S. D. M., 1985, ApJ, 292, 371
Davis, M., Summers, F. J. & Schlegel, D. 1992, Nature, 359, 383
Davis, R. L., Hodges, H. M., Smoot, G. F., Steinhard P. J. & Turner, M. S. 1992, Phys. Rev. Lett., 69, 1856
de Bernardis, P. et al. 1993, ApJ Lett, submitted
de Bernardis, P., de Gasperis, G. C., Masi, S. & Vittorio, N. 1994, ApJ Lett, submitted
de Bernardis, P., Masi, S., Melchiorri, B., Melchiorri, F., & Vittorio, N., 1992, ApJ, 396, L57
de Gasperis, G. C., Muciaccia, P. F., & Vittorio, N. 1993, ApJ, submitted
Dodelson, S., & Jubas, J. M., 1993, Phys. Rev. Lett., 70, 2224
Dragovan, M. et al. 1993, preprint
Duus, A. & Newell, B. 1977, ApJ Supp, 35, 209
Edge, A. C., Stewart, G. C., Fabian, A. C., & Arnaud, K. A. 1990, MNRAS, 245, 559
Efstathiou, G., Bond, J. R., & White, S. D. M., 1992, MNRAS, 258, 1P
Gaier, T., Schuster, J., Gundersen, J., Koch, T., Meinhold, P., Seiffert, M., & Lubin, P., 1992, ApJ, 398, L1
Górski, K. M., Stompor, R., & Juszkiewicz, R. 1993, ApJ, 410, L1
Górski, K. M., Silk, J., & Vittorio, N., 1992, Phys. Rev. Lett., 68, 733
Gunderson, J. O., Clapp, A., Devlin, M., Holmes, W., Fischer, M., Meinhold, P., Lange, A., Lubin, P., Richards, P., & Smoot, G., 1993, ApJ, 413, L1

Gush, H. P., Halpern, M., & Wisshnow, H., 1990, Phys. Rev. Lett., 65, 537
Henry, J. P., Gioia, I. M., Maccacaro, T., Morris, S. L., Stocke, J. T., & Wolter, A. 1991, preprint
Holtzmann, J. 1989, ApJ Supp, 71, 1
Kofman, L. & Starobinskii, A. A., 1985, Sov. Astron. Lett., 11, 271
Kowalski, M. P., Ulmer, M. P., Cruddace, R. G., & Wood, K. S. 1984, ApJ Supp, 56, 403
Liddle, A. R. & Lyth, D. H. 1993, MNRAS, in press
Lucchin, F. & Matarrese, S., 1985, Phys. Rev. D, 32, 1316
Lucchin, F., Matarrese, S., & Mollerach, S. 1992, ApJ, 401, L49
Maddox, S. J., Efstathiou, G., & Sutherland, W. J. 1990, MNRAS, 242, 43P
Matarrese, S., Pantano, O., & Saez, D. 1993, preprint
Mather, J. C. et al. 1990, ApJ, 354, L37
Meinhold, P., Clapp, A., Devlin, M., Fischer, M., Gunderson, J., Holmes, W., Lange, A., Lubin, P., Richards, P., & Smoot, G., 1993, ApJ, 409, L1
Muciaccia, F., Mei, S., de Gasperis, G., & Vittorio, N., 1993, ApJ, 410, L61
Peebles, P. J. E. 1987, Nature, 327, 210
Penzias, A. A. & Wilson, R. W. 1965, ApJ, 142, 419
Press, W. H. & Schechter, P. 1974, ApJ, 187, 425
Readhead, A. C. S., Lawrence, C. R., Meyers, S. T., Sargent, W. L. W., Hardebeck, H. E., Moffet, A. T., 1989, ApJ, 346, 566
Rowan-Robinson, M. et al. 1990, MNRAS, 247, 1
Sachs, R. K. & Wolfe, M. A., 1967, ApJ, 147, 73
Scaramella, R. & Vittorio, N. 1990, ApJ, 353, 372
Scaramella, R. & Vittorio, N. 1991, ApJ, 375, 439
Scaramella, R. & Vittorio, N. 1993a, ApJ, 411, 1
Scaramella, R. & Vittorio, N. 1993b, MNRAS, 263, L17
Schuster, J., Gaier, T., Gundersen, J., Meinhold, P., Koch, T., Seiffert, M., Wuencshe, C. A., & Lubin, P., 1993, ApJ, 412, L47
Silk, J. & Juszkiewicz, R. 1991, Nature, 353, 386
Smoot G. F. et al. 1992, ApJ, 396, L1
Sugiyama, N., Silk, J., & Vittorio, N. 1993, ApJ, in press
Tegmark, M., Silk, J. & Blanchard, A. 1993, preprint
Visnhiac, E. T. 1987, ApJ, 322, 597
Vittorio, N. & Silk, J., 1992, ApJ, 385, L9

## DISCUSSION

**F. Stecker**: I was just wondering if you are planning to test the two component dark matter model as well.

**Vittorio**: I have very preliminary results. Let me show them to you just for reference. I am leaving $n$ as a free parameter. I have $b$ here and this model has 10% of the critical density in massive neutrinos. The point is that if you gain something in the agreement between the 1° experiments and COBE, although you pay that by having a very large $b$, you are barely able to recover the situation with the peculiar velocities. This is something which makes me worry, because if that is confirmed it means the whole family of flat universes are in trouble.

**P. Lubin**: You showed the prediction for the 90 arcminute ACME experiment. Do you have the equivalent at 0.5° to be compared to the MAX balloon experiment and to the Goddard experiment?

**Vittorio**: Not really. I don't have the numbers yet, in terms of prediction at different angular scales. I have some graphs I can show you.

**M. Rees**: Just a technical question about your assumptions on the gravitational wave contribution. I thought that in Starobynski's paper, he predicted a gravitational wave contribution even when $n = 1$, whereas, as I understood it, you did not get gravitational waves except in tilted models.

**Vittorio**: Yes. I have followed essentially Lucchin et al., in which they use a power law inflation model. In this kind of model, the gravitational background vanishes for $n = 1$. Clearly if you add that contribution, you are changing the results quite drastically.

**M. Rees**: I just wanted to emphasize that your assumptions were very sensitive to details of the inflation model.

**Vittorio**: Absolutely.

**J. Mould**: Nicola, I think you tended to minimize the conflict between anisotropy and peculiar velocity results by adopting the Bertschinger et al. large scale peculiar velocity. If you were to adopt, instead, the results we get from Tully-Fischer's analysis of galaxy clusters or the Lauer and Postman's results on the brightest cluster members, then the conflict would be worse.

**Vittorio**: I agree completely.

**Q.**: Yes, I think there is some confusion about the inflationary models, about what is meant by $n = 1$. In Starobinski's paper, for example, what is important is the nature of the inflationary field and of the potential; even things which follow the Zel'dovich scaling solution closely are, in fact, not exactly $n = 1$, because there is a logarithmic term that you can approximate by a power-law over the range of scales relevant to the COBE small scales. In other words, the steeper the logarithmic term over that range, the more important is the gravitational radiation. So, looking at your numbers, I think they were quite similar, at least in the range of 0.8 or 0.9, to what you would get from Starobinski's results.

**J. Peacock**: Nicola, the measurement of the spectral index that you are making out of the peculiar velocity field presumably refers to the slope of the power spectrum at wavelengths around 100 Mpc. I think there is good evidence that the spectrum is still curved there, so you might well expect to find something which is $n < 1$, which is still turning over.

**Vittorio:** Sorry, that is the primordial spectral index.

**J. Peacock:** But you have to assume a transfer factor.

**Vittorio:** Cold dark matter. That was inside the family of cold dark matter models.

**J. Peacock:** Okay, but that does not fit even at 100 Mpc wavelength.

**Vittorio:** In the sense that you have extra power, I admit it, but the point is, if you take these models and you try to ask which is the most probable value that you may have, it turns out that you have, until 100 Mpc, $n = 0.4$.

**J. Peacock:** This is a very specific CDM model with a particular density. I think, perhaps, it is better to be empirical and say that the power we observe in the density field on a few hundred Mpc wavelengths sets a limit to how tilted the spectrum can be, and probably no more than a couple of tenths deviation from $n = 1$.

**Vittorio:** I agree completely. I think that a phenomenological approach should be complemented in all this kind of stuff.

**S. White:** I have a question about the velocity constraints which is related to your comment about the cosmic variance for the correlation function. We only have one measurement of the microwave background dipole velocity, and two measurements of peculiar velocity that are obtained for spheres centered on us. So these three measurements that you have, the microwave dipole and the two velocities from Bertschinger, are actually strongly correlated. The IRAS dipoles that you used were not, in fact, velocity data; they are telling us something about the distribution in space of the galaxies, but they are not true velocity data. I think perhaps it might be fairer to ask what are the constraints if we discard the IRAS dipoles and take into account the correlation between those three things. Are they really so stringent anymore?

**Vittorio:** The answer is: I made a joint probability analysis on the three data sets so I am taking into account the full correlation matrix of the data sets, namely the correlation between the dipole and the back motions and the fact that the back motions are not uncorrelated. I am taking also into account the fact that the IRAS data are not velocity data; you can correlate the amount of power that is required to fit the IRAS dipole and the amount of power which is required to fit the back velocity. Now, I do not have the single 95% confidence level so I am going by heart. If I remember well, it is only when you take the three data sets together that you have a strong constraint. When you use only dipoles or only back velocity, or only IRAS data, the 95% confidence level is just a bit larger and intersects the $n = 1$ case. So, the point is that you may find a fit for a single observable, but when you try to fit all the observables, which is relevant for constraining the power spectrum over a wide range of wave numbers, then you are in trouble.

**S. White:** My point was that the dipole data from IRAS were a different kind of data and make a different assumption about how galaxies formed, and how galaxies trace the mass distribution. So, perhaps you might want to separate that particular constraint from the true velocities constraint.

**Q.:** Just a sort of follow-up on that. I was wondering if you considered in your analysis the cosmic variance that you would expect in the velocities. I was not sure whether that is what you were saying or not.

**Vittorio:** The answer is yes. Because I am taking the full distribution of peculiar velocity over the ensemble with the full correlation matrix.

**Q.:** Could I ask a question about the spectral constraints on the index $n$, and whether those might be improved to provide a usable test? This is something that John Mather mentioned, so maybe John or Nicola or Ruth can answer.

**R. Daly:** I think there is very little possibility for improving those constraints. The $y$ or $\mu$ parameters would have to come down by orders of magnitude to improve that. So I do not anticipate any better constraint.

# THE RADIO BACKGROUND EMISSION—
# THE LONG AND SHORT OF IT

M. S. Longair
Cavendish Laboratory
Madingley Road
Cambridge CB3 0HE
England

**Abstract.** A brief survey is presented of the extragalactic background radiation in the radio, centimetre and millimetre wavebands, excluding the Cosmic Microwave Background Radiation. Little progress has been made in the study of the long wavelength radio background radiation over recent years. At millimetre wavelengths, the possibility of searching for the strongly redshifted far-infrared luminosity associated with the first generations of star formation in young galaxies is discussed.

## 1. SALUTATIONS

Let me begin with two items which are the source of great pleasure in attending this symposium in honour of Riccardo Giacconi. The first is the obvious one that we should celebrate Riccardo's enormous contributions to the undoubted success of the Space Telescope Science Institute. This has been a complex and difficult task but one which is now showing the just rewards for a really huge effort on the part of many people.

The second great pleasure is the fact that Rashid Sunyaev has been able to attend this symposium. It was more than 20 years ago that Rashid and I produced our spectrum of the extragalactic background radiation at all accessible electromagnetic wavelengths and I reproduce our spectrum in Fig. 1. The remarkable thing about Fig. 1 is that it is still a reasonably accurate representation of the overall spectrum of the background radiation. The reason for this is that in those wavebands in which the background is reasonably easy to detect, background radiation was amongst the earliest observations to be made by telescopes with low angular resolution. In contrast, in those wavebands in which the background was swamped by the contribution of discrete sources, this has by and large remained the same.

I must confess that my heart sank somewhat when I saw that the topic I have been asked to discuss is one of the least popular and exciting aspects of the background radiation—*The Radio Background—Observations*. Essentially nothing has happened in this area for about 25 years and most of the exciting topics will be covered by John Peacock who will discuss the interpretation of the background radiation and, in

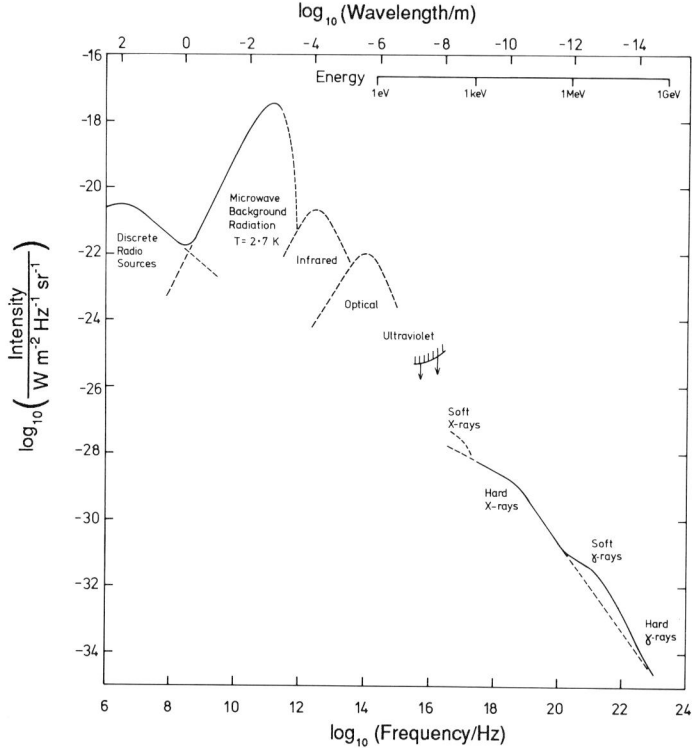

**Figure 1.** *The spectrum of the extragalactic background radiation as is was known in 1969 (Longair and Sunyaev 1971). The solid lines indicate regions of the electromagnetic spectrum in which extragalactic background radiation had been measured. The dashed lines are theoretical estimates of the background radiation due to discrete sources and should not be taken very seriously.*

particular, what we can learn about the cosmological evolution of the radio source and quasar populations.

I will therefore do two things. First, I will talk about what I am meant to talk about—the classical extragalactic radio background radiation and then I will turn to some new work which Andrew Blain and I have been doing concerning the millimetre and sub-millimetre background radiation which we believe can tell us a lot about the early evolution of galaxies.

## 2. THE RADIO BACKGROUND RADIATION

The story begins in 1933 with Jansky's discovery of the radio emission from the Galaxy. It was immediately apparent that, on large angular scales, the radio sky is dominated by diffuse Galactic emission. As is well known, this great discovery caused little stir in the astronomical community and it was only after the Second World War that the nature and origin of the radio background emission became the object of astronomical interest. By the late 1940s, the emission mechanism was identified as synchroron radiation and, at about the same time, the first of the discrete radio sources was identified.

At that period, one of the principal motivations for attempting to extract the diffuse extragalactic component of the radio background radiation was related to the question of the distances and luminosities of the discrete radio sources which continued to be discovered as the sky surveys discovered more and more faint sources. The argument is a well-known one and goes as follows. Suppose the sources have typical luminosities $L_\nu$ and space densities $\rho_L$. Then the diffuse background emission due to a uniform cosmological distribution of these sources is

$$I_\nu \propto \rho_L L_\nu$$

On the other hand, if we also measure the number of sources brighter that a given flux density $S$, $N(\geq S)$, that number is given by

$$N(\geq S) \propto \rho_L L_\nu^{3/2}$$

Since the observed background intensity $I_\nu$ is an upper limit to the integrated intensity, and $N(\geq S)$ is fixed, we can find a lower limit to $L_\nu$. This was the argument used by Martin Ryle to demonstrate reasonably convincingly that the bulk of the discrete radio sources had to be distant extragalactic objects. It was also the motivation for attempting to disentangle the intensity of the isotropic radio background emission from the anisotropic Galactic radio emission which was much more intense. This was a very difficult observational programme and several generations of Cambridge research students were almost broken in attempting to find a credible result.

The problem is that the radio sky is dominated by the synchrotron emission of our own Galaxy as is beautifully demonstrated by the map of the whole sky due to Glyn Haslam and his colleagues at the Max Planck Institute for Radio Astronomy at Bonn. As a result, wherever one looks in the sky, there is always intense radiation in the far out sidelobes of the radio telescope. The best one can do is to map the sky at different wavelengths with *geometrically scaled antennae* so that although the sidelobe problem is not eliminated, at least it should be the same at different frequencies. What one observes on the sky is

$$I_\nu(\alpha, \delta) = I_{\text{gal}}(\alpha, \delta) + I_0(\nu)$$

where the first term on the right-hand side represents the anisotropic component associated with the Galaxy and the term $I_0(\nu)$ represents the isotropic extragalactic component. The procedure is then to map the sky at different frequencies, assume that the anisotropic component has the same radio spectrum in all directions and then find $I_0(\nu)$. The procedure only works because the Galactic continuum spectrum is different from that of the diffuse extragalactic component, specifically, the spectrum of our Galaxy having the form $I_\nu \propto \nu^{-0.4}$ at frequencies less than about 200 MHz whereas the extragalactic sources have much steeper spectra.

The best results are still those presented by Bridle (1967). It is convenient to express the results in terms of the brightness temperature of the radiation $T_b = (\lambda^2/2k)I_\nu$. At the traditional wavelength of 178 MHz, the frequency of the revised 3C Catalogue, the results are a follows. The minimum sky temperature at 178 MHz is about 80 K and includes both the minimum Galactic component as well as the isotropic component. As the errors build up, it is not possible to determine both the intensity and spectrum of the extragalactic component and so the isotropic component is extracted assuming different values for the radio spectral index. If $\alpha = 0.75$, the isotropic background temperature is $30 \pm 7$ K; if $\alpha = 0.9$, the intensity corresponds to $15 \pm 3$ K. The typical spectral index of radio sources at 178 MHz is about $\alpha = 0.8$.

These figures should be compared with the brightness temperature found when the source counts are integrated to the lowest flux densities observed. The integrated background emission to sub-millijansky levels corresponds to about 20 K. It is interesting to identify the principal contributors to the discrete source background on the basis of modelling the source counts. If we simply adopt the local radio luminosity function for extragalactic radio sources and assume that there was no evolution of the population with cosmic epoch, we would expect a radio background at 178 MHz of only about 1–2 K. When the effects of strong evolution of the source population is taken into account, the background emission from the evolving component of the population increases to about 16–19 K. To these components we have to add the contribution of normal galaxies which amounts to about 4 K and the low luminosity 'starburst' galaxies which probably contribute a further few K to the total background.

Thus, it seems that virtually all the radio background emission can be attributed to discrete sources and there is not much room left for any other contribution to the background radio emission at low frequencies. One contribution of possible cosmological interest is the upper limit to the intensity of intergalactic bremsstrahlung which would have a flat radio spectrum, $I_\nu \propto \nu^0$. As a result, the best limit comes from observations at about the minimum of the radio background emission which occurs at about 400 MHz because at higher frequencies, the Cosmic Microwave Background Radiation becomes the dominant component. Once the discrete source component of the background and the Cosmic Microwave Background Radiation are removed, the upper limit to any residual diffuse component would amount to about $T_{400 \text{ MHz}} \leq 0.1$ K.

What all of this means is not my job—John Peacock will take up the story of the astrophysical and cosmological implications of these observations. I will end this story with two footnotes. The first is the touching story reported by Jasper Wall at the 1989 Heidelburg meeting on the Galactic and Extragalactic Background Radiation (Wall 1990). In 1964, Jasper and Donald Chu were attempting to measure the background radiation at a frequencies of 320 and 707 MHz. They found to their distress that they could not obtain the 'right' answer—their background spectrum was too flat (Wall, Chu and Yen 1970). As research students, the tacit assumption was made that they had simply made some error in the calibration of their experiment. Only in the next year was the discovery of the Cosmic Microwave Background Radiation reported which accounted for their excess antenna temperature.

The second footnote concerns the extragalactic background emission at very long wavelengths, $1 - 10$ MHz. This is an even more unfashionable waveband because the observations are very difficult to make because of ionospheric absorption and refraction. However, at certain locations in the auroral zone, it is possible to observe the sky at 10 MHz. In the 1960s Chris Purton and Alan Bridle did as good a job as could be done at that time at these very low frequencies from the Penticton Radio Observatory (Bridle and Purton 1968). The sky is still dominated by the synchrotron emission of the Galaxy but, because of the differences in spectral indices, the extragalactic component is relatively more important. The process which becomes important at these low frequencies is bremsstrahlung absorption so that, at 10 MHz, the Galactic plane is observed in absorption (Purton 1966). As observations are made at frequencies less than 10 MHz, the distance at which the bremsstrahlung optical depth becomes unity decreases. The spectrum of the background radiation in the region of the Galactic pole has, however, been determined from the Canadian RAE1 satellite and the shape of the extragalactic component of the background was determined (Clark et al. 1970). Evidence was found that the extragalactic spectrum showed a cut-off at low frequencies, $\nu \leq 3$ MHz (Fig. 2).

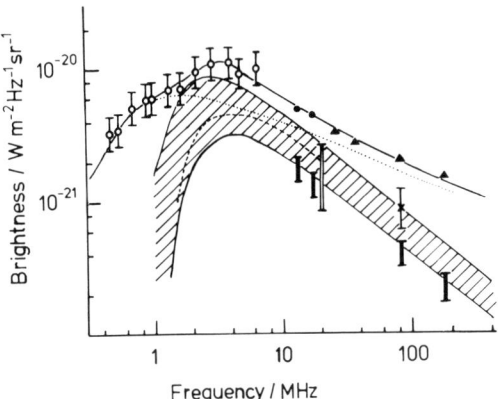

**Figure 2.** *The spectrum of the radio sky in the direction of the 'north halo minimum'. The solid line shows the best fit to the total background. The dotted line shows the Galactic contribution and the dashed line the estimated extragalactic contribution, the shaded region indicating the uncertainties in the latter estimate. Independent estimates of the extragalactic background are also shown. (From Simon 1977)*

The origin of this behaviour was discussed by Simon (1977). The obvious interpretation of the cut-off is that it is associated with synchrotron self-absorption in the discrete sources which make up the background. She studied the predicted spectra of a complete sample of 3CR radio sources to very low frequencies for which detailed radio structural information was available. Compact components and hot spots become synchrotron self-absorbed at frequencies $\nu \geq 100$ MHz and the only components which contribute to the $1 - 10$ MHz background radiation are the most diffuse components. Because of the strong inverse correlation between diffuse structure and radio luminosity, the greatest contributions to the background in the 1–10 MHz waveband come from relatively low luminosity sources (Fig. 3). Simon evaluated the predicted background spectrum when account was taken of the cosmological evolution of these sources and found that she could account quite naturally for the inferred turn-over in the isotropic radio background spectrum.

## 3. THE BACKGROUND RADIATION AND GALAXY FORMATION REVISITED

I will now change tack completely and look at some aspects of galaxy evolution and in particular at the very beautiful analysis by Lilly and Cowie (1987) of the constraints on metal production in large redshift galaxies. Let me first repeat their argument.

The analysis begins with the observation that a prolonged burst of star formation in a galaxy has a remarkably flat emission spectrum out to the Lyman limit at 91.2 nm. This is nicely illustrated by the model star-bursts synthesised by Bruzual and presented by White (1989) (Fig. 4). These spectra show the integrated spectra of the starburst galaxy at different ages assuming that the star formation rate is a constant and that the

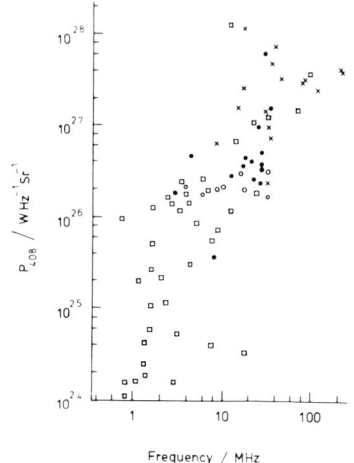

**Figure 3.** *The relation between radio luminosity at 408 MHz and the frequency at which the radio source is expected to exhibit synchrotron self-absorption. The radio sources form a representative sample of the radio sources in the 3CR catalogue. (From Simon 1977).*

stars are formed with the same Salpeter mass function. The flatness of the spectrum is due to the fact that, although the most luminous blue stars have short lifetimes, they are constantly being replenished by new stars. Furthermore, the intensity of the flat part of the spectrum is directly proportional to the rate of formation of heavy elements since their energy is primarily derived from the conversion of hydrogen into helium which is the first stage in the formation of the heavy elements—these are only formed in stars with mass greater than about $4M_\odot$. From the model starbursts and from simple physical arguments, it can be shown that the intensity of the flat spectrum region of the spectrum is related to the mass of heavy elements produced by the simple relation

$$L_\nu = 2 \times 10^{22} \left( \frac{\dot{M}Z}{1 M_\odot \text{ year}^{-1}} \right) \text{ W Hz}^{-1}$$

at all wavelengths longer than the Lyman continuum edge at 91.2 nm. $\dot{M}Z$ is the rate of formation of heavy elements. It is a simple calculation to work out the background intensity due to such sources and, provided the Lyman limit is not redshifted into the observing waveband, Lilly and Cowie show that the intensity expected for a given amount of element formation is independent of the cosmological model. Specifically, the background intensity due to the formation of a density of the heavy elements $\rho_m$ is

$$I_\nu = 7.5 \times 10^{-25} \left( \frac{\rho_m}{10^{-31} \text{ kg m}^{-3}} \right) \text{ W m}^{-2} \text{ Hz}^{-1} \text{ sr}^{-1}$$

Notice that the density used in this relation is the density of heavy elements observed at the present epoch and that a density of $10^{-31}$ kg m$^{-3}$ of heavy elements would correspond roughly to $Z = 0.01$ in an $\Omega = 0.01$ Universe. The beautiful thing about this relation is that, by inserting the intensity of the background radiation originating

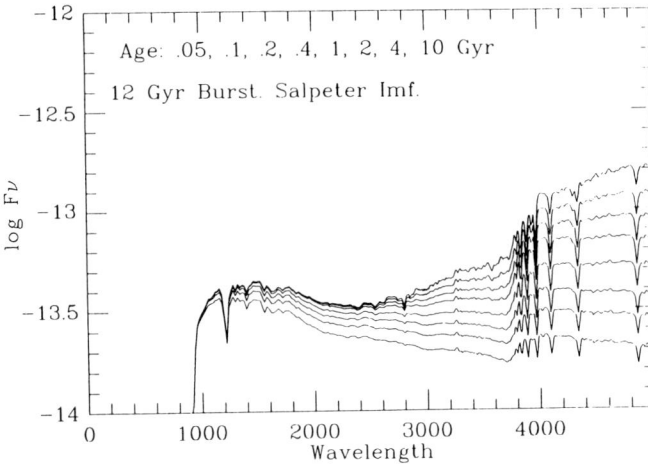

**Figure 4.** *Synthetic spectra for a region with constant star formation rate at the ages indicated. A Salpeter initial mass function has been assumed with cut-offs at 75 and 0.08 $M_\odot$. The spectra were generated by Gustavo Bruzual from a recent version of his evolutionary synthesis programmes (from White (1989).*

in a particular redshift interval $\Delta z$, we can immediately read off the density of metals synthesised in that interval.

Cowie and Lilly have observed a class of flat spectrum objects in their very deep optical surveys. Originally, it was though that these objects lay at large redshifts but it is now believed that they have redshifts roughly one. The background intensity due to such objects amounted to about $6.6 \times 10^{-25}$ W m$^{-2}$ Hz$^{-1}$ sr$^{-1}$. Lilly and Cowie interpret this result as meaning that a significant fraction of the heavy elements must have been produced about a redshift of one. In fact, this heavy element abundance is significantly less than the maximum permissible. If we were to assume that $H_0 = 50$ km s$^{-1}$ Mpc$^{-1}$, the upper limit to the baryon density in the Universe as determined by the need to produce at least the observed abundance of deuterium is about 0.1. If a maximum metal abundance of $Z = 0.03$ is adopted, the total background intensity due to metal formation could be up to about 30 or 40 times the intensity already detected by Lilly and Cowie. In fact such an intensity would exceed the upper limit to the background intensity reported by Toller (1990).

Now, it is well known that galaxies undergoing bursts of star formation are not only sources of ultraviolet continuum radiation but also are strong emitters in the far infrared waveband because of the presence of dust in the star forming regions. According to Weedman (1993), in a sample of star forming galaxies studied by the IUE, most of the galaxies emit much more of their luminosities in the far infrared rather than in the ultraviolet region of the spectrum. As a result, it is quite possible that most of the radiation associated with the formation of the heavy elements is not radiated in the ultraviolet-optical region of the spectrum but in the far infrared region and would permit a higher abundance of the elements as compared with the existing optical and ultraviolet limits.

This was one of the motivations for undertaking a study of the feasibility of detecting the far-infrared emission from star-forming galaxies at large redshifts in the submillimetre waveband.

## 4. SUBMILLIMETRE COSMOLOGY

Andrew Blain and I have been carrying out some computations of the expected source counts and background emission expected from star-forming galaxies at large redshifts in the submillimetre and millimetre wavebands (Blain and Longair 1993). Until recently, the prospects for making surveys of sources in the submillimetre waveband have not been very encouraging because of the lack of array detectors which would allow a significant region of sky to be surveyed. The situation will change dramatically in the near future with the introduction of submillimetre bolometer array detectors on telescopes such as the James Clerk Maxwell Telescope. Specifically, the Submillimetre Common User Bolometer Array (SCUBA) currently being completed for that telescope will enable the mapping of regions of the sky in these wavebands to be carried out about 10,000 times faster than is possible with the current generation of single element detectors.

It might be thought that the detection of star-forming galaxies at cosmologically interesting distances would be very difficult because nearby examples of these types of galaxy are only weak submillimetre emitters. This problem is, however, more than offset by the enormous far infrared luminosities of these galaxies which are redshifted into the submillimetre waveband at redshifts greater than about 1. Specifically, the far infrared spectra of IRAS galaxies peak about 100 $\mu$m and have very steep spectra, $I_\nu \propto \nu^\alpha$ where $\alpha$ is about 3–4. As a consequence, the 'K-corrections' are very large and negative at submillimetre wavelengths. The result is that, at redshifts greater than 1, the flux density of a standard IRAS galaxy is more or less independent of redshift until the far infrared maximum is redshifted through the submillimetre wavebands. This is illustrated in Fig. 4 which shows the expected flux density-redshift relations for a galaxy emitting $10^{13}$ $L_\odot$ with a standard dust emission spectrum at temperatures of 30 and 60 K as observed at 450 and 1100 $\mu$m. Correspondingly, the counts of submillimetre sources show a remarkable behaviour at those flux densities at which the 'coasting phase' in the flux density-redshift relation is reached. The predicted differential number counts for a single luminosity class of source at different wavelengths and for different assumed temperatures of the dust grains are shown in Fig. 5. These differential counts have to be convolved with the luminosity function of the sources and this can be found from the IRAS luminosity function derived by Saunders et al. (1990). The differential source counts for a uniform population of sources is shown in Fig. 6 in which it can be seen that there is an enormous excess over the expectations of a 'Euclidean' model. It must be emphasised that these computations are carried out for a *uniform* world model and that the apparent 'excess' is entirely due to the large and negative K-corrections. If the effects of cosmological evolution are included, an even more remarkable excess of faint sources and extraordinarily steep source counts are predicted. Fig. 7 shows the results of incorporating the effects of luminosity evolution of the form $L \propto (1+z)^3$ in the redshift interval $0 \leq z \leq 2$ and a constant value at larger redshifts, $L = 27L_0$ where $L_0$ is the luminosity of sources at zero redshift; according to Peacock (1993), this form of evolution can account not only for the radio and optical counts of quasars and radio sources but also for the counts of IRAS galaxies. In this case, there would be very large

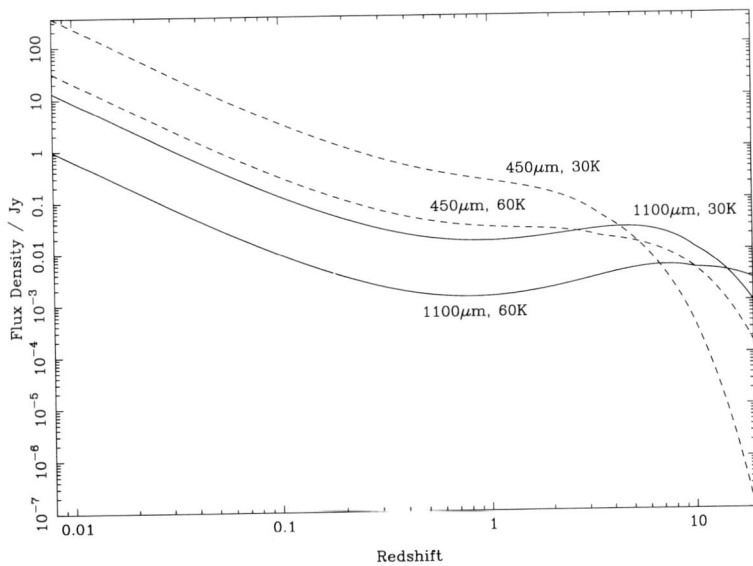

**Figure 5.** *The flux density-redshift relations for a standard dust emission spectrum from a source of far infrared luminosity $10^{13} L_\odot$ evaluated for dust temperatures of 30 and 60 K and for wavelengths of 450 and 1100 µm (Blain and Longair 1993).*

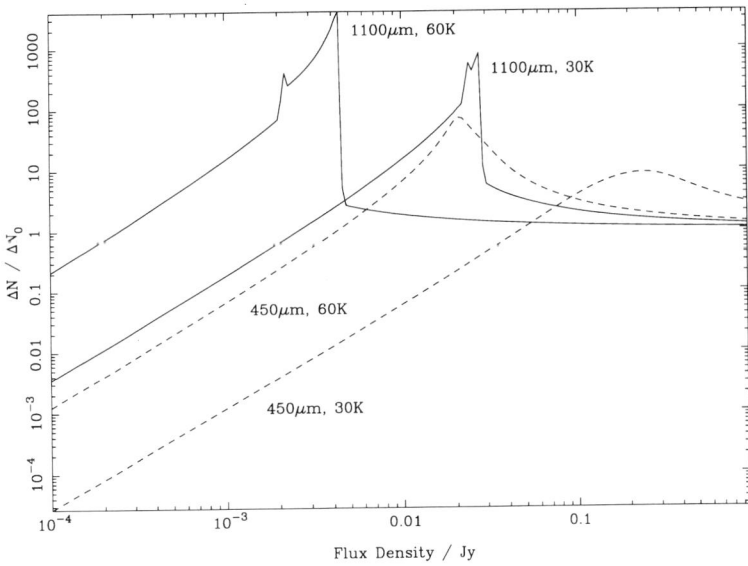

**Figure 6.** *Differential source counts normalised to the expectations of a Euclidean world model for a uniform distribution of standard dust sources at temperatures of 30 and 60 K as observed at wavelengths of 450 and 1100 µm. The bolometric luminosity of the dust source is assumed to be $10^{13}$ $L_\odot$ (Blain and Longair 1993).*

surface densities of submillimetre sources at flux densities which will be accessible to instruments such as SCUBA.

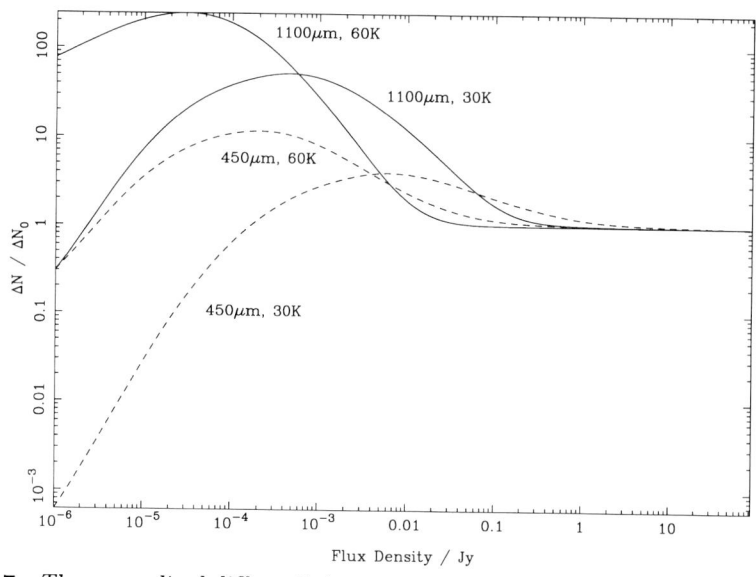

**Figure 7.** *The normalised differential source counts of all IRAS galaxies at wavelengths of 450 and 1100 µm for assumed dust temperatures of 30 and 60 K. It is assumed that the comoving number densities and luminosities of the sources are unchanged with cosmic epoch. (Blain and Longair 1993).*

## 5. THE BACKGROUND RADIATION AND GALAXY FORMATION

As part of our analysis, we have investigated the feasibility of distinguishing different models of galaxy formation by submillimetre observations. The expectations of Hot Dark Matter or "pancake" models in which large scale structures form first at relatively late epochs can be well approximated by the evolution models described at the end of the last section and result in very steep number counts of submillimetre sources. In fact, it is a quite general result that any model in which the bulk of the star and galaxy formation takes place at redshifts of the order 2–5 results in large number densities of sources at accessible submillimetre flux densities. The background radiation from such populations are strongly constrained by the fact that the spectrum of the Cosmic Microwave Background Radiation is known to be very precisely of black-body form in the wavelength interval 2500 to 500 µm. The problem can be alleviated if it is assumed that the dust grains in the large redshift star-forming galaxies are at a higher temperature, say, 60–80 K.

The currently favoured picture of galaxy formation involving Cold Dark Matter and hierarchical clustering of galaxies can be modelled using the Press-Schechter formulation for the mass function of galaxies as a function of cosmic epoch (Press and Schechter 1974). We have converted this function into a rate of coalescence of small galaxies into larger ones and assumed that each time this occurs a fixed fraction of the mass involved in the collision results in star formation with the standard dust emission spectrum. As expected, the number counts in these models are not nearly as remarkable at those expected in the strong evolution models because the galaxies are built up gradually over a long time period and become more rather than less luminous at later epochs. The number densities of submillimetre sources are expected to be much smaller in these models at the same flux density.

The millimetre background radiation from these models is, however, of considerable interest. The results of computations of the background intensity expected from these models is shown in Fig. 8 and compared with the current upper limits to the deviations of the spectrum of the Cosmic Microwave Background Radiation from a pure black-body spectrum. It can be seen that the upper limits are precisely parallel to the upper limits to the contribution which such sources could make to the background radiation. It can be seen that these models are already constrained by the remarkable precision with which the spectrum of the Cosmic Microwave Background Radiation is known to be of black body form. The point of special interest is the fact that the predicted background radiation spectrum extends well into the millimetre waveband. This background is associated with star formation in coalescing galaxies at redshifts of the order of 10 or more. The intriguing point is that these galaxies must be far infrared emitters due to the star formation which must occur as the small galaxies collide to form larger ones. It is apparent that the precise spectrum is sensitive to the exact assumptions made about the amount of star formation associated with each coalescence but, quite independent of these predictions, the millimetre and submillimetre background radiation provide a direct measure of the rates of star formation as a function of cosmic epoch.

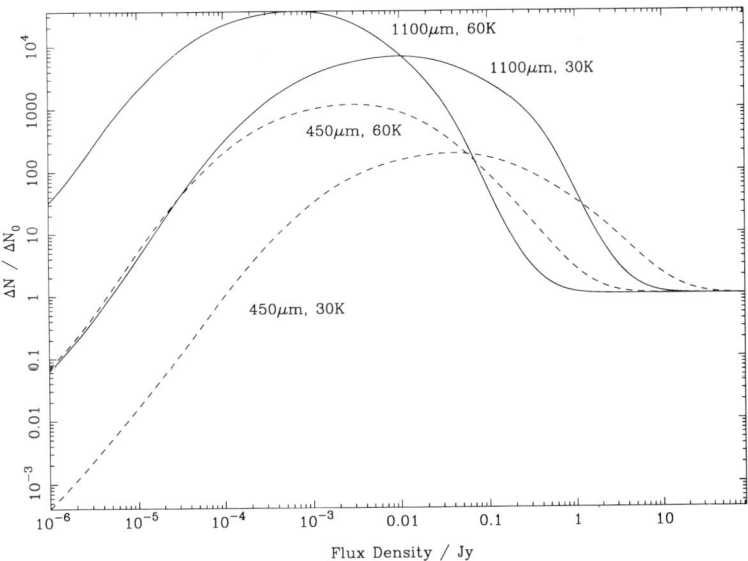

**Figure 8.** *The normalised differential source counts of all IRAS galaxies at wavelengths of 450 and 1100 μm for assumed dust temperatures of 30 and 60 K. It is assumed that the comoving number densities are unchanged with cosmic epoch but that the luminosities of the sources evolve as $(1+z)^3$ in the redshift interval $0 < z < 2$ and remain constant at 27 times the local luminosity at all redshifts greater than 2. (Blain and Longair 1993).*

The prediction of this analysis is that, at some sensitivity level, it must be possible to detect the integrated emission from star formation in young galaxies and that by measuring precisely the spectrum of the background due to these galaxies in the mil-

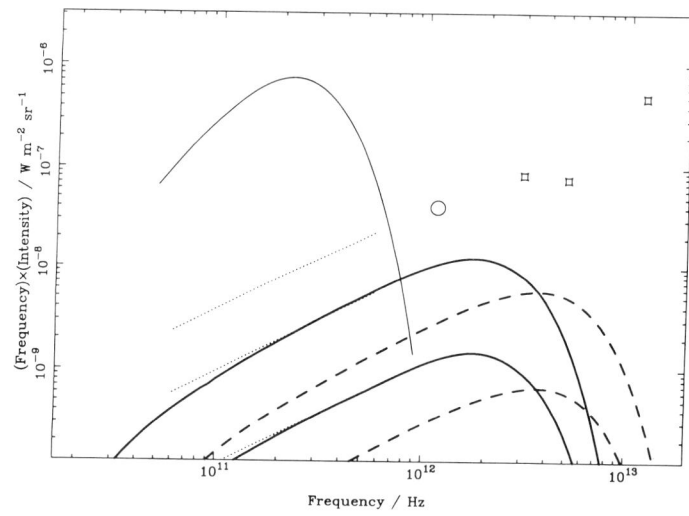

**Figure 9.** *Comparison of the integrated background emission expected from dusty merging galaxies with the intensity of the Cosmic Microwave Background Radiation and upper limits to the far infrared background radiation from the IRAS survey. The assumed temperatures of the dust grains are 30 K (solid lines) and 60 K (dashed lines). In each case, the upper curve has been normalised to the maximum allowable density of metals at the present epoch. The lower curve corresponds to about one tenth that density of metals. The dotted lines correspond to 1%, 0.25% and 0.1% of the maximum intensity of the Cosmic Microwave Background Radiation. (Blain and Longair 1993).*

limetre and submillimetre wavebands, the rate of star formation at very large redshifts can be read off directly. These observations would be of the utmost cosmological importance. The magnificent COBE spectrum of the background is already constraining these models but with further increase in sensitivity, the young galaxies must make their presence known.

## REFERENCES

Blain, A. W. and Longair, M. S. (1993). *Mon. Not. R. astr. Soc.*, (in press).
Bridle, A. H. (1967). *Mon. Not. R. astr. Soc.*, **136**, 219.
Bridle, A. H. and Purton, C. (1968). *Astron. J.*, **73**, 717.
Clark, T. A., Brown, L. W and Alexander, J. K. (1970). *Nature*, **228**, 847.
Cowie, L. (1988). In The Post-Recombination Universe, (eds N. Kaiser and A.N. Lasenby), 1. Dordrecht: Kluwer Academic Publishers.
Lilly, S. J. and Cowie, L. L. (1987). In *Infrared Astronomy with Arrays*, (eds. C. G. Wynn-Williams and E. E. Becklin), 473. Honolulu: Institute for Astronomy, University of Hawaii Publications.
Longair, M. S. and Sunyaev, R. A. (1971). *Uspekhi Fiz. Nauk*, **105**, 41.
Peacock, J. A. (1993). In *The Nature of Compact Objects in Active Galactic Nuclei*, NATO Advanced Study Institute, Cambridge 1992 (in press).
Press, W. and Schechter, P. (1974). *Astrophys. J.*, **1187**, 425.

Purton, C. (1966). Ph.D. Dissertation, Cambridge University.

Saunders, W., Rowan-Robinson, M., Lawrence, A., Efstathiou, G., Kaiser, N., Ellis, R. S. and Frenk, C. S. (1990). *Mon. Not. R. astr. Soc.*, **242**, 318.

Simon, A. C. (1978). *Mon. Not. R. astr. Soc.*, **180**, 429.

Toller, (1990). In *The Galactic and Extragalactic Background Radiation*, (eds. S. Bowyer and C. Leinert), 21. Dordrecht: Kluwer Academic Publishers.

Wall, J. V. (1990). In *The Galactic and Extragalactic Background Radiation*, (eds. S. Bowyer and C. Leinert), 327. Dordrecht: Kluwer Academic Publishers.

Wall, J. V., Chu. T. Y. and Yen, J. L. (1970). *Aust. J. Phys.*, **23**, 45.

Weedman, D. (1993). *The Physics of Active Galaxies*, (ed. G. Bicknell), First Stromlo Symposium, Canberra, (in press).

White, S. D. (1989). In *The Epoch of Galaxy Formation*, (eds. C. S. Frenk, R. S. Ellis, T. Shanks, A. F. Heavens and J. A. Peacock), 1. Dordrecht: Kluwer Academic Publishers.

## DISCUSSION

**B. Wang:** Malcolm, we did a calculation on the IR background radiation with evolution which has been published in *ApJ* a couple of years ago. A lesson you can learn from this exercise, I think, is that the evolution is usually coupled to the temperature. Typically the dust temperature is determined by the flux density from star formation. So if we want to put evolution in, you usually shift, I agree with you completely, the temperature from 30 K to 60 K. On the other hand, for simplest models you couple those two things and you do not have this freedom to arbitrarily adjust the evolution and the temperature.

**Longair:** I think my concern was that, in many of the models in the literature, it is difficult to disentangle all the effects that are going on simultaneously. The reason for carrying out our calculations with a single temperature was simply to make it very clear what was going on. It is very important to know the range of dust temperatures which are present in these objects. I suspect it is quite complicated and it is indeed naive to assume we can model these sources by a single dust temperature.

**C. Lonsdale:** I assume you can fit the 60 micron source counts with your merging model.

**Longair:** This is a programme which we are actively pursuing at the moment. I believe we can obtain a satisfactory fit without stretching the parameters but we will have to wait and see.

**P. Lubin:** These are all assuming a completely transparent atmosphere. What would you expect to see from Mauna Kea?

**Longair:** We have carried out all our calculations for the submillimeter windows at

450 and 1100 μm. The latter window is very good from Mauna Kea and the shorter wavelength window has 40% transparency under good observing conditions. We know that the programme is feasible, because we have already detected sources down to about the 0.1 Jansky level at 350 μm with single element bolometers. We are expecting to do even better with SCUBA. For example, we hope to be able to compensate in real time for the atmospheric fluctuations with the 91 pixel array.

**P. Lubin**: Which opacity do you expect to have at, say, 400 μm for your observing?

**Longair**: In the best weather conditions we get about 50% opacity in the 450 μm and 350 μm wavebands. The advertised figures are that about 40% transparency is obtained about 30% of the time in these wavebands. We would probably start the programme at 1 mm, where the transparency is about 95% if the sky is clear.

**S. White**: Just a comment. I believe there are several other objects at high redshifts that have recently been detected in the submillimeter including some of the highest redshift quasars. I have heard of one detection from IRAM by MacMahon and, I think, also Andreani *et al.* have detected several sources at IRAM. So it is clear you can find very high redshift things in this waveband.

# THE RADIO BACKGROUND: RADIO-LOUD GALAXIES AT HIGH AND LOW REDSHIFTS

J. A. Peacock
Royal Observatory
Blackford Hill
Edinburgh EH9 3HJ
UK

**Abstract.** This paper is in two unequal halves. After dealing with the possibility of a genuine continuum background at $\lambda \gtrsim 1$ cm, and showing that it is unlikely to arise in interesting circumstances, the remainder of the discussion is devoted to discrete radio sources, and their consequences for cosmology. Three main issues are considered: (i) what makes a galaxy radio loud?; (ii) what do we know about how the population of radio-loud galaxies has changed with epoch?; and (iii) what can observations of high-redshift radio galaxies tell us about general questions of galaxy formation and evolution? The main conclusion is that radio galaxies are remarkably ordinary massive ellipticals. The high-redshift examples are generally old and red and do not make good candidates for primaeval galaxies.

## 1. INTRODUCTION

The purpose of this review is to see what facts of cosmological interest can be dredged from wavelengths of above a few cm. In order to deal with modern research, rather than ancient history, it will be necessary to cheat a little and concentrate on the discrete-source population, rather than genuine smooth backgrounds—a strategy adopted by many other speakers at this meeting. However, to do duty to the advertised title, we begin with a few comments about what a non-discrete background might actually mean, were it to exist. Following this, the concentration will be on radio galaxies: why are they active, and how has the degree of activity changed with redshift? The final sections attempt to liberate us from the shackle of the radio waveband altogether, and to ask what general conclusions may be drawn about stellar evolution and galaxy formation from optical/IR data on high-redshift radio galaxies.

Notation: the Hubble constant, where quoted explicitly, is given in the form $h = H_0/100$ kms$^{-1}$Mpc$^{-1}$. If not otherwise specified, $\Omega = 1$ and $h = 0.5$ are assumed.

## 2. THE SMOOTH RADIO BACKGROUND

Malcolm Longair has described how the Cavendish Laboratory spent the 1960s practicing human sacrifice in order to determine the extragalactic radio-source background, with the following approximate result:

$$I_\nu \simeq 6000 \left(\frac{\nu}{1\text{ GHz}}\right)^{-0.8} \text{ Jy sr}^{-1},$$

to within an uncertainty of about 20% in amplitude and 0.1 in spectral index. This background dominates over the CMB for $\lambda \gtrsim 1$ m, and is consistent with the integrated contribution of discrete sources.

On the other hand, it is also not ruled out that a genuine continuum background might exist at up to 10% or so of the above level. What would this mean if it was really so? The hope would be to learn something about diffuse intergalactic gas, and there are two standard emission mechanisms to which we might appeal: synchrotron radiation and bremsstrahlung. The parameters available are the density of the emitting plasma, parameterised by its contribution to $\Omega$ (in the case of synchrotron radiation, the electrons would have an assumed power-law energy distribution), plus the local value of either the magnetic field, $B$ or temperature $T$—both of which should scale as $(1+z)^2$. The resultant background can then be worked out in the standard way (see Longair 1978). For synchrotron radiation, we get

$$I_\nu \simeq 10^{14} \,\Omega h \left(\frac{B}{\text{nT}}\right)^{1.8} \left(\frac{\nu}{1\text{ GHz}}\right)^{-0.8} \text{ Jy sr}^{-1}.$$

What is a plausible value for the intergalactic magnetic field? It is worth recalling that magnetic fields are very much a skeleton in the closet of cosmology, since we cannot easily rule out rather large values—which would significantly change our ideas about structure formation, for example. A nice review of the issue is given by Coles (1992); he argues that $B$ could be as large as $10^{-4}$ nT. This would allow observed magnetic fields in astrophysical sources to be made via compression, rather than dynamo effects, and would greatly alter the progress of galaxy clustering. For such a field, the observed background would be produced with $\Omega h \sim 10^{-3}$. This is an implausibly high density for a plasma with fully relativistic electrons, but it is perhaps surprising that the effect is this close to being interesting.

Turning to bremsstrahlung, one can simply try scaling old solutions for the X-ray background in which a 'low'-energy flux of around $10^{-3}$ Jy sr$^{-1}$ is produced by models with $T \simeq 10^8$ K and $\Omega h^2 \simeq 0.1$. Since bremsstrahlung emissivity scales as $T^{-1/2}$, this implies

$$I_\nu \simeq 10^3 \,(\Omega h^2)^2 \left(\frac{T}{1\text{K}}\right)^{-1/2} \text{ Jy sr}^{-1}.$$

If we ignore the difficulty in keeping plasma at such temperatures ionized, this seems the closest that the radio background is likely to get to setting constraints on Cold Dark Matter...

## 3. WHAT MAKES A RADIO-LOUD GALAXY?

Turning now to the infinitely more interesting issue of the population of discrete radio sources, we first review what is known about the causes that lead to enhanced radio emission.

For orientation, it is convenient to give a sketch of the population, ordered according to output. Define $P \equiv \log_{10} L_{21\,\mathrm{cm}}/\mathrm{WHz}^{-1}\mathrm{sr}^{-1}$; at $P \gtrsim 24$ we find the classical radio galaxies and quasars, conventionally divided roughly into Fanaroff-Riley (1974) FRII objects like Cygnus A and compact sources such as 3C273. At intermediate powers $23 \lesssim P \lesssim 24$, we find FRI sources: twin-jet objects such as 3C31, often lurking in clusters. At $P \lesssim 24$, we find all the rest of astronomy: 'radio-quiet' quasars, starburst galaxies and normal galaxies. We shall be concerned here with the bona-fide radio galaxies having $P \gtrsim 23$.

Two outstanding systematics of such galaxies have long been known: they are virtually without exception associated with elliptical/S0 galaxies, and moreover with the massive members of this class. This strong tendency for the probability of strong radio activity to increase with optical luminosity is illustrated in Figure 1. It is however interesting that this figure conceals a more complex behaviour noted by Owen & White (1991). They showed that the more powerful FRII sources are actually *less* likely in the most massive galaxies—*i.e.* the FR transition shifts to higher radio power at higher optical power. This may indicate an influence of the local density on the ability of a radio jet to remain stable (see Prestage & Peacock 1989). Nevertheless, the increased tendency of more massive galaxies to produce sources of FRI output or above is not in conflict with this interesting discovery.

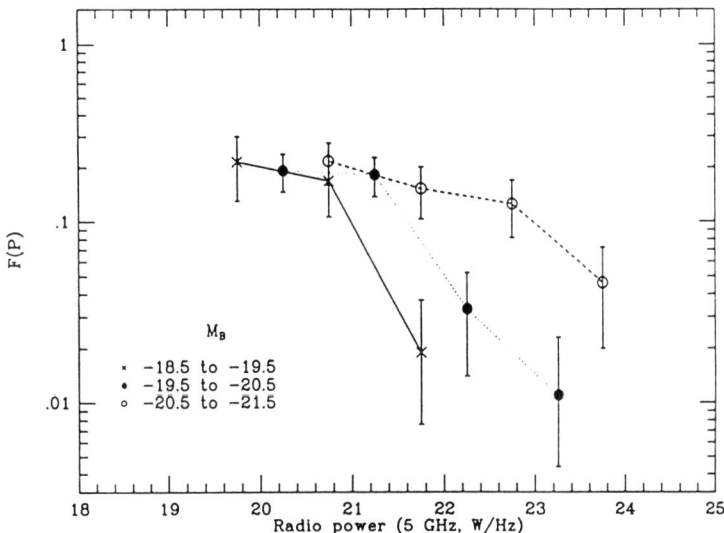

**Figure 1.** *The differential probability distribution of radio power (for $h = 1$), for different bins in optical luminosity, taken from Sadler, Jenkins & Kotanyi (1989). Above $P = 21 - 22$, the probability of radio emission is a very rapidly rising function of optical luminosity.*

The rather narrow spread in stellar luminosity for radio galaxies has long been known, and is perhaps best illustrated in the infrared Hubble diagram (Lilly & Longair 1984), which displays an rms of only 0.4 mag. The average absolute magnitude is somewhat brighter than for normal ellipticals. The most direct way of demonstrating this is not to rely on local samples, where powerful radio galaxies are rare, but to turn to a direct comparison at intermediate redshifts. Aragón-Salamanca et al. (1991) give $K$-band data on ellipticals in the A370 cluster at $z = 0.37$, from which a Schechter $K^* = 16.3$ for the ellipticals can be determined. Lilly & Longair give $K = 15.2$ for the mean radio galaxy at this redshift, but this is in a $10''$ diameter aperture, whereas the cluster data are in $4''.8$ apertures. The aperture correction at this radius is well modelled by $L \propto r^{0.5}$, which introduces a small (0.4 mag.) correction, and leads to the conclusion

$$\langle L_{\rm RG} \rangle \simeq 1.9 L_{\rm E}^*.$$

Both the large size of this mean luminosity and its small dispersion may be understood quantitatively as empirical manifestations of the strong trend of radio activity with optical luminosity. If we say that the probability of a galaxy hosting a strong radio AGN is $P \propto L_{\rm opt}^\beta$, then multiplying this rising power law into the exponential truncation of a Schechter function for the elliptical population as a whole gives roughly the observed mean luminosity and scatter if $\beta \simeq 4 - 5$. This may also go a good deal of the way towards explaining the dominance of elliptical hosts: the $L*$ values for elliptical galaxies tend to be a few tenths of a magnitude brighter than for spirals, a gap which may be stretched to as much as a magnitude if we allow for typical bulge-to-disk ratios to obtain the $L^*$ ratios between ellipticals and spiral bulges. We would then predict $N_{\rm E}/N_{\rm S} \simeq 2.5^\beta \sim 10^2$. In other words, massive ellipticals dominate powerful radio sources because only they have the exceptionally deep potential wells needed for the most active radio AGN. This is far from the whole story: first, any possible spiral identifications for powerful radio sources must be more at the $\lesssim 10^{-3}$ level; second, the whole reasoning rests on the strong $L^\beta$ trend which remains unexplained. There is still a major puzzle here.

Are there any other distinguishing features of radio galaxies? Almost all other peculiarities can either be traced directly to the effect of the AGN (such as the strong narrow emission lines), or to the peculiarity of high mass already discussed. It would be important to know if there were any systematic differences between those galaxies that turn on a radio active nucleus and those that do not—but there is no strong evidence for any such difference. Various suggestions have been made, but these have usually turned out to be small and subtle effects, whose reality generates controversy.

For example, about a decade ago it was suggested that radio-loud galaxies were redder by about 0.03 mag. in $B - V$ (Sparks 1983), rounder (Disney & Sparks 1984), more rapidly rotating (Jenkins 1984) and in denser environments (Sparks et al. 1984) than their radio-quiet counterparts of the same optical luminosity. Sparks et al. (1984) argued that these trends could be understood within a single picture of fuel gathering in potential wells, with the deeper wells being more successful at generating radio activity. However, in subsequent years the picture has become somewhat more complicated as further data have accumulated. For example, Heckman et al. (1985) found that the suggestion of excess rotation was due to a few incorrect measurements in the compilation used by Jenkins. Smith & Heckman (1989) found a normal distribution of axial ratios and claimed that galaxies were bluer—sometimes by as much as 0.2 mag. in $B - V$. Finally, Smith & Heckman (1990) found environments consistent with those of radio-quiet ellipticals. Part of the problem here may be that any peculiarity may be a function of radio power, so that different studies can yield different answers unless they use the same definition of radio-loudness. Also, the range of properties of radio-quiet ellipticals is large and diverse; misleading conclusions may be reached unless there is a large and complete comparison sample. What is needed is a large sample of radio-loud ellipticals whose properties can be compared to a radio-quiet set matched in optical luminosity and redshift.

In the meantime, claimed peculiarities of radio ellipticals need to be treated with caution. Two properties which are presently in this provisional class are the suggestion that radio ellipticals have low-level isophote distortions indicative of merging (Smith & Heckman 1989), and the question of dynamics. Smith, Heckman & Illingworth (1990) found that radio-loud ellipticals lie on the 'fundamental plane' in size/luminosity/velocity dispersion space, but there are some suggestions that they may occupy a different region of the plane—being brighter at a given velocity or size (Sansom, Wall & Sparks 1987; Romanishin & Hintzin 1989). It will be interesting to see how these suggestions hold up. We certainly badly need some clear set of clues as to what triggers these objects.

## 4. LUMINOSITY FUNCTIONS

Now consider what we know empirically about the abundance of radio AGN at high redshift, and what constraints this information may set on models of structure formation.

### 4.1 Observational results

No significant new datasets relevant to the luminosity function of powerful radio sources have been published since the study of the RLF published by James Dunlop and myself in 1990. This was based on nearly-complete redshift data on roughly 500 sources down to a limit of 100 mJy at 2.7 GHz, plus fainter number-count data and partial identification statistics.

The main conclusions of this study were firstly to affirm long-standing results (Longair 1966; Wall, Pearson & Longair 1980) that the RLF undergoes differential evolution: the highest luminosity sources change their comoving densities fastest. Nevertheless, because the RLF curves, the results can be described by a model of pure luminosity evolution for the high-power population, in close analogy with the situation for optically-selected quasars (Boyle et al. 1987). The characteristic luminosity in this case increases by a factor $\simeq 20$ between the present and a redshift of 2. Similar behaviour applies for both steep-spectrum and flat-spectrum sources, which provides some comfort for those wedded to unified models for the AGN population. There is a remarkable similarity here to the evolution of 'starburst' galaxies, distinguished by blue optical-UV continua and strong emission from dust which make them very bright in the IRAS 60-$\mu$m band. It has been increasingly clear since the work of Windhorst (1984) that such galaxies make up a substantial part of the radio-source population below $S \simeq 1$ mJy. The evolution of these objects at radio wavelengths and at 60 $\mu$m is directly related because there exists an excellent correlation between output at these two wavebands. Rowan-Robinson et al. (1993) have exploited this to investigate the implications of IRAS evolution for the faint radio counts. They find good consistency with the luminosity evolution $L \propto (1+z)^3$ reported for the complete 'QDOT' sample of IRAS galaxies by Saunders et al. (1990).

Were it not for the fact that some populations of objects show little evolution (e.g. normal galaxies in the near-infrared: Glazebrook 1991), one might be tempted to suggest an incorrect cosmological model as the source of this near-universal behaviour. The alternative is to look for an explanation which owes more to global changes in the

Universe than in the detailed functioning of AGN. One obvious candidate, long suspected of playing a role in AGN, is galaxy mergers; Carlberg (1990) suggested that this mechanism could provide evolution at about the right rate (although see Lacey & Cole 1993). Why the evolution does not look like density evolution is still a major stumbling block, but it seems that we should be looking at this area quite intensively, given that mergers have been implicated in both AGN and starbursts, and that there may be some evidence for their operation from the general galaxy population (Broadhurst, Ellis & Glazebrook 1992).

However, it is unclear how much emphasis should be placed on this apparent universality; particularly, limited statistics make it uncertain just how well luminosity evolution is obeyed. For example, Goldschmidt et al. (1992) have produced evidence that the PG survey is very seriously incomplete at $z \lesssim 1$; if confirmed, this would imply that the evolution of quasars of the very highest luminosities is *less* than for those a few magnitudes weaker. Furthermore, the QDOT database was afflicted by an error in which 10% of the galaxies were assigned incorrectly high redshifts (Lawrence, private communication); this will probably weaken the IRAS degree of evolution. It may well be that the degree of unanimity described above will prove spurious, and that we will be left with the unsurprising situation that a complex phenomenon like AGN evolution can only be described simply when the samples are too small to show much of the detail.

### 4.2 Redshift cutoff and interpretation

At higher redshifts, the uncertainties increase as the data thin out, but there is evidence that the luminosity function cannot stay at its $z = 2$ value at all higher redshifts. The form of this 'redshift cutoff' is uncertain: we cannot at present distinguish between possibilities such as a gradual decline for $z > 2$, or a constant RLF up to some critical redshift, followed by a more precipitous decline. We therefore present a 'straw man' model designed to concentrate the minds of observers, in which the luminosity evolution goes into reverse at $z \simeq 2$ and the characteristic luminosity retreats by a factor $\simeq 3$ by $z = 4$ (Figure 2).

This model predicts the following fraction of objects at $z > 3.5$ as a function of 1.4-GHz flux-density limit: 0.5% at 100 mJy; 3% at 1 mJy. Without some form of cutoff, these numbers would be about a factor of 5 higher. The reason for the increased ease of detecting a cutoff at low flux density is that the RLF is rather flat at low powers; for $\rho \propto P^{-\beta}$ and $S \propto \nu^{-\alpha}$, we expect $dN/dz \propto (1+z)^{-\beta(2+\alpha)-1/2}$. Steep spectra and a steep RLF thus discriminate against high redshifts, but at low powers the flatter RLF helps us to see whatever high-$z$ objects there are more easily. It should be relatively easy to test for the presence of a cutoff on the basis of these predictions. This is especially true at low flux densities (see Figure 3). Here, we still sample the flat portion of the RLF even at high redshift, and so the predicted numbers of high-redshift sources is large without a cutoff—around 15% at $z > 4$ for a sample at 1 mJy.

Whether or not the redshift cutoff is real, we seem to have direct evidence that the characteristic comoving density of radio galaxies has not altered greatly between $z \simeq 4$ and the present. Integrating to 1 power of 10 below the break in the RLF, we find

$$\rho \simeq 10^{-6} h^3 \text{ Mpc}^{-3}.$$

Is this a surprising number? In models involving hierarchical collapse, the characteristic mass of bound objects is an increasing function of time. At high mass, the abundance

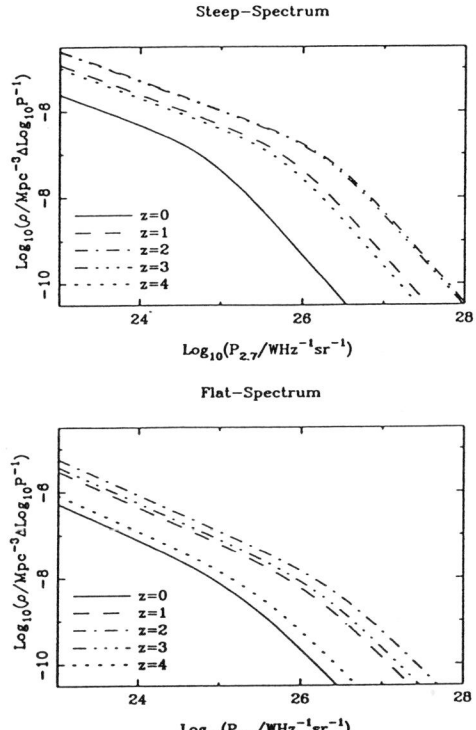

**Figure 2.** *The evolving RLF, according to the pure luminosity evolution model of Dunlop & Peacock (1990). The main features are a break which moves to higher powers at high redshift, but which declines slightly at $z \gtrsim 2$. the strength of the break and the rate of evolution are comparable for both radio spectral classes.*

of objects falls exponentially if the statistics of the density field are Gaussian. Clearly, a model such as CDM (which falls in this class) will be embarrassed if the density of massive objects stays high to indefinite redshifts. The analysis of this problem, using the Press-Schechter mass-function formalism (Press & Schechter 1974) was first given by Efstathiou & Rees (1988) for optically-selected quasars.

There are two degrees of freedom in the analysis: what mass of object is under study, and what are the parameters of the fluctuation power spectrum? For the first, Efstathiou & Rees had to construct a long and uncertain chain of inference leading from quasar energy output, to black-hole mass, to baryonic galaxy mass, to total halo mass. For radio galaxies, things are much simpler, because we can see the galaxy directly. Infrared observations imply that, certainly up to $z = 2$, the stellar mass of radio galaxies has not changed significantly. At low redshift, there is direct evidence that the mass of radio galaxies exceeds $10^{12}$ $M_\odot$, so it seems reasonable to adopt this value at higher redshift. Figure 4 shows the Press-Schechter predictions for two COBE-normalized CDM models. The low-$h$ model which fits the shape of the galaxy-clustering power spectrum (Peacock 1991) intersects the observed number density at low-ish redshifts (7–8), whereas the 'standard' $h = 0.5$ model with its higher degree of small-scale power predicts many more objects. This is clearly only a suggestive coincidence at present, but

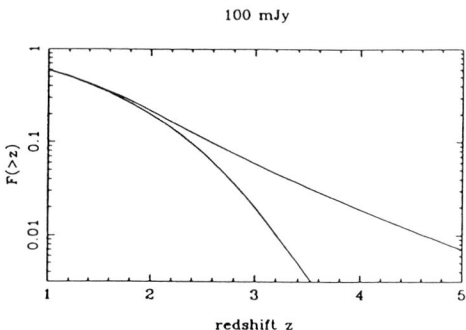

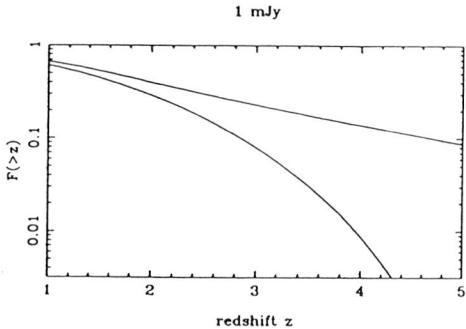

**Figure 3.** *A plot of the integral redshift distributions predicted for two samples limited at 1.4-GHz flux densities of 100 mJy and 1 mJy. The upper line shows a prediction for a luminosity function which is held constant for $z \gtrsim 2$; the lower line shows the prediction of the 'negative luminosity evolution' model of Dunlop & Peacock (1990).*

it is clearly interesting that the model which most nearly describes large-scale structure also predicts that the formation of massive objects should occur near the point at which we infer a lack of high-$z$ AGN.

### 4.3 Black-Hole abundances

In the spirit of this meeting, it is probably important to concentrate on integrated properties of the radio-source population. One important feature of this sort is the relic density of black holes deposited by the work of past AGN. This is something which has been discussed extensively for radio-quiet quasars, but which has not been given so much attention in the radio waveband alone. The advantage of doing this is that, as discussed above, we have a rather good idea of which galaxies host radio-loud AGN, and therefore we know where to look for any debris from burned-out AGN. The basic analysis of this problem goes back to Soltan (1982). He showed that the relic black-hole density may be deduced observationally in a model-independent manner, as follows.

The mass deposited into black holes in time $dt$ by an AGN of luminosity $L_\nu$ is

$$d[M_\bullet c^2] = \epsilon^{-1} g\,[\nu L_\nu]\,dt,$$

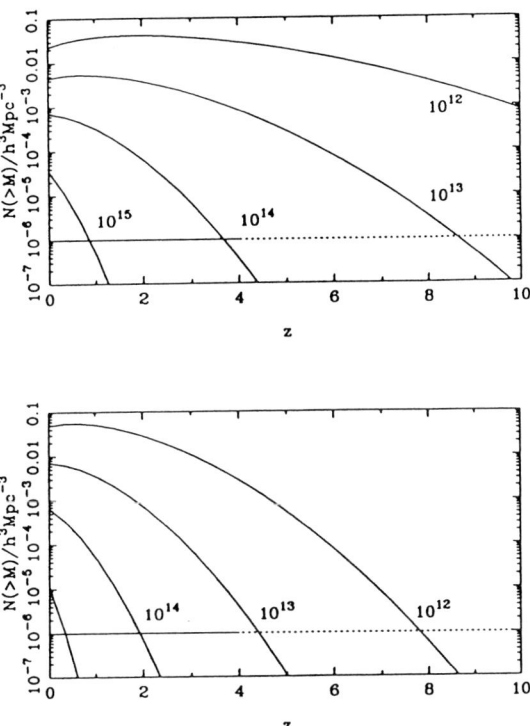

**Figure 4.** The epoch dependence of the integral mass function in CDM, calculated using the Press-Schechter formalism as in Efstathiou & Rees (1988). The normalization is to the COBE detection of CMB fluctuations. Results are shown for two Hubble constants: the 'standard' $\Omega h = 0.5$ (upper panel) and $\Omega h = 0.3$ (lower panel). Here, $\Omega h$ is merely a fitting parameter used to describe the shape of the power spectrum, and it does not presuppose a true value of the Hubble constant. The vertical scaling of density with $h$ is given explicitly, and the mass values assume $h = 0.5$. The extra small-scale power in the former case means that many more massive hosts than the observed radio-galaxy number (horizontal line) are predicted, even at $z \gtrsim 10$.

where $\epsilon$ is an efficiency, and $g$ is a bolometric correction. To obtain the total mass density in black holes, we have to multiply the above equation by the luminosity function (which already gives the comoving density, as required) and integrate over luminosity. The integral can be converted to one over redshift and flux density, and the integrand depends of the observable distribution of redshifts and flux densities, so the answer is model dependent. Doing this for the Radio LF gives a much lower answer than for optically-selected QSOs, which have a much higher density:

$$\rho_\bullet = 10^{11.7}\, \epsilon^{-1} g\ M_\odot \mathrm{Gpc}^{-3} \quad \text{(QSO)}$$
$$\rho_\bullet = 10^{9.0}\, \epsilon^{-1} g\ M_\odot \mathrm{Gpc}^{-3} \quad \text{(Radio)}$$

Since we know rather well the present density of massive elliptical galaxies (e.g. Loveday et al. 1992), we may distribute half the above radio mass into ellipticals above the median radio-galaxy luminosity, with the following result for the mean hole mass:

$$\langle M_\bullet \rangle \simeq 2000\, \epsilon^{-1} g\, h^{-3}\ M_\odot.$$

What is the bolometric correction for radio galaxies? We know that the total output generally peaks in the IRAS wavelength regime, with an effective $g \sim 100$ (Heckman, Chambers & Postman 1992); this gives

$$\langle M_\bullet \rangle \simeq 2 \times 10^5 \, \epsilon^{-1} \, h^{-3} \, M_\odot,$$

which paints a rather less optimistic prospect for detection than studies based on the output of QSOs. This is because, even with such a large $g$, the actual energy radiated by radio galaxies is rather low, and this is not compensated for fully by the relative rareness of the host galaxies. The above figure is not easy to reconcile with large black-hole masses suggested for some radio AGN. For example, Lauer et al. (1992) suggest a central mass of $M_\bullet \simeq 3 \times 10^9 \, M_\odot$ for M87. Without suggesting that M87 is greatly atypical, this can always be made consistent by assuming a low enough efficiency. However, this would not fit well with the view that radio galaxies are powered via electrodynamic extraction of black-hole rotational energy (e.g. Blandford 1990); here the efficiency can be up to $\epsilon = 1 - 2^{-1/2}$. If masses of order $10^9 M_\odot$ are substantiated in several radio galaxies or radio-quiet massive ellipticals, this would be quite a puzzle. Probably the simplest solution would be to suggest that the total energy was higher than suggested by the above sum—perhaps because radio ellipticals spend part of their lives as QSOs, where the total energy output would be considerably higher for a given radio power.

## 5. HI SEARCHES FOR HIGH-REDSHIFT GALAXIES

We now turn to the question of what the radio waveband has to say about the properties of galaxies seen at high redshifts. One unique capability of radio astronomy for cosmology is the detection of neutral hydrogen via the 21 cm line. This tends to receive most attention at low redshifts via the Tully-Fisher relation and the studies of the distance scale and peculiar velocities. However, it also gives a unique way of detecting neutral gas at high redshift—even beyond the limit of $z \simeq 5$ where quasar absorption-line studies can probe. Particularly motivated by early 'pancake' theories of galaxy formation in which purely baryonic models give a supercluster-scale coherence length to the mass distribution, there have been a number of attempts over the years to use low-frequency observations to detect neutral hydrogen at high redshifts (e.g. Davies, Pedlar & Mirabel 1978 [$z = 3.3$ & $4.9$]; Bebbington 1986 [$z = 8.4$]; Uson, Bagri & Cornwell 1991 [$z = 3.3$]; Wieringa, de Bruyn & Katgert 1992 [$z = 3.3$]). These are sensitive only to rather large structures: for a Gaussian velocity dispersion $\sigma_v$, the expected flux density is

$$\frac{S}{\mathrm{mJy}} = 19.9 \left(\frac{M_{\mathrm{HI}}}{10^{14} \, M_\odot}\right) \left(\frac{\sigma_v}{10^3 \, \mathrm{km \, s^{-1}}}\right)^{-1} \frac{h^2}{D^2 \, (1+z)},$$

where $D$ is comoving distance divided by $c/H_0$—e.g. $D = 2(1 - [1+z]^{-1/2})$ in an $\Omega = 1$ model. Since sensitivities of a few mJy are typically attained, the experiment is sensitive to masses in the range $10^{14} - 10^{15} \, M_\odot$.

Most such experiments have yielded only upper limits, but the VLA experiment of Uson, Bagri & Cornwell (1991) claimed the detection of a resolved object with a peak flux density of 10 mJy. The inferred parameters of their object were

$$z = 3.397$$
$$M_{\rm HI} \simeq 10^{14}\, h^{-2}\, M_\odot$$
$$\theta \simeq 5' \simeq 1\, h^{-1}\ \text{proper Mpc}$$
$$\sigma_v \simeq 77\ \text{km s}^{-1}.$$

This experiment caused much debate, particularly the authors' claim that this was an example of a Zeldovich pancake. The characteristics of the emission are certainly hard to understand in any other way. The gas mass and size of object, together with the effective volume of space surveyed, are about right for a rich cluster of galaxies. However, in addition to the minute velocity dispersion, one would also not expect to find intracluster gas in a neutral state. In hierarchical models, it is continually shock heated by new infalling clumps of mass as structure grows. The only neutral gas would be associated with individual galaxies, producing much less massive neutral condensations (*e.g.* Subrahmanian & Padmanabhan 1993). The only possibility might be a group of unusually neutral-rich galaxies resembling the damped Lyman-$\alpha$ absorption systems seen in quasar spectra; in this context, it is worth noting that Wolfe (1993) has shown these to lie in regions of high density (at least in terms of cross-correlation with weaker Lyman-$\alpha$ emitters). In any case, it would still be necessary to appeal to the coincidence of seeing a cluster close to its turn-round time to explain why the velocity dispersion is so small (and even this does not solve things completely, since there will be a dispersion associated with substructure).

Only in models with an initial coherence length does the gas have time to cool and regain its neutrality following heating at the initial collapse of the cluster. Without attempting to turn history back to a time before dark matter, perhaps the least radical modification would involve warm dark matter with a coherence length of a few Mpc. This would in any case lead to the usual 'top-down' chain of events for galaxy formation. Since we believe that objects of cluster mass in fact mainly formed relatively recently (Lacey & Cole 1993; see also the contribution to this volume by S. White), this would have important implications for the ages of galaxies. For this reason, it is vital that the Uson *et al.* object be either confirmed or shown not to exist. Van der Kruit (private communication) suggests that the Westerbork group have indeed failed to detect it, which may cause some relief to those distressed by the above discussion. Whatever the eventual outcome, such observations will continue with increasing sensitivity and will be capable of setting interesting constraints on conditions at high redshift.

## 6. STELLAR POPULATIONS AT HIGH REDSHIFT

### 6.1 The golden age

Finally, for a line of argument that turns out to lead in completely the opposite direction—*i.e.* to galaxy formation at rather high redshift—we turn to the stellar populations in high-redshift radio galaxies. Most of the 1980s constitute a vanished age of innocence for the radio cosmologists: at this time, they were the only ones able to find galaxies at $z \gtrsim 1$ in any sort of numbers. A series of investigations established several interesting properties for these objects, in particular

(i) The well-defined $K$–$z$ relation for 3CR radio galaxies, consistent with purely passive evolution of their stellar populations, and producing 1 mag. of brightening by $z \simeq 1$ (Lilly & Longair 1984).

(ii) The large scatter in the optical-IR (Lilly & Longair 1984) and optical (*e.g.* Spinrad & Djorgovski 1987) colours of 3CR galaxies, which was interpreted as reflecting the occurrence of bursts of star formation in otherwise passively evolving objects.

(iii) Lilly (1989) argued for a two-component model in which a bluer component was superimposed onto a rather red underlying galaxy. To reproduce the spectral energy distribution of the red component (and thus of the reddest radio galaxies) required ages greater than 1 Gyr, pointing to high formation redshifts (Lilly 1989, Dunlop et al. 1989b, Windhorst, Koo & Spinrad 1986), although there was some controversy over the model dependence of the exact ages (Chambers & Charlot 1990).

(iv) Perhaps the high-water mark of this period was the discovery by Lilly (1988) of 0902+34 at $z = 3.4$ (at a time when the galaxy redshift record was 1.8). The apparently red colours of this object argued for a large enough age that the bulk of the stars must have formed at $z \gtrsim 6$—an inference of enormous importance for models of galaxy formation.

However, over the last few years a revisionist tendency has appeared—leading to all the above achievements being questioned. Even at the time, there was some doubt whether we could be sure that the above behaviour was representative of all galaxies. Fears of a radio-induced bias appeared well founded with the discovery of what has become known as the 'alignment effect': the realisation that at large redshifts ($z \gtrsim 0.8$) the optical and radio axes of many of the most powerful radio galaxies are aligned (McCarthy et al. 1987; Chambers, Miley & van Breugel 1987). Near-IR images of 3CR galaxies appeared to confirm that the infrared morphologies of these objects were in general just as peculiar as their optical morphologies (Chambers, Miley & Joyce 1988; Eisenhardt & Chokshi 1990; Eales & Rawlings 1990). These discoveries provide direct evidence of radio-induced 'pollution' of the UV-optical light of radio galaxies, and this led some authors to suggest that these sources are thus useless as probes of galaxy evolution in general (*e.g.* Eisenhardt & Chokshi 1990).

Furthermore, it has become apparent that Lilly's galaxy 0902+34 does not have the properties initially claimed. The $K$ flux is rather lower than Lilly's measurement, and a large fraction of this smaller total is contributed by the [OII] 3727Å line, which is redshifted into the $K$ window. The result is that the galaxy in fact looks very young: nearly flat-spectrum with no evidence for the presence of an old component. On this basis, and considering other similar objects at extreme redshifts, Eales et al. (1993) have argued that radio galaxies at $z \gtrsim 2$ are in effect protogalaxies observed in the process of formation.

Before accepting this remarkable reversal of conventional wisdom, however, it is worth bearing in mind that the galaxies under discussion are among the most luminous few dozen radio AGN in the entire universe (inevitably: they are the high-redshift members of bright samples with $S \sim 1-10$ Jy). In order to draw any general conclusions about galaxy formation, it is necessary to understand the effect the AGN has on the optical/IR properties of the galaxy within which it is embedded.

## 6.2 Alignments as a function of power

What is required is to be able to study the properties of galaxies with a wide range of radio powers, and this is what James Dunlop & I have attempted in some recent work (Dunlop & Peacock 1993). In order to eliminate possible confusion with any epoch dependence, we worked with a redshift band around $z \simeq 1$. At this redshift, it is relatively easy to select samples unbiased by optical selection, and the objects are bright enough that high-quality data can be obtained. We considered galaxies from two catalogues: 19 high-power 3CR galaxies; 14 low-power comparison galaxies with $S_{2.7\,\mathrm{GHz}} > 0.1$ Jy from the Parkes Selected Regions (PSR) (Downes et al. 1986; Dunlop et al. 1989a). The PSR galaxies are a factor $\simeq 20$ less radio luminous than their 3CR counterparts. Radio luminosity is the only significant difference between the radio properties of the two samples.

Our principal dataset on these galaxies is deep infrared images, taken with the $62 \times 58$ pixel InSb array camera IRCAM, on the 3.9m United Kingdom Infrared Telescope (UKIRT), with the camera operating in the 0.62-arcsec/pixel mode. From these images, we investigated the extent of the the alignment effect at $z \simeq 1$. To avoid subjective factors, the infrared position angles were determined automatically by using the moments of the sky-subtracted flux within some circular aperture. We decided to vary the diameter of the aperture to adapt to the size of the radio source, because there are virtually no examples of optical or IR emission extending beyond the radio lobes. If the diameter of the radio source lay between 5 and 8 arcsec, an aperture equal in diameter to the radio source was used. If the radio source was greater than 8 arcsec in diameter, an 8 arcsec diameter was used (larger apertures generally contain foreground objects). If the radio source was smaller than 5 arcsec in diameter, a 5-arcsec diameter was used.

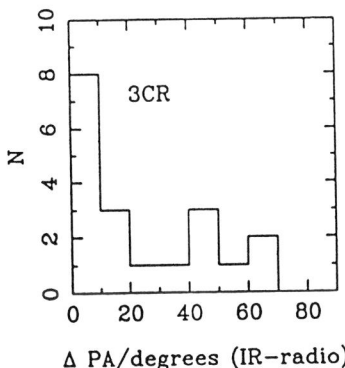
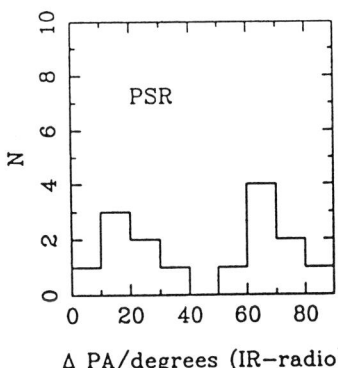

**Figure 5.** Histograms of (IR Radio) position angle differences for the 3CR and PSR samples. The clear difference seen here is completely robust to different methods for determining position angles. It is related to the fact that the PSR galaxies are also rounder, and generally lacking in an extended aligned component of blue light.

Figure 5 shows the resulting IR-radio alignment histogram for the 3CR and PSR subsamples. The infrared alignment effect is extremely obvious in our data for the 3CR galaxies, which appears to contrast with the conclusions of Rigler et al. (1992). Much of the apparent discrepancy arises from the fact that we have a larger sample. Position angles for objects in common generally agree well, but with some exceptions which are due to different methods of analysis; Rigler et al. (1992) sometimes use a large aperture where their position angle is affected by companion objects. In contrast to the 3CR sub-sample, there is no evidence of any significant alignment between the infrared and radio morphologies of the PSR galaxies. This result is very robust and quite obvious given the images: the PSR galaxies are rounder, with generally little sign of the disturbance evident in many of the 3CR images.

This argues in favour of the two-component model advanced by Lilly (1989) and Rigler et al. (1992). In this, the underlying galaxy is round, but there is a component of variable amplitude which is elongated along the radio axis, and it is this which leads to the alignment. Our data demonstrate that the strength of this component correlates well with radio power, as is perhaps not so surprising in retrospect. Certainly, several models for the production of this light exist that predict a correlation with radio power (scattering, induced star formation, inverse Compton emission—see e.g. Daly 1992 for a review). We shall not be concerned here with having to plump for a specific model, but it is worth noting that evidence is starting to mount in favour of the explanation in terms of scattering from a hidden blazar. The main argument in this direction is the measurement of polarization with E-vector perpendicular to the radio axis. The first measurements of this effect gave very low percentage polarizations, implying that this could not be the dominant mechanism. However, with better resolution, imaging polarimetry is now producing polarized fractions of $\gtrsim 20\%$ in the outer parts of strongly aligned galaxies (Jannuzi & Elston 1991; Tadhunter et al. 1992; Cimatti et al. 1993). Given geometrical dilution, it now seems plausible that the aligned component results from scattering in at least some objects.

## 6.3 Colours and ages of radio galaxies

Having seen that the extent of the aligned component scales so dramatically with radio power, we now look for other optical/infrared properties which correlate with power. Given that the aligned component is often bluer than the nucleus of the galaxy, we should certainly expect to see some correlation between colour and power. A useful way of quantifying the degree of UV activity was introduced by Lilly (1989). He assumes that the observed spectrum of a radio galaxy arises from a combination of two distinct components—an 'old' population with a well-developed 4000Å break, and a 'young' flat-spectrum component. This simple model can be fitted to the observed colours by varying one parameter. This is $f_{5000}$: the fractional contribution of the flat-spectrum component to the galaxy light at a rest wavelength of 5000Å. This method can also be used with some success to estimate the redshift for objects which lack spectroscopy (see Lilly 1989; Dunlop & Peacock 1993). Some of the PSR objects had their redshifts estimated in exactly this way: the redder objects with low $f_{5000}$ also have low levels of emission-line activity and so are of course the hardest spectroscopic targets.

This procedure is illustrated in Figure 6. For the 'old' or 'red' component we chose to adopt a spectrum capable of producing the reddest colours seen in radio galaxies at $z \simeq 1$ (e.g. 3C65); in practice this was achieved using the spectrum produced by a stellar population of age 10 Gyr in an updated version of the models of Guiderdoni & Rocca-Volmerange (1987). For the 'young' or 'blue' component, we decided to adopt a power-law spectrum ($f_\nu \propto \nu^{-\alpha}$) with a spectral index $\alpha = 0.2$. This choice of spectrum can be justified at two different levels. First, the exact value of $\alpha$ was chosen in the spirit of scattered quasar light; Barvainis (1990) concluded that the mean value for the optical spectral index in high luminosity quasars (i.e. those whose optical spectra are essentially uncontaminated by a host galaxy contribution) is $\alpha = 0.2$. Second, empirically, this form of spectrum is an excellent representation of the approximately flat $f_\nu$ optical-UV continuum actually observed in high-redshift radio galaxies.

In Figure 7 we show the quantitative relation between this definition of UV activity and radio power. Radio power and $f_{5000}$ appear to be strongly correlated (no PSR

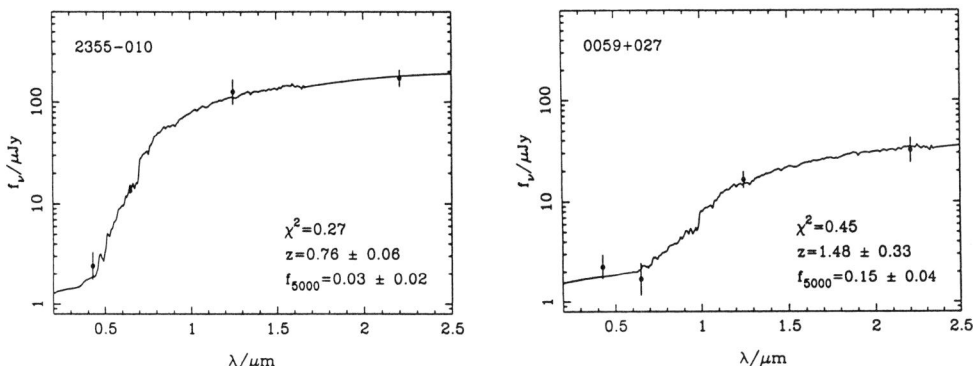

**Figure 6.** Two examples of the spectral fitting used to determine estimated redshifts and $f_{5000}$, the relative contribution of the flat-spectrum component. The 'red' component is the spectrum produced by a 1-Gyr 'Burst' model of galaxy evolution at an age of 10 Gyr. The blue component is a power-law with spectral index $\alpha = 0.2$ ($f_\nu \propto \nu^{-\alpha}$), the mean optical spectral index found for quasars by Barvainis (1990). 2355-010 is a red radio galaxy with only a very small value of $f_{5000}$, while 0059+027 is one of the bluer galaxies in the PSR sample.

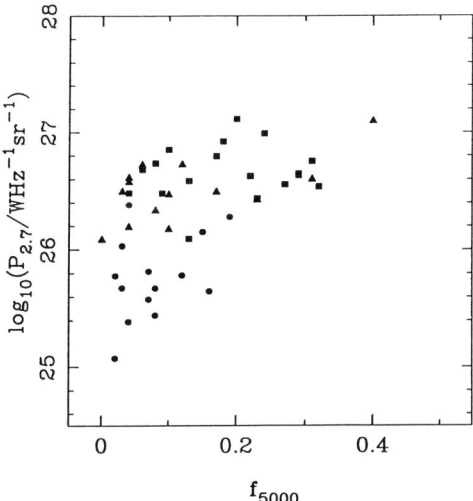

**Figure 7.** Radio power, $P_{2.7\,GHz}$, versus $f_{5000}$ for the combined 46-source 3CR/1-Jy/PSR sample. 3CR sources are shown as squares, 1-Jy sources as triangles, and the PSR sources as circles. Notice that the correlation is mainly in the sense of setting an upper limit to $f_{5000}$ at given power.

galaxy has $f_{5000} > 0.19$ whereas more than half the 3CR galaxies have $f_{5000} > 0.20$). This result contrasts sharply with that of Lilly (1989), who reported that in his combined 3CR and 1-Jy sample there was no significant correlation between $f_{5000}$ and $P_{408\,MHz}$. The origin of the difference appears to be an error in Lilly's calculation of radio luminosity. An interesting aspect of the relation with power is that all sub-samples appear to possess a range of $f_{5000}$ values, but with power apparently setting the upper limit in $f_{5000}$. This suggests the existence of a second parameter which determines the actual level of UV light—see Dunlop & Peacock (1993) for further discussion.

For the present, the point to emphasise is that this diagram provides a quantitative definition of a radio-quiet galaxy. At least at $z \simeq 1$, any galaxy with $P_{2.7} \lesssim 10^{25.5}$ WHz$^{-1}$sr$^{-1}$ (for $h = 1/2$) has a negligibly small level of UV activity. There have been some suggestions that UV activity and alignments are functions specifically of redshift, but there is little evidence that this is anything other than a reflection of the above trend in a flux-limited sample. Until proven otherwise, the natural null hypothesis is that galaxies below this power level at higher redshifts also reflect the properties of the general population of massive ellipticals.

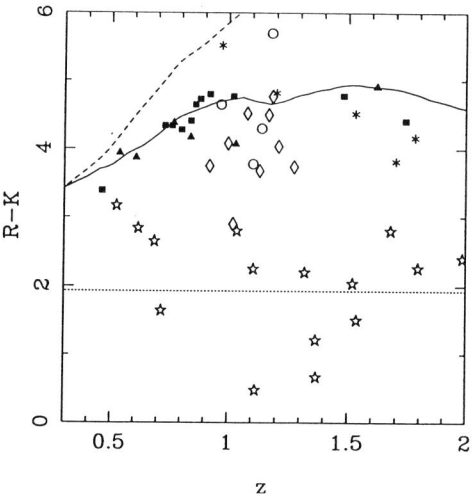

**Figure 8.** Comparison of the $R - K$ colours of the PSR galaxies (solid squares and triangles) and the 3CR galaxies in the subsample (open circles and diamonds). PSR galaxies with measured redshifts are denoted by solid triangles, those with estimated redshifts by solid squares. 3CR galaxies whose K-band morphologies are aligned with $15°$ of the radio axis are denoted by diamonds, and the remainder by open circles. Also shown are five 1-Jy galaxies (from Lilly 1989) (asterisks), and all spectroscopically confirmed quasars with $0.5 < z < 2.0$ in the PSR sample (stars). The dashed line shows the effect of simply k-correcting the spectrum. The solid line shows a very old ($z_f = 50$, $\Omega_0 = 0$, $H_0 = 50$ kms$^{-1}$ Mpc$^{-1}$) UV-hot model of elliptical galaxy evolution (Rocca-Volmerange 1989).

In Figure 8 we compare the $R - K$ colours of the PSR and 3CR galaxies. Several other objects which are not part of our PSR and 3CR subsamples have been included here for comparison purposes. These are (i) the very red 3CR galaxy 3C65, (ii) the five 1-Jy galaxies with measured redshifts for which $r - K$ colours are given by Lilly (1989), and (iii) all spectroscopically confirmed quasars with $0.5 < z < 2.0$ in the Parkes Selected Regions sample for which $R - K$ colours exist (Dunlop et al. 1989a).

This diagram displays a number of important features. First, with the obvious exception of 3C65, the PSR galaxies are consistently redder than the 3CR galaxies; moreover, the PSR galaxies display remarkably little dispersion in their optical-infrared colours. This is well consistent with the findings of Rixon, Wall & Benn (1991) at lower redshift: they found the rest-frame colours of radio ellipticals at $z < 0.3$ to be constant to within a few hundredths of a magnitude. In contrast, the 3CR galaxies scatter downwards from the well-defined PSR locus towards the region of colour space occupied by the PSR quasars (the very red galaxies 3C65 and 1129+37 appear to be exceptional). Of the six 3CR galaxies with $R - K \leq 4.0$, all but one (3C252) have $K$-band morphologies clearly aligned with their radio axes.

The homogeneity of the PSR galaxies, along with the lack of any dramatic alignment effect in the redder galaxies, suggests instead that the true optical-infrared colour of a radio-quiet elliptical at $z \simeq 1$ is actually $R-K \simeq 4.8$. Values of $f_{5000} \simeq 0.05$ might be a feature of most elliptical galaxies at $z \simeq 1$. This is certainly consistent with the results of Aragón-Salamanca et al. (1993). From optical/IR photometry of clusters of galaxies up to $z = 0.8$, they conclude that ellipticals (mainly radio-quiet) in the highest-redshift clusters are slightly bluer than present-day ellipticals. On the assumption that these galaxies formed in a single burst, their data allow the epoch of formation to be as low as $z = 2$. However, the radio-selected samples extend the range still further. Although the above discussion has concentrated on the situation at $z \simeq 1$, the PSR sample contains a number of galaxies inferred from colour-estimated redshifts and from the $K-z$ relation to have $z \simeq 2$. These also are apparently old and red, with $R - K \simeq 4 - 5$. If this is taken to imply a minimum age of $1h^{-1}$ Gyr, the formation redshift is pushed out to between 3.3 and 7.2, depending on $\Omega$. Note that this is the epoch at which the whole galaxy must be assembled: ellipticals cannot have been assembled from many small clumps after star formation had ceased (Bower, Lucey & Ellis 1992). It will be fascinating to pursue this line of argument in mJy samples, where we may hope to find 'normal' radio galaxies at $z > 3$. If these are still red, the consequences for galaxy formation models will be radical indeed.

## 7. CONCLUSIONS

This review has given a brief summary of the properties of galaxies as viewed in the radio background. In conclusion, it is worth emphasising three points:

(i) Although some factors such as galaxy mass and Hubble type strongly dispose a galaxy to host a radio-loud AGN, we still have no definite understanding of why this should be so. Other 'distinguishing marks' of radio galaxies might be helpful in this process, but few if any are clearly established.

(ii) With certain exceptions (such as the situation at $z \gtrsim 2$), we have a good statistical description of how the abundance of radio AGN evolves. Again, though, we are very far from understanding why active nuclei found it so much easier to function at high redshift.

(iii) High-redshift radio galaxies should probably not be thought of as in any way primaeval. If we ignore the few dozen most luminous sources in the universe, then the optical/IR properties of high-redshift radio galaxies are consistent with those of radio-quiet ellipticals. They appear to be red and old: theories in which most massive galaxies complete their star formation at $z \gtrsim 4$ are required.

# REFERENCES

Aragón-Salamanca, A., Ellis, R.S. & Sharples, R.M., 1991. *Mon. Not. R. astr. Soc.*, **248**, 128.
Aragón-Salamanca, A., Ellis, R.S. Couch, W.J. & Carter, D., 1993. *Mon. Not. R. astr. Soc.*, **262**, 764.
Barvainis, R., 1990. *Astrophys. J.*, **353**, 419.
Bebbington, D.H.O., 1986. *Mon. Not. R. astr. Soc.*, **218**, 577.
Blandford, R.D., 1990. *Active galactic nuclei*, 20th SAAS-FEE lectures.
Bower, R.G., Lucey, J.R. & Ellis, R.S., 1992. *Mon. Not. R. astr. Soc.*, **254**, 601.
Boyle, B.J., Shanks, T. & Peterson, B.A., 1987. *Mon. Not. R. astr. Soc.*, **235**, 935.
Broadhurst, T.J., Ellis, R.S. & Glazebrook, K., 1992. *Nature*, **355**, 55.
Carlberg, R.G., 1990. *Astrophys. J.*, **350**, 505.
Coles, P., 1992. *Comments on Astrophys.*, **16**, 45.
Chambers, K.C. & Charlot, S., 1990. *Astrophys. J.*, **348**, L1.
Chambers, K.C., Miley, G. & Joyce, R., 1988. *Astrophys. J.*, **329**, L75.
Chambers, K.C., Miley, G.K. & van Breugel, W.J.M., 1987, *Nature*, **329**, 624.
Cimatti, A., di Serego Alighieri, S., Fosbury, R.A.E., Salvati, M. & Taylor, D., 1993. *Mon. Not. R. astr. Soc.*, in press.
Daly, R.A., 1992. *Astrophys. J.*, **399**, 426.
Davies, R.D., Pedlar, A. & Mirabel, I.F., 1978. *Mon. Not. R. astr. Soc.*, **182**, 727.
Disney, M.J. & Sparks, W.B., 1984. *Mon. Not. R. astr. Soc.*, **206**, 899.
Dunlop, J.S., Peacock, J.A., Savage, A., Lilly, S.J., Heasley, J.N. & Simon, A.J.B., 1989a. *Mon. Not. R. astr. Soc.*, **238**, 1171.
Dunlop, J.S., Guiderdoni, B., Rocca-Volmerange, B., Peacock, J.A. & Longair, M.S., 1989b. *Mon. Not. R. astr. Soc.*, **240**, 257.
Dunlop, J.S. & Peacock, J.A., 1990. *Mon. Not. R. astr. Soc.*, **247**, 19.
Dunlop, J.S. & Peacock, J.A., 1993. *Mon. Not. R. astr. Soc.*, in press.
Eales, S.A. & Rawlings, S., 1990. *Mon. Not. R. astr. Soc.*, **243**, 1P.
Eales, S., Rawlings, S., Puxley, P., Rocca-Volmerange, B. & Kuntz, K., 1993. *Nature*, **363**, 140.
Efstathiou, G. & Rees, M.J., 1988. *Mon. Not. R. astr. Soc.*, **230**, 5P.
Eisenhardt, P. & Chokshi, A., 1990. *Astrophys. J.*, **351**, L9.
Fanaroff, B.L. & Riley, J.M., 1974. *Mon. Not. R. astr. Soc.*, **167**, 318.
Glazebrook, K., 1991. PhD thesis, Univ. of Edinburgh.
Goldschmidt, P., Miller, L., La Franca, F. & Cristiani, S., 1992. *Mon. Not. R. astr. Soc.*, **256**, 65P.
Guiderdoni, B. & Rocca-Volmerange, B., 1987. *Astr. Astrophys.*, **186**, 1.
Heckman, T.M., Illingworth, G.D., Miley, G.K. & van Breugel, W.J.M., 1985. *Astrophys. J.*, **299**, 41.
Heckman, T.M., Chambers, K.C. & Postman, M., 1992. *Astrophys. J.*, **391**, 39.
Jannuzi, B.T. & Elston, R., 1991. *Astrophys. J.*, **366**, L69.
Jenkins, C.R., 1984. *Mon. Not. R. astr. Soc.*, **207**, 445.
Lacey, C. & Cole, S., 1993. *Mon. Not. R. astr. Soc.*, **262**, 627.
Lauer, T.R. *et al.*, 1992. *Astr. J.*, **103**, 703.
Lilly, S.J. & Longair, M.S., 1984. *Mon. Not. R. astr. Soc.*, **211**, 833.
Lilly, S.J., 1988. *Astrophys. J.*, **333**, 161.

Lilly, S.J., 1989. *Astrophys. J.*, **340**, 77.
Longair, M.S., 1966. *Mon. Not. R. astr. Soc.*, **133**, 421.
Longair, M.S., 1978. *Observational cosmology*, 8th SAAS-FEE lectures.
Loveday, J., Peterson, B.A., Efstathiou, G.P. & Maddox, S.J., 1992. *Astrophys. J.*, **390**, 338.
McCarthy, P.J., van Breugel, W., Spinrad, H. & Djorgovski, S., 1987. *Astrophys. J.*, **321**, L29.
Owen, F.N. & White, R.A., 1991. *Mon. Not. R. astr. Soc.*, **249**, 164.
Peacock, J.A., 1991. *Mon. Not. R. astr. Soc.*, **253**, 1P.
Press, W.H. & Schechter, P., 1974. *Astrophys. J.*, **187**, 425.
Prestage, R.M. & Peacock, J.A., 1989. *Mon. Not. R. astr. Soc.*, **230**, 131.
Rigler, M.A., Lilly, S.J., Stockton, A., Hammer, F. & Le Fèvre O., 1992. *Astrophys. J.*, *385*, 61.
Rixon, G.T., Wall, J.V. & Benn, C.R., 1991. *Mon. Not. R. astr. Soc.*, **251**, 243.
Rocca-Volmerange, B., 1989. *Mon. Not. R. astr. Soc.*, **236**, 47.
Romanishin, W. & Hintzin, P., 1989. *Astrophys. J.*, **341**, 41.
Rowan-Robinson, M., Benn, C.R., Broadhurst, T.J., Lawrence, A. & McMahon, R.G., 1993. *Mon. Not. R. astr. Soc.*, **263**, 123.
Sadler, E.M., Jenkins, C.R. & Kotanyi, C.G., 1989. *Mon. Not. R. astr. Soc.*, **240**, 591.
Sansom, A., Wall, J.V. & Sparks, W.B., 1987. *Structure & dynamics of elliptical galaxies*, IAU Symp. no 127, ed T. de Zeeuw (D. Reidel), p429.
Saunders, W., Rowan-Robinson, M., Lawrence, A., Efstathiou, G., Kaiser, N., Ellis, R.S. & Frenk, C.S., 1990. *Mon. Not. R. astr. Soc.*, **242**, 318.
Smith, E.P. & Heckman, T.M., 1989. *Astrophys. J.*, **341**, 658.
Smith, E.P. & Heckman, T.M., 1990. *Astrophys. J.*, **348**, 38.
Smith, E.P., Heckman, T.M. & Illingworth, G.D., 1990. *Astrophys. J.*, **356**, 399.
Soltan, A., 1982. *Mon. Not. R. astr. Soc.*, **200**, 115.
Sparks, W.B., 1984. *Mon. Not. R. astr. Soc.*, **204**, 1049.
Sparks, W.B., Disney, M.J., Wall, J.V. & Rodgers, A.W., 1984. *Mon. Not. R. astr. Soc.*, **207**, 445.
Spinrad, H. & Djorgovski, S., 1987. in Hewitt A., Burbidge G., Fang L.Z., eds, Proc. IAU Symp. 124, Observational Cosmology. Reidel, Dordrecht, p129.
Subramanian, K. & Padmanabhan, T., 1993. *Mon. Not. R. astr. Soc.*, in press.
Tadhunter, C.N., Scarrott, S.M., Draper, P. & Rolph C., 1992. *Mon. Not. R. astr. Soc.*, **256**, 53P.
Uson, J.M., Bagri, D.S., & Cornwell, T. J., 1991. *Phys. Rev. Lett.*, **67**, 3328.
Wall, J.V., Pearson, T.J. & Longair, M.S., 1980. *Mon. Not. R. astr. Soc.*, **193**, 683.
Wieringa, M.H., de Bruyn, A.G. & Katgert, P., 1992. *Astr. Astrophys.*, **256**, 331.
Windhorst, R., 1984. PhD thesis, Univ. of Leiden.
Windhorst, R.A., Koo, D.C., Spinrad, H., 1986. in Madore B.F., Tully R.B., eds, NATO & Advanced Research Workshop, Galaxy distances and Deviation from Universal Expansion. Reidel, Dordrecht, p197.
Wolfe, A.M., 1993. *Astrophys. J.*, **402**, 411.

## DISCUSSION

**Van der Kruit**: Several months ago, De Bruyn tried to confirm the Uson detection at Westerbork, but he could not. De Bruyn's data were more sensitive than Uson's original observations, so I think the signal-to-noise ratio was better. I do not remember all of the details, but in Uson's field there was an absorption, which was confirmed at Westerbork, but there was also an emission, which definitely was not there—or, at least, it was well below the limits.

**Peacock**: I was not aware of these observations. Thank you.

**R. Daly**: I just wanted to comment on the alignment effect that you briefly touched on. Chambers *et al.* have HST images of 4C41.17 and the optical continuum is sitting right on the regions of radio emission, suggesting a process of inverse Compton scattering. In addition, although polarization has been detected in the extended optical emission for a few objects, in most of the sources where polarization has been detected, it is in the nuclear regions.

**Peacock**: No, that is not true! There are a number of examples with imaging polarimetry (*e.g.*, the work of Scarrott, Tadhunter). A lot of polarimetry gives a total measurement only, but that is far from proving that the polarized flux comes from the nucleus. Where extended polarization has been sought, it has been detected in the cases examined to date.

**R. Daly**: Less than half a dozen.

**Peacock**: How many do you want? That is six out of six.

**R. Daly**: Anyway, I just wanted to comment that scattering from AGN light is probably important, but other processes should also be considered. There is a lot of evidence that they are important.

**Peacock**: That is true—and what you say about inverse Compton, of course, is more important at high redshifts.

**V. Khersonsky**: I want to add a couple of points about Uson's object. Last fall, as far as I know, Frank Briggs from Pittsburgh University tried to observe this object at Arecibo. He confirmed the absorption features but he did not confirm the emission.

**F. Stecker**: If I put some of your conclusions together (let me know if I am right on this), it seems that if you take your high-$z$ cutoff $z = 2$ from your previous work with Dunlop, and combine it with some of these very old radio ellipticals, then it takes about a Gyr for an elliptical to turn on and that turning-on has nothing to do with mergers. Is that correct?

**Peacock**: Right. This is the question of relating the ages of things and their abundance at high redshift. We know roughly that the density of these objects is $10^{-6} \, h^3$ Mpc$^{-3}$ at $z = 2$. The evidence, at least in the model I presented, is that at high redshift the number density remains the same, but with a lower characteristic power. The sort of question you would like to ask is whether at high redshift there are enough objects with the mass of a galaxy to go around. This is related to the calculations of Efstathiou & Rees for explaining the abundance of high redshift quasars. They had to go through a rather uncertain chain of arguments to get the sort of mass needed for the host galaxy. How much mass do you need for a quasar? For the case of radio galaxies at low redshift, the answer can be obtained directly from observations, and infrared data imply that the stellar mass of radio galaxies has not changed up to $z = 2$. But where do you reach the point when there are not enough objects with $M \sim 10^{12} M_\odot$ to go around? The answer depends on the galaxy-clustering power spectrum. I will just briefly advertise a new compilation I have made which puts together a number of data sets into a picture of remarkable consistency. This says that you observe at large (100 Mpc) wavelengths an exact extrapolation of the power that can be read off from COBE assuming scale invariance, although you get high apparent values for the Hubble Constant or for $\Omega$. There is too much power at small scales by comparison with standard CDM, but a value $\Omega h \simeq 0.25$ fits the data. With the Press-Schechter mass-function formalism and a standard CDM you find that you hit the critical limit $10^{-6} h^3$ Mpc$^{-3}$ for the density at a redshift of, say, 7 to 8. But if you take the standard CDM with a COBE normalization, then there would be no conflict with the data until a much higher redshift. So, what I am saying is that maybe if massive objects formed around these redshifts, there is a consistency between the galaxy-clustering power spectrum, the ages of these galaxies and their abundance. I hope that this answers your question.

**F. Stecker**: Are you saying they do not turn on till $z \sim 2$?

**Peacock**: That is probably about right: the elliptical AGNs have finished turning on by a redshift of two, but that process can take a while.

**S. White**: I was going to make a comment on the first point of your conclusions. It does not seem to me very puzzling that radio sources occur in elliptical galaxies. After all, one of the main problems in accounting for an energetic phenomenon is getting a source of fuel, and the obvious source of fuel would be stellar mass loss. This will produce many of the observed properties. First of all, in a large galaxy it is much more likely that the stellar winds will direct the mass loss towards the center rather than outside, because of the large escape velocity. That will give you a strong blast towards the bright galaxies. Secondly, you have to accumulate a certain amount of fuel in the center of the galaxy before it gets unstable enough to fuel the nucleus. So, the nucleus will only be switched on some fraction of the time, and this would explain why some of the galaxies are radio loud and some other are not. The same phenomenon would also explain why the host galaxy is old, since the radio source may turn on very long after the galaxy is formed. Thirdly, if you use the standard stellar initial mass function, the mass loss rate from the stars is much higher at redshift $z = 1$ than it is today, so it is obvious why it is easier to get radio sources at $z = 1$.

**Peacock**: Well, I was trying to emphasize a general result for galaxy evolution, and the fact that this is done using radio galaxies is entirely irrelevant. I think there is a great consistency with optical data, certainly out to moderate redshifts. When you consider cluster ellipticals and try to see how their colors evolve as a function of redshift, you detect a slight amount of bluening at high redshift. But certainly you would need a formation redshift way above $z = 1$, in case of a single burst. At redshift 1, I have been arguing that the low power radio galaxies are effectually radio quiet and that is confirmed, in the sense that this agrees where it overlaps, by the pure optical data. The point is that radio galaxies allow us to make the same statement for galaxies at redshift 2: they are still old, and I do not believe that this is influenced by the fact that they are AGN.

**S. White**: But what I was saying is there is a natural explanation for the radio properties even if things were formed at high redshift.

**Peacock**: I am trying to get away from radio astronomy here. Many of the arguments you suggest have been put forward in the past (*e.g.*, Bailey & Macdonald); they are plausible, but far from being tested in any way. I am still struck by how inactive radio galaxies appear. For example, look at the results people like Gus Evrard obtain from detailed simulations of galaxy formation. I have seen him make pictures of how clusters of galaxies should appear at $z \simeq 1$, according to some variant of CDM. The thing which is sitting in the center of the clusters, which had better be an elliptical, is not as red as the observations imply. According to the simulations, it has lots of star formation in the recent past, and that is because gas has been raining in quite recently. This gas flow might very well be expected to power an active nucleus – but it would not make a radio galaxy as optically inactive as the ones we see.

**J. Wall**: John, some years back, when I was studying nearby bright ellipticals, it seemed to me and some colleagues of mine, including Bill Sparks, that the environment is very important in the decision of a galaxy whether or not to become a radio galaxy. I do not think you mentioned the environment. Would you like to comment on that?

**Peacock**: I mentioned environment in the context of Frazer Owen's correlation of Fanaroff-Riley class with galaxy mass. Given that the active nucleus in the galaxy is switched on, the environment clearly has some influence to play on what sort of radio source is produced. However, if you take for example a large-scale measurement of the environment, something like the cross-correlation amplitude or the richness of the cluster the radio source sits in, you find clusters around radio-loud galaxies of exactly the same strength as you would expect around radio-quiet ellipticals of the same stellar luminosity. There must be some low-level differences, but they are fairly subtle. It is surprising how lacking this thing is in gross systematics given what is going on in the middle.

**C. Lonsdale**: You said the evidence now is that these objects are old and red. Red may be, but could they be red because of reddening dust, rather than age?

**Peacock**: They could be. In fact, some probably are. The sharp eyed amongst you might have noticed on this diagram of a wonderfully precise colour-redshift track for old boring ellipticals, two objects which are really quite startlingly red. One of those is 3C 65; why it is that red, nobody knows. One possibility is that it has dust in it, and another possibility is that you are seeing a little bit of an AGN kicking through at long wavelengths. But what this diagram emphasizes is the overall homogeneity. If these objects have the colors they do for reasons other than stellar ages, then they have got their acts synchronized to a remarkable extent. Age is the simplest explanation.

# CONFERENCE SUMMARY

Martin J. Rees
Institute of Astronomy
Madingley Road
Cambridge CB3 0HA, UK

The talks we've heard over the last three days have themselves been comprehensive reviews. I shall therefore not attempt a complete summary, but will merely offer a few concluding impressions. This meeting is being held in honour of Riccardo Giacconi. It is therefore appropriate to start with the X-ray background, discovered by him and his colleagues more than 30 years ago. This classic work was based on only a few minutes of data from a sounding rocket. We now know that the background is mainly sources, many hundreds per square degree. The focus has now shifted, therefore, to the nature of these sources, and, in particular, why the background displays spectral features such as a "bump" at 30 keV. This feature must be sufficiently conspicuous in the spectrum of a typical source that it is not smeared out by the range in redshift of all the objects contributing the background.

What physics naturally gives rise to a temperature such that $kT = 30(1+z)$ keV? Some authors attribute the feature to a so-called "reflection bump." The idea here is that the albedo of a cold slab is depressed at high energies because of Compton recoil, and at low energies, below 20 keV, by photoelectric absorption. If half or more of the primary radiation from the source suffers internal reflection, the resultant spectrum will have a conspicuous bump in roughly the right place, if the typical redshift is $z \simeq 1-2$. Alternative ideas involve transmission or Comptonization by a thermal plasma. Radiation transmitted through a slab displays an exponential cutoff for $kT$ above $\frac{mc^2}{\tau}$. A suitable range of optical depths $\tau$ can then yield an appropriate integrated spectrum. A third possibility is that the primary emission mechanism involves a thermal pair plasma. This has a characteristic temperature which depends inversely on the compactness parameter. Provided that this parameter does not span a broad range, a standardized spectrum again results.

Until recently, spectra of individual AGNs were only available at relatively low X-ray energies. However, Ginga spectra of several objects revealed not only evidence of fluorescent Fe emission, but also a flattening towards high energies indicative of a hard X-ray "bump." Further information has come more recently from the OSSE experiment on GRO. This suggests that typical AGNs (in other words those that are not blazars) have X-ray spectra which cut off remarkably steeply above 50 keV: the cutoff is either exponential, or, if a power law, is definitely steeper than 2.2. We now know, therefore, that there are some individual AGNs whose spectra, over the entire range from 1 or 2 keV up to more than 100 keV, mimic that of the X-ray background. It is important to note, however, that, even if we believe that the X-ray background is due to sources over that entire range, there was no *a priori* reason to expect that any single source would have the same spectrum as the background. It could perfectly well have turned out that a very different mix of objects was responsible in different parts of the X-ray band. Indeed, to some extent this is true, because hard X-ray surveys are now revealing a population of sources which would appear very weak below 10 keV.

The X-ray background obviously serves as a probe of clustering on large scales at redshifts of order 2: the isotropy below a few keV already sets constraints on the number of sources and how they are clustered. It would also be important to improve the precision of the earlier HEAO-1 measurement of the dipole in the X-ray background. Provided that the microwave dipole is indeed due to our local velocity, rather than being an intrinsic effect on a cosmological scale, we would expect the X-ray dipole to point in more-or-less the same direction. Its amplitude should be similar: indeed it should be somewhat larger because, in addition to the Doppler contribution, there will also be an enhancement due to the excess of sources pulling us in that direction.

The X-ray background has been at the forefront of X-ray astronomy over the last 30 years. As Malcolm Longair reminded us, this contrasts with the situation in the radio band, where the background has been frustratingly hard to measure and has attracted relatively little interest. There are two reasons for this difference. First, in the radio band, the foreground galactic contribution is strong, rendering it difficult to isolate the extragalactic component. Second, because of differences in techniques, the strongest extragalactic radio sources were identified early on, whereas it took several years after the discovery of the X-ray background before any individual extragalactic sources were reliably identified.

In this latter respect, the radio background resembles the gamma-ray background. It now seems clear that the dominant high energy gamma-rays come from an atypical subset of AGNs whose brightest members are already identified—the so-called 'blazars,' with relativistic jets aimed almost directly towards us. The EGRET experiment has provided important data on this in the last year. It seems that the relativistic jets from blazars could be dumping most of their energy into high energy photons, with a spectrum which may extend up to TeV energies. An estimate of the total contribution of such sources to the background depends on an uncertain extrapolation of the counts towards fainter objects. However, it may well turn out that such objects are the primary contributors to the background, at least above 10 MeV.

The background in the optical band, amounting to almost 100 times the energy density in the X-ray background, is due primarily to galaxies. The UV background due to massive stars has a flat spectrum, and, as several speakers have reminded us, relates directly to the amount of heavy elements produced. Associated with the integrated light from observed galaxies, down to 25th magnitude or thereabouts, was the synthesis of a substantial fraction of the heavy elements now observed. Therefore, unless there are a lot of hidden heavy elements, most of the total UV and optical background must be accounted for by faint observed galaxies. Moreover, there cannot be a much larger background in the infrared due to starlight reprocessed by dust.

There have been heroic attempts over the years, as Dick Henry reminded us, to make direct measurements of the 1000–2000 Å UV background. However, the shorter-wavelength XUV, composed of photons above the Lyman limit in energy, is not directly observable. Nonetheless, it is worth pointing out that this particular background has very important and readily observable indirect effects. Observations of high redshift quasars tell us that intergalactic hydrogen was already ionized by $z = 5$. This requires that at least one XUV photon must already have been produced for every atom in the IGM. What are the likely sources? The first is early galaxy formation. O or B stars generate at least $10^4$ photons, mainly in the XUV, for every baryon in them. Therefore only a small fraction of nucleosynthesis would need to occur at redshifts above 5 in order to provide enough photons to ionize the IGM. The second candidate sources are the high redshift AGNs themselves. These give a hard spectrum, able to ionize helium as well as hydrogen. It is, I believe, a tenable conjecture that H is ionized by UV from OB stars in small galaxies at redshifts well above 5, but HeII is ionized much later, maybe at redshifts of 3 or 4, by AGNs. Such conjectures are testable by more careful study of the absorption spectra along the lines of sight to the high redshift quasars. The XUV background has important effects on the Lyman alpha forest, and on the so-called 'inverse effect.' There could even be interesting feedback effects from the intensity of the background on the rate of star formation.

The final band that has been discussed extensively at this meeting is the far infrared and microwave part of the spectrum, where recent advance has been spectacular. However, before turning to this it is perhaps worth mentioning that a mystery faces us as we push towards zero frequencies. Below the radio band we approach the zero frequency limit, the "DC" component of the electromagnetic background—in other words cosmic magnetic fields. These are ubiquitous, but it is still a mystery how they get started. Dynamo amplification is believed to operate on several scales, but a so-called 'seed field' is required. The origin of this 'seed' is uncertain, and it would be important in various astrophysical contexts, for instance high redshift absorption lines and the IMF in newly-formed galaxies, to have a clearer idea of how quickly magnetic fields can amplify to the stage when they become dynamically important.

We have heard the remarkable new results from the FIRAS experiment, revealing that the microwave background displays no deviations from a black body spectrum, even at levels of order $10^{-4}$. This sets strong limits on the amount of Comptonization in a hot IGM, thereby ruling out a diffuse medium that could contribute significantly to the X-ray background. There are also constraints on very early heat ejection into the universe even before recombination, as well as on far infrared emission from dust due to Population III stars, etc. It is particularly fortunate that Rashid Sunyaev is here, for it was he, with Zeldovich, who wrote most of the classic papers quantifying the microwave background distortions predicted by various cosmological effects.

We also had an update on the DMR experiment, whose first year of data revealed temperature fluctuations on scales of more than $10°$, at the level of $10^{-5}$. We also heard from Phil Lubin the preliminary results of other experiments on smaller angular scales, which may soon reveal fluctuations at similar levels. Such data provide key evidence on a phase of cosmic evolution when the density perturbations were still of small amplitude. As Niccola Vittorio explained, they already provide severe constraints on theories, when combined with the evidence on large-scale structure and streaming motions.

I find it interesting, incidentally, to compare and contrast the communal reaction to the results from these two magnificent experiments on the COBE satellite. The data from FIRAS represent an improvement of much more than a factor 100 on the precision with which the shape of the millimetre background spectrum is known. Moreover, the smallness of the distortions has important implications, ruling out many widely-discussed hypotheses about the early universe. The precision with which the present background spectrum fits a black body is surprising to many people, and certainly could not have been confidently predicted. The DMR results, in contrast, represented an improvement by a factor of 2 or 3 on the sensitivity levels achieved by earlier experiments. Moreover, the level of the detected fluctuations was not specially surprising. (Indeed, what would have been remarkable would have been the lack of fluctuations at within a factor 2 of the measured level. It would then have been hard to understand the present largescale structures in the universe, unless these had been pulled together by some process more efficient than gravitational clustering.) In this perspective, it's curious that it should have been the DMR result that attracted far greater publicity. Media interest is certainly capricious, and I hope very much that the FIRAS results will get a due share of acclaim, so that the balance is redressed.

To conclude, I would like to look forward, and highlight three areas where we can expect exciting developments in the next few years.

The first of these will undoubtedly be the further exploration of the angular fluctuations in the microwave background as a function of $\theta$. We can clearly expect either positive results, or upper limits which are low enough to be a severe constraint on theories, on a variety of angular scales. The scales of $1°$ or less are harder to interpret, because these are sensitive to the thermal history of the intergalactic medium and the nature of the last scattering. However, it is these scales which relate most directly to the clusters and superclusters which are directly observed, and which should help most to discriminate among the rival theories for galaxy formation.

The second growth area will undoubtedly be early galactic evolution. We need better data on the high redshift galaxies; we can expect better statistics on AGNs out to redshifts of 5, and even further if they exist at earlier times. We also want to know what happened during the 'dark age' between recombination and the epoch when the first bound systems formed. Clues to this will come from the optical, infrared, and X-ray background, and from indirect inferences about the XUV band. In parallel with observational progress, there will be great enhancements in the scope and realism of numerical simulations. (Indeed, as computers advance from gigaflops towards teraflops, people like myself whose cerebration is done without electronic aids at about a milliflop (or maybe two if not jetlagged) will feel increasingly marginalised unless we can change our ways.)

A third area, central to extragalactic astrophysics but which also relates to the subject of this meeting, is understanding the spectrum of different classes of AGNs. We know that these emit in all bands from the radio up to TeV gamma-rays, and probably dominate the high energy background. This is an enterprise where observations are needed using a variety of techniques, and where data from the refurbished HST is eagerly awaited.

# POSTER PAPERS

# THE LARGEST STRUCTURES: LIMITS FROM THE RADIO-SOURCE BACKGROUND

C. R. Benn
Isaac Newton Group of Telescopes
Apartado 321
38780 Santa Cruz de La Palma, Tenerife
Spain

J. V. Wall
Royal Greenwich Observatory
Madingley Road
Cambridge CB3 0EZ
UK

**Abstract.** Using Voronoi-foam modelling together with estimates of the luminosity function and its epoch dependence, we estimate the scatter which cellular structure of various scales imposes on radio source counts. Comparison with observed isotropy of source counts places a limit of $\sim 100$ Mpc on the largest scale of cellular superclustering in the galaxy distribution.

Galaxy redshift surveys provide our current of view of the universe of galaxies arranged in cellular structure, with luminous matter defining thin ($\sim 5h^{-1}$ Mpc) sheets and filaments bounding near-spherical voids of sizes up to $\sim 100h^{-1}$ Mpc. Structures on such large scales, consistent with the COBE fluctuations (Smoot et al. 1992), are hard to reconcile with the simplest Cold Dark Matter models with $\Omega = 1$. The scales are comparable to the depths of the surveys themselves; the 2-D distributions of galaxies from Schmidt surveys (*e.g.*, Maddox *et al.* 1990) sample greater volumes economically, but are at the mercy of large-scale systematic effects. In contrast, extragalactic radio-source surveys typically sample scales comparable with the size of the observable universe, and are largely free from systematic effects such as calibration and obscuration. We show here how they may be used to determine the scale of the largest structures.

## 1. MODELLING

Voronoi tessellations have proved to be useful models of the actual distribution of galaxies (*e.g.*, Icke & van de Weygaert 1991) particularly on the larger scales; the cell walls, edges and vertices of the tessellation represent sheets, strings and Abell clusters of galaxies. Thus a realistic modelling of the effect of the cellular structure on source counts may be obtained by sprinkling radio sources on the surfaces of a Voronoi tessellation in numbers approximating the estimated mean density as a function of radio luminosity and redshift, and then calculating the source counts derived from a number of randomly-positioned pencil beams through it. Such calculations (Benn 1993) indicate (Fig. 1) that there is an optimum angular diameter of survey area for seeking anisotropy: the area-to-area anisotropy will increase with the number of detected radio sources per intercepted supercluster, and for very small survey angle, this number increases with opening angle because of the increasing area of intercepted cell wall, while the number of supercluster cells intercepted increases very slowly. At larger opening angles the number of sources rises at the same rate as the number of intercepted cells, so that a large survey area gives no extra signal-to-noise. Signal-to-noise increases with angle until the survey diameter is $\sim$ cell size at the redshift $z_0$ contributing the bulk of the sources. For $z_0 \sim 1$, $100h^{-1}$ Mpc corresponds to a subtended angular diameter of $\sim 3°$, close to the angular diameter of the 5C surveys carried out with the Cambridge One Mile Telescope.

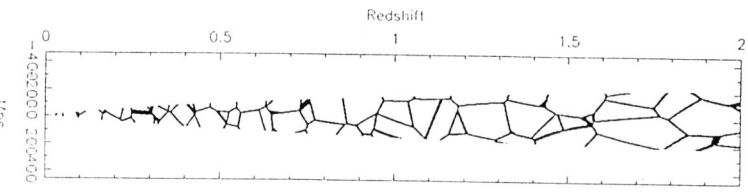

**Figure 1.** *Voronoi-foam simulation of a pencil-beam survey $5°$ in diameter, mean cell radius R = 120 Mpc, $H_0$ = 50 km/s/Mpc.*

## 2. OBSERVATIONS

Away from the Galactic plane, extragalactic radio sources show little evidence of anisotropy (*e.g.,* Webster 1976), with the exception of the few brightest and nearest ($z < 0.02$) sources which favour the plane of the local supercluster (Shaver & Pierre 1989). Indeed the modelling shows that the absence of significant anisotropy or clustering in surveys covering *large areas of sky* to flux-density limits $\sim 0.3 - 1$ Jy implies $R < 250h^{-1}$ Mpc. Grueff (1988) found hints of anisotropy in the distribution of B3 sources on scales of about 10 deg; such anisotropy would be consistent with $R \sim 100h^{-1}$ **Mpc**. A number of large surveys reaching to limits 30–100 mJy have become available over the last few years, including an NRAO 1.4-GHz survey (Condon & Broderick 1985, 1986) to 100 mJy of $\sim$ 30000 sources; and an NRAO 4.85-GHz survey to 25 mJy of $\sim$ 55000 sources (Condon, Broderick & Seielstad 1989). Our preliminary analyses show no evidence for anisotropy, in agreement with prediction.

With regard to *pencil-beam surveys* the published 5C surveys reach a surface density of $\sim 10^5$ sources sr$^{-1}$ and catalogue a total of $\approx$ 3000 sources with $S_{408} > 10$ mJy in 13 areas of $\sim 0.004$ sr distributed over $\approx 2$ sr of the Northern hemisphere. Two, 5C10 and 5C12, show marginally significant departures from randomness; the deviations are spread uniformly over the survey fields, and the optical identifications indicate that the deviations are not due to 'local' groupings of nearby ($z < 0.1$) objects. These deviations suggest $R \sim 100h^{-1}$ Mpc; otherwise the 5C data constrain $R < 100h^{-1}$ **Mpc**. The redshift distribution for some 40 objects in the 5C12 region (Wall, Rixon & Benn 1993) shows clumpiness similar to that in galaxy redshift surveys, associations of two to four objects with $\sim$ identical redshifts from 0.1 to 0.3 indicating structures on scales **10 to 30$h^{-1}$ Mpc**.

At yet higher source densities, source counts from three deep surveys carried out with the Westerbork Synthesis Radio Telescope at 1.4 GHz which reach a surface density $\sim 4 \times 10^5$ sources sr$^{-1}$ were compared by Oort (1987). For $S_{1.4} > 0.5$ mJy, no significant area-to-area differences in source counts was found, and there was but weak evidence for variation of the angular correlation function from one area to another. Comparison with the foam model again supports $R < \mathbf{100}h^{-1}$ **Mpc**.

## 3. CONCLUSIONS

Modelling with a Voronoi-foam universe anticipates the *general isotropy of the radio source counts* despite the cellular structure, and shows how this is a consequence of the small mean number of bright sources per supercluster volume. At fainter flux densities, however, traces of the galaxy anisotropy are expected to induce *small area-to-area variations in the source counts*. Moderately deep pencil-beam surveys are shown to provide the strongest constraints on the largest scale of superclustering. In fact the observed limits to isotropy of the 5C and $\mu$Jy source counts set an upper limit of $R = \sim 100h^{-1}$ Mpc to the diameter of the cell size. A deep and uniform VLA 'tiling' of the northern sky will provide ideal data to determine the scale of the largest structures in the universe.

## REFERENCES

Benn, C. R., 1993, in *Observational Cosmology*, eds. Chincarini, G. *et al.*, in press
Condon, J. J. & Broderick, J. J., 1985, *AJ*, **90**, 2540; 1986, *AJ*, **91**, 1051
Condon, J. J., Broderick, J. J. & Seielstad, G. A., 1988, *AJ*, 97, 1064
Grueff, G., 1988, *A&A*, **193**, 40
Icke, V. & van de Weygaert, R., 1991, *Q. Jl. R. astr. Soc.*, **32**, 85
Maddox, S. J., Esfstathiou, G., Sutherland, W. J. & Loveday, J., 1990, *MNRAS*, **242**, 43P
Oort, M. J. A., 1987, PhD thesis, University of Leiden
Shaver, P. A. & Pierre, M., 1989, *A&A*, 220, 35
Smoot, G. F. *et al.* 1992, *ApJ Lett.*, **396**, L1
Wall, J. V., Rixon, G. T. & Benn, C. R., 1993, in *Observational Cosmology*, eds. Chincarini, G., *et al.*, in press
Webster, A. S., 1976, *MNRAS*, **175**, 71

# BIASSES IN THE ESTIMATION OF THE UV BACKGROUND STRENGTH

Brian Espey
Department of Physics & Astronomy
University of Pittsburgh
Pittsburgh, PA 15260
USA

**Abstract.** We have examined the possibility that current estimates of measuring the strength of the UV background by means of the so-called 'inverse' or 'proximity' effect may be in error due to various systematic biasses. We conclude that the largest source of uncertainty in current estimates comes from our lack of knowledge of the systemic velocity of the host QSOs. Bases on current estimates of the mean blueshift of the emission lines of Ly$\alpha$ 1216Å and CIV 1549Å observed in samples of luminous QSOs we conclude that current data is consistent with QSOs accounting for at least 25% of the background intensity. It is possible that QSOs may account for the whole of the UV background if current models of QSO evolution and revised intergalactic opacities are used.

## 1. UV BACKGROUND ESTIMATES

Current estimates of the extraglactic UV background at intermediate and high redshifts are based upon analysis of the decrease in the number of Ly$\alpha$ absorption clouds close to the redshift of a QSO. It is generally believed that the decrease in absorbing systems is due to the increasing photoionization of the gas clouds close to the QSO continuum source and so it is possible to predict the strength of the UV background from observations of the ionizing spectrum of a QSO and the distance of each absorbing cloud from it. We have re-analysed the data of Lu *et al.* (1991) to check whether systematic errors in the determination of either the QSO ionizing spectrum, or the distance estimates are sufficient to bias current background estimates. We conclude that aside from the uncertainties in the number and spectral shape of QSOs at high redshift as well as the optical depth of the Universe to ionizing radiation, the biggest uncertainty is in our estimates of the location of Ly$\alpha$ clouds from the continuum source of an individual QSO.

We have taken the Lu *et al.* data sample and calculated how much current estimates might be in error depending on the magnitude of the blueshift of the high ionization lines of Ly$\alpha$ 1216Å and CIV 1549Å with respect to the systemic velocity—this is best illustrated by reference to Figure 1. We note that it is possible for measurements of the UV background to be in error by anywhere between $\approx 0\%$ and $\approx 72\%$ relative to the values deduced without allowance for velocity shifts. Our best-guess estimate of the velocity shift in the Lu et al. data places it somewhere in the shaded region which is equivalent to a *decreased* background estimate of 72% relative to the zero shift case. This now places the 'QSO only' models within a factor of 3–4 and possibly even within a factor of 2 of accounting for all of the ionizing photons required in an optically thick Universe (Madau 1992, Meiksin & Madau 1993). More complete details are provided in Espey (1993).

## REFERENCES

Espey, B.R. 1993. *Ap. J. (Letters)*, **411**, L59.
Lu, L., Wolfe, A.M. & Turnshek, D.A. 1991, *Ap. J.*, **367**, 1.
Madau, P. 1992, *Ap. J.*, **389**, L1.
Meiksin, A. & Madau, P. 1993, *Ap. J.*, **412**, 34.

## TABLE 1
### UV BACKGROUND STRENGTH & SOME SYSTEMATIC ERRORS

| Source of error | $J_\nu$ correction | for change of... |
|---|---|---|
| Revised Lyman limit fluxes | −11% | +18% increase in fluxes |
| Galactic extinction | +26% | $\overline{E(B-V)} = 0.103^1$ |
| Velocity shifts | −72% | $\bar{v} \approx 1450 \text{ km s}^{-1}$ |
| Choice of fitting function index | −64% | $\alpha = 0.7 \rightarrow 0.4$ |

1. Values ranged between $0.0 \rightarrow 0.48$ mags.

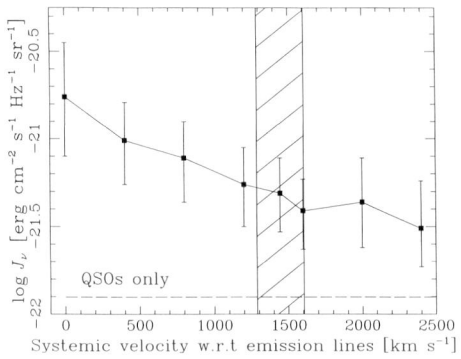

**Figure 1.** *The effect of a change in the value of the assumed offset of the emission lines from the systemic velocity. The 'best-guess' velocity lies somewhere in the shaded region which was determined from measurements of QSOs of a similar luminosity to those used in the Lu et al. sample. The horizontal dashed line marked 'QSOs only' indicates the value of $\log(J_\nu)$ expected for a Universe which is optically thick near the Lyman limit and in which QSOs are the only sources of ionizing photons (Madau 1992).*

# THE CASE FOR AN EXTRAGALACTIC ORIGIN FOR THE HIGH GALACTIC LATITUDE DIFFUSE ULTRAVIOLET BACKGROUND

Richard C. Henry and Jayant Murthy
Center for Astrophysical Sciences
Henry A. Rowland Department of Physics and Astronomy
The Johns Hopkins University
Baltimore, MD 21218

**Abstract.** A simple model is presented which can be used to predict the diffuse background that is to be expected at high galactic latitudes due to the light of galactic plane OB stars scattered from high latitude dust. When combined with a recent and highly reliable determination of the scattering pattern of interstellar grains in the far ultraviolet, this model indicates that such scattering cannot account for the continuous spectrum that is seen at high latitudes by all observers. The spectrum of the extragalactic background is $\sim$ flat longward of 1216 Å and is undetected shortward of that wavelength, suggesting an origin in hydrogen recombination radiation, although the intensity is higher than is expected in common cosmological models (Henry 1991).

There is considerable dust at high galactic latitudes; for example Hauser et al. (1984) report, from their study of IRAS cirrus observations, that $A_V = 0.1$ mag at high latitudes. (We use that value, which is probably too high by at least a factor of two, in order to be conservative in our conclusions.) Also, there are many bright OB stars in or near the galactic plane (Fig. 1) For our simple model for the scattered light of these stars, we integrate the Henyey-Greenstein (1941) scattering function

$$H(\theta) = (1 - g^2)/4\pi \; (1 + g^2 - 2g\cos\theta)^{-3/2}$$

over the back-scattering directions, $\pi/2$ to $\pi$, obtaining

$$B = 1/2 - 1/(2g) + (1 - g^2) \Big/ \left(2g\sqrt{(1+g^2)}\right)$$

for B, the fraction of scattered light that is backscattered. Our model is, then, that at high latitudes we expect a scattered intensity $S = B\,G\,a\,\tau$, where G is the local far-ultraviolet interstellar radiation field ($\sim$ 10,000 photons cm$^{-2}$ s$^{-1}$ sr$^{-1}$ Å$^{-1}$ ("units"), Henry, Anderson, and Fastie 1980), $a$ is the grain albedo ($\sim$ 0.65 at 1500 Å, Witt et al. 1992), and $\tau$ is the far-ultraviolet optical thickness of the high-galactic-latitude scattering layer. With $\tau_\lambda = 0.921\,A_\lambda$ and with $E_{1500-V}/E_{B-V} = 5.3$ (Bless and Savage 1972, for $\zeta$ Oph) and $A_V = 3E_{B-V}$, we obtain $\tau = 0.255$ (we emphasize again that this is surely much too high a value).

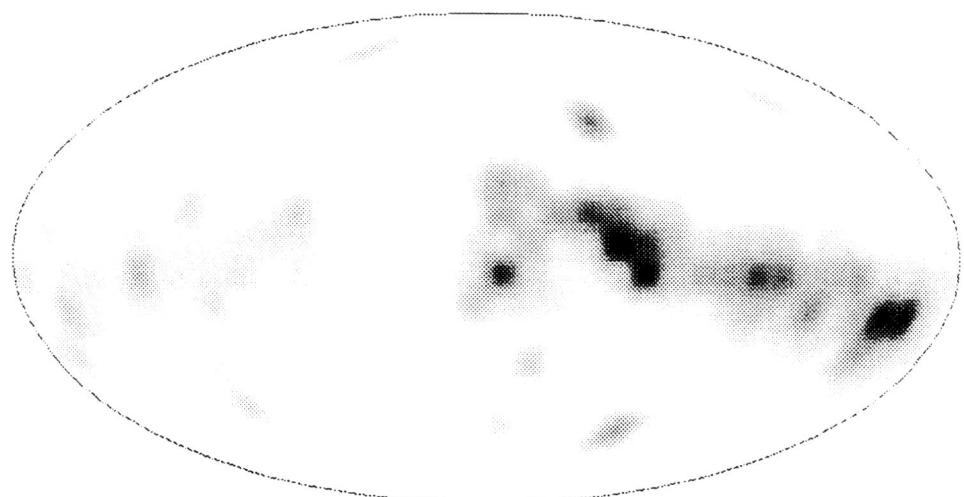

**Figure 1.** A linear, just saturated, "photograph" of the sky at 1565 Å constructed from the TD-1 observations. This image contains only the light of the stars; that is, of the source function for scattering. The north galactic pole is at the top, and the galactic center is at the center. Just more than 78% of the source function originates between galactic longitudes 180° and 360°.

For detailed study of scattered light at any particular region of the sky, one unquestionably wants to use a detailed model. However, such models are complex and not generally available. The only competing simple model is that of Jura (1979), which predicts the scattered light as a function of five variables: the source function in the disk (roughly our G), $\tau_0$ ($\sim$ 0.85, Joubert et al. 1983) the optical thickness of the galaxy in the ultraviolet, $a$, $g$, and the galactic latitude $b$. Use of Jura's model is illustrated nicely in Joubert et al. 1983. We prefer our model: because of its simplicity (no evaluation of an exponential integral is required); because it is valid for all values of $g$ (Jura's model fails for large values of g); and because it does *not* give a galactic latitude dependence: the source function of Fig. 1 shows clearly that asymmetry in galactic longitude is very strong (Henry 1977), equalling that in latitude (just more than 78% of the source function originates at $|b < 21°|$). If more than a crude estimate of the flux expected at high latitudes is desired, use of a detailed model is a necessity.

A crude estimate using our model is, however, very revealing. In Table 1 we present as a function of the Henyey-Greenstein scattering parameter $g$, the predicted high galactic latitude flux, from our model and from that of Jura (for $b = 90°$), using the values that were specified above for the necessary parameters.

The observed level of cosmic background reported at moderate and high latitudes by large numbers of observers is about 300 units (Henry 1991). A glance at Table 1 shows that if the value of $g$ in the ultraviolet is, say, 0.6 or greater (keeping in mind the conservative values of our chosen parameters), then the cosmic high-latitude background is not scattered starlight and is presumably extragalactic.

An important new measure of $g$ in the ultraviolet has been obtained on the *Astro* mission by Witt et al. 1992. Their value is $\sim$ 0.75, and they emphasize that their result is model-independent. Results similar to those of Witt et al. for $a$ and $g$ were also

## Table 1.
### Backscattered Light $S$ (units) at High Latitudes as a Function of $g$

| g | B | S(present) | S(Jura) |
|---|---|---|---|
| 0.98 | 0.004 | 7 | (−53) |
| 0.90 | 0.023 | 38 | (7) |
| 0.80 | 0.051 | 84 | 81 |
| 0.75 | 0.067 | 110 | 118 |
| 0.70 | 0.084 | 139 | 155 |
| 0.60 | 0.124 | 205 | 229 |
| 0.40 | 0.225 | 372 | 378 |
| 0.30 | 0.286 | 473 | 452 |
| 0.00 | 0.500 | 828 | 675 |
| −0.90 | 0.977 | 1619 | 1342 |

obtained by Henry (1981). We conclude that it appears likely that the high latitude diffuse background is extragalactic in its origin.

## REFERENCES

Bless, R. C., & Savage, B. D. 1972, *Ap.J.*, **171**, 293

Hauser, M. G., Gillett, F. C., Low, F. J., Gautier, T. N., Beichman, C. A., Neugebauer, G., Aumann, H. H., Baud, B., Boggess, N., Emerson, J. P., Houck, J. R., Soifer, B. T., & Walker, R. G. 1984, *Ap.J.*, **278**, L15

Henry, R. C. 1977, *Ap.J. (Suppl.)*, **33**, 451

Henry, R. C. 1981, *Ap.J.*, **244**, L69

Henry, R. C. 1991, *ARA&A*, **29**, 89

Henry, R. C., Anderson, R. C., & Fastie, W. G. 1980, *Ap.J.*, **239**, 859

Henyey, L. G., & Greenstein, J. L. 1941, *Ap.J.*, **93**, 70

Joubert, M., Masnou, J. L., Lequeux, J., Deharveng, J. M., & Cruvellier, P. 1983, *A.&A.*, **128**, 114

Jura, M. 1979, *Ap.J.*, **227**, 798

Witt, A. N., Petersohn, J. K., Bohlin, R. C., O'Connell, R. W., Roberts, M. S., Smith, A. M., & Stecher, T. P. 1992, *Ap.J.*, **395**, L5

# ON THE ABSORPTION OF UV RADIATION IN LYα CLOUDS

Valery Khersonsky and David Turnshek
Department of Physics and Astronomy
University of Pittsburgh
Pittsburgh, PA 15260

Neutral hydrogen clouds distributed throughout the Universe form a class of objects which have an important influence on the absorption of UV radiation at different cosmological epochs. A general understanding of the neutral hydrogen clouds might be achieved through a better understanding of the Lyα forest clouds seen in quasar spectra. However, the origin and structure of the putative *primordial* Lyα clouds are poorly understood up to now. A promising model of the Lyα clouds in general was developed by Rees (1986) [see also Ikeuchi (1991), Miralda-Escudé and Rees (1993)]. In this model 'minihalos' of dark matter (DM) are postulated to provide the potential wells required to stabilize the Lyα clouds. Minihalos are natural products of the cold dark matter (CDM) scenario. Therefore, investigations of Lyα clouds in the framework of this model allow us to obtain some parameters of the DM at different epochs in the evolution of the Universe. An understanding of the Lyα clouds would provide the basis for a correct approach to the problem of their optical depth, $\tau^C_\lambda(z, z_e)$. Here $\lambda$ is the observed wavelength at an epoch with redshift $z$ and $z_e$ is the redshift at the emission epoch. In this contribution we wish to briefly discuss two points:

(a) *The statistical approach to absorber description.* The optical depth can be estimated using the equation

$$\tau^C_\lambda(z, z_e) = \int_z^{z_e} dz' \int_0^\infty dN \left\{1 - exp\left[-N \sum_X x_X(\sigma_X^{ph}(\lambda') + \sigma_X^{res}(\lambda'))\right]\right\} \left(\frac{d^2 \mathcal{N}_C(z', N)}{dz' dN}\right). \quad (1)$$

Here $\sigma_X^{ph}(\lambda')$ is the photoionization cross section, $\sigma_X^{res}(\lambda')$ is the cross-section for the resonance transition, $\lambda'$ is the wavelength of radiation which is absorbed at epoch $z'$, $\lambda' = \lambda/(1+z')$, $X$ = HI, HeI and HeII, $N$ is a neutral hydrogen column density of a cloud, and $x_X$ is the fractional abundance of atoms or ions with respect to the neutral hydrogen atoms. The distribution of the number of $Ly\alpha$ clouds in column densities and redshifts, $d^2 \mathcal{N}_C/dN dz'$, is invariant with respect to the epoch of the observation. Therefore we can consider it with respect to the current epoch. To connect this distribution with the properties of $Ly\alpha$ clouds we take into account that a line of sight has a probability $2p(dp/dN)dN$ ($p$ is impact parameter) to be associated with an HI column density between $N$ and $N + dN$. This probability depends on the set of cloud properties which we designate as $(\beta)$. Then the probability under discussion can be presented as $2p(N, \beta(z))(dp(N, \beta(z))/dN)dN$ where $\beta(z)$ expresses the evolution of the cloud properties. We can write that,

$$\frac{d^2\mathcal{N}_C(z,N)}{dzdN} = -\frac{2c}{H_0(1+z)^2\sqrt{1+\Omega_0 z}} \int_{(\beta)} n_C(z,\beta(z)) \frac{dp(N,\beta(z))}{dN} p(N,\beta(z)) d\beta(z). \quad (2)$$

The integral over $(\beta)$ means averaging over the properties of the clouds with the distribution function $n_C(z_e,(\beta))$, which is simply the number density of clouds with given properties. In turn, the properties of this distribution are connected *with the distribution and properties of DM minihalos*. Thus, the integral in equation (2) in fact means averaging over the properties of DM minihalos with the distribution function $n_{DM}(z_e,(\beta))$, which is simply the number density of minihalos with given properties at redshift $z_e$. Equations (1) and (2) provide a complete statistical description of the contribution of Ly$\alpha$ clouds to the absorption of UV radiation in the DM minihalo model.

(b) *The expected properties of the absorption profiles*. The characteristics of Ly$\alpha$ clouds in hydrostatic equilibrium in a DM potential well are strongly dependent on the properties of the potential well and the flux of ionizing radiation coming into the cloud from the intergalactic and intercloud medium. The distribution of gas in the DM potential well must be strongly inhomogeneous (Rees 1988) and we can expect higher density gas in the central parts of the well. Since a medium with inhomogeneous density is usually thermally unstable, this might lead to the formation of a small cold region. If such were the case, a Ly$\alpha$ cloud might consist of a region with two phases: hot ($T > 10,000K$) and warm ($T \approx 5000 - 7000K$). The actual equation of state will depend on the flux of UV ionizing radiation. In particular, since the background UV ionizing flux is much less at lower redshift, we might expect low-redshift clouds to more easily form a multi-component structure with the warm and cold regions of neutral gas. If such a region were seen in absorption in a background quasar spectrum, evidence for a narrow component in the spectral line profile might be seen. Our illustrative calculations show that such a line profile would be composed of a narrow Doppler core (having a Doppler width of about $10\ km\ s^{-1}$ if the temperature of the dense warm part is $T \approx 5000\ K$) and wide wings corresponding to absorption from the tenuous hot parts of the cloud. Recently, Pettini et al. (1991) published evidence for the presence of such narrow lines in the spectrum of the QSO 2206-199N. Their results have come under close scrutiny (see Webb and Carswell 1991). However, if their results are confirmed, that would mean that warm gas has been detected in Ly$\alpha$ clouds even at redshift $z > 2$.

Ideally, one would like results from a statistical study of many different Ly$\alpha$ line profiles. This might provide information on the distribution of parameters for DM wells. Such data could be combined with statistical investigations of $d^2\mathcal{N}_C(z,N)/dzdN$, which depends on the distribution of DM wells in space. By making some assumptions about the distribution of DM wells in space and making fits to observational data, we could expect to determine some of the parameters which define the distribution of CDM fluctuations.

Thus, by combining results on the incidence of Ly$\alpha$ lines as a function of redshift together with results from high-resolution studies of line profiles, we might gain a better understanding of the nature of the DM and at the same time correctly account for the absorption of UV radiation from QSOs and star forming galaxies by Ly$\alpha$ clouds.

## REFERENCES

Ikeuchi, S. 1991, *Adv. Space. Res.*, **11**, No. 2, 245

Miralda-Escudé, J., and Rees, M. J. 1993, *MNRAS*, **260**, 617

Pettini, M., Hunstead, R. W., Smith, L. J., and Mar, D. P. 1991, *MNRAS*, **246**, 545

Rees, M. J. 1986, *MNRAS*, **218**, 25P

Rees, M. J. 1988, in QSOs Absorption Lines: Probing the Universe, eds. Blades, Turnshek, and Norman, (CUP), p. 107

Webb, J. K., and Carswell, R. F. 1991, *Proc. of the ESO Mini-Workshop on Quasar Absorption Lines*, eds. Shaver, Wampler and Wolfe, (ESO), p. 3

# ON THE CONTRIBUTION OF STAR FORMING GALAXIES TO THE UV BACKGROUND RADIATION

Valery Khersonsky and David Turnshek
Department of Physics and Astronomy
University of Pittsburgh
Pittsburgh, PA 15260

The contribution of star forming galaxies (SFGs) to the UV background radiation at different cosmological epochs is discussed in the framework of a model in which the galaxy mass spectrum (GMS) evolves via mergers. Mergers initiate global star formation processes in galaxies. We have performed illustrative calculations of this process by considering mergers of galaxies as a coagulation of particles. Using data about the spectral energy distributions (SEDs) of SFGs and mass/luminosity ratios, the total UV emissivity at different redshifts can be estimated and these results can be compared with the contribution from QSOs.

## BASIC ASSUMPTIONS

1. *Mergers of galaxies dominate the process of forming the GMS during the evolution of the Universe.* Evolution of the GMS starts from an initial mass spectrum, $n(x, z_0)$, at epoch $z_0$. The parameters of the initial mass spectrum and redshift $z_0$ have to be selected by trying to obtain the best fit to the GMS at the current epoch which can be calculated from the Schechter function.

2. *Galaxies contain approximately 30% baryonic matter*; $\Omega = 1$ and $h_{100} = 0.55$ apply.

3. *Global star formation on galactic scales in galaxies is initiated by interactions and mergers.* Local star formation is not considered. The duration of star formation is $10^8$ years.

4. *The number density of SFGs is assumed to be zero at redshift $z = z_0$.*

5. *During star formation the normalized SED changes in the manner described by Bruzual (1983a, 1983b).* The SED is the same for all SFGs. The contribution of different galaxies to the background radiation emissivity is determined by their total luminosities or, equivalently, their total masses.

## MAIN CONCLUSIONS

1. The evolution of the GMS via mergers starts approximately at redshift $z = 11$. The initial GMS contains mostly dwarf galaxies. The total number density of these dwarfs is six orders of magnitude larger than at the current epoch (Figure 1).

2. The mean time between a collision or merger of two galaxies is larger for dwarf galaxies than for massive galaxies because of the weaker gravitational interaction of partners (Figure 2). This time is less than the current Hubble time $t_h(z)$. The frequency of collisions of galaxies of intermediate and large mass is larger than for small massive galaxies. These galaxies can undergo strong star formation, rapid evolution of chemical composition and structure, etc. Galaxies of small mass experience collisions relatively rarely and evolve slowly. They can form a population of dwarf galaxies with small metallicity, for example, extragalactic HII regions.

3. Quantitatively almost all galaxies of intermediate and large mass are SFGs at large redshift (Figure 3). At small redshift the frequency of collisions is small and the relative fraction of SFGs, $q(M, z)$ is also small. The mass spectrum of SFGs, $n_*(M, z)$, at all redshifts contains mostly dwarfs galaxies, but at small redshift the contrast between the number density of dwarf and giant galaxies is less than at large redshift (Figure 4).

4. The spectral emissivity of UV radiation is maximal at large redshift (Figure 5). The contribution of SFGs to the UV background radiation can be comparable to the contribution from QSOs. Estimates of the contribution from QSOs are discussed, for example, by Miralda-Escude and Ostriker (1990).

## REFERENCES

Bruzual, G. A. 1983a, *Ap.J. Suppl.*, **53**, 497
Bruzual, G. A. 1983b, *Ap.J.*, **273**, 105
Miralda-Escude, J. and Ostriker, J. P. 1990, *Ap.J.*, **350**, 1

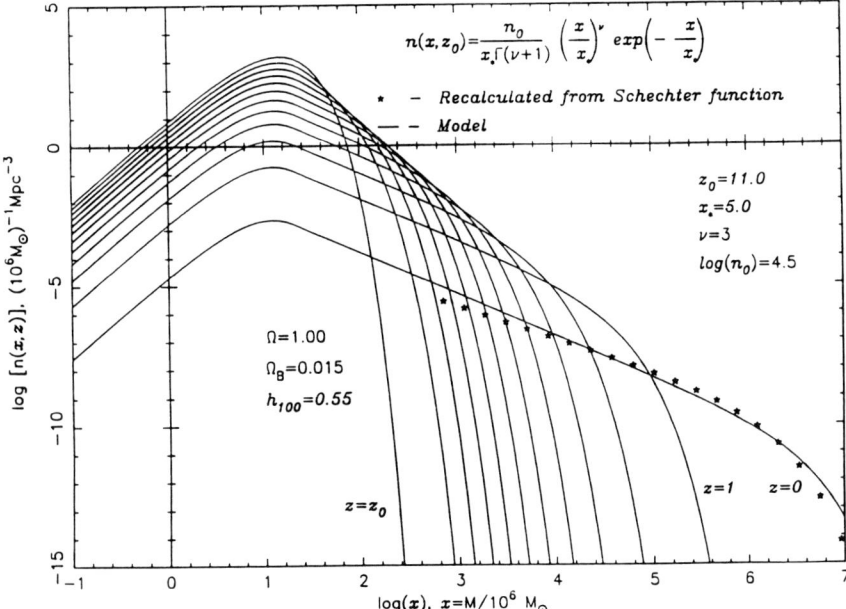

**Figure 1.** *Galaxy mass spectrum evolution.*

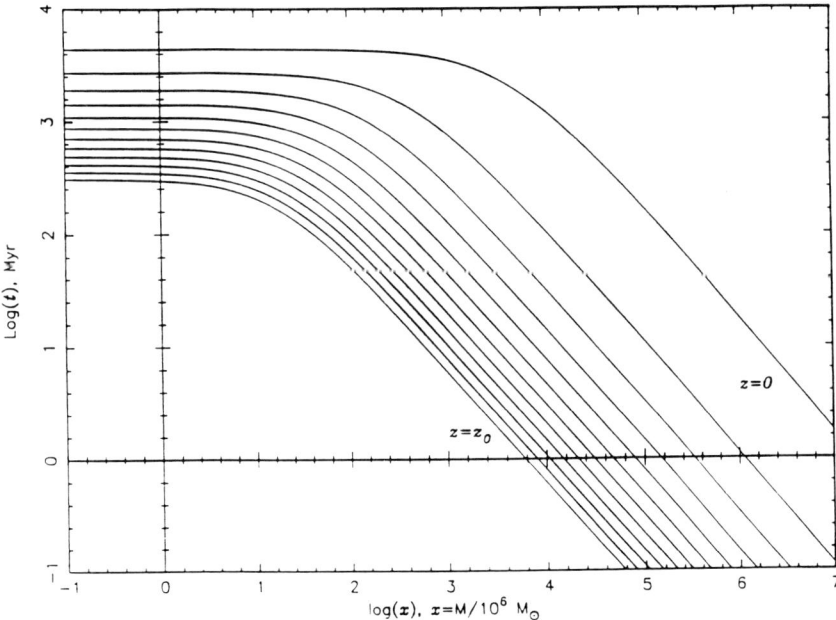

**Figure 2.** *Time between collisions for galaxy with mass M.*

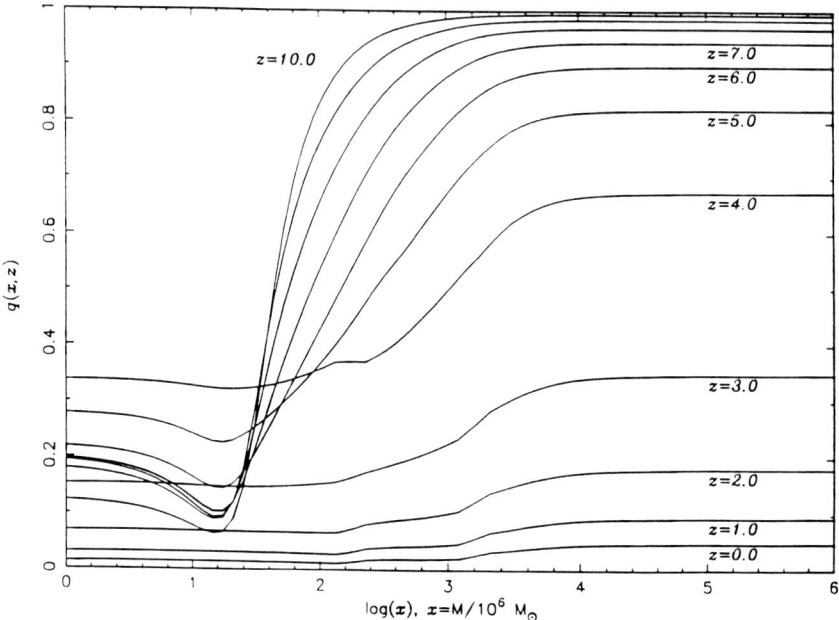

**Figure 3.** *Fraction of star forming galaxies,* q(x,z).

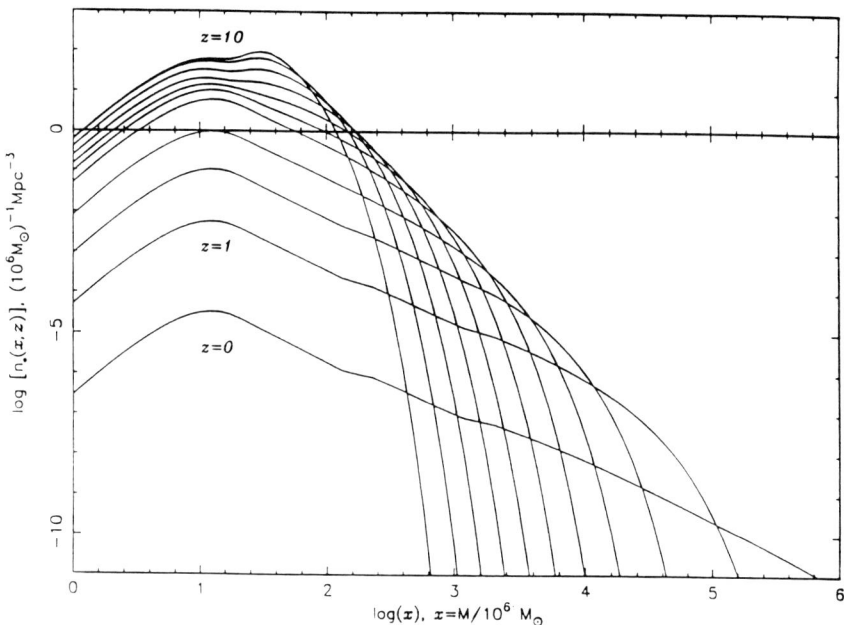

**Figure 4.** *Mass spectrum of star forming galaxies,* $n_*(x,z)$.

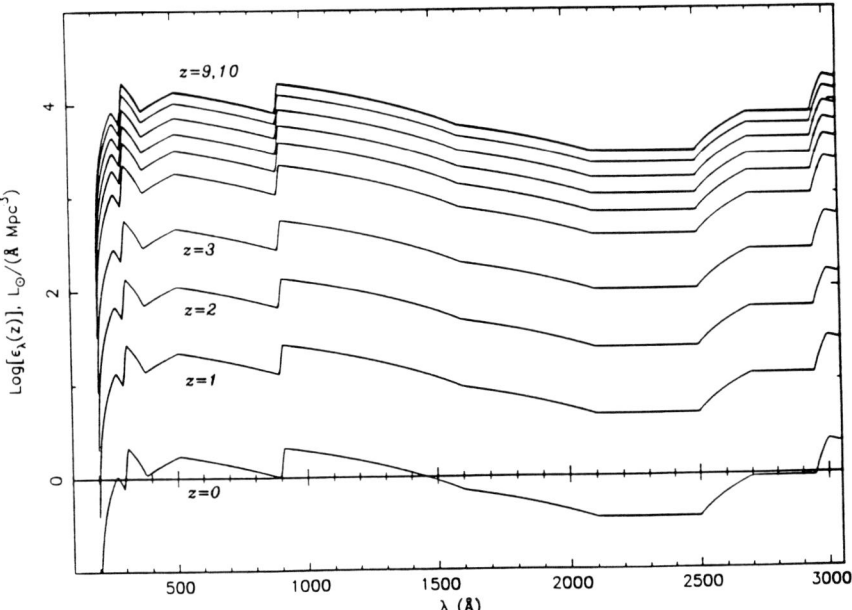

**Figure 5.** *Spectral emissivity of SFGs at different redshifts.*

# HITCHHIKER, DWARFS AND THE EBL

I. Morgan and S. P. Driver
Department of Physics and Astronomy
University of Wales College of Cardiff
P.O. Box 913
Cardiff, CF2 3YB, Wales, UK

**Abstract.** We discuss here the first results from the Hitchhiker camera. These consist of galaxy number counts in a deep field, using four filters B, V, R and I. We present an extreme dwarf dominated model which could account for the number counts and galaxy colours. We also discuss the contribution to the EBL from the detected galaxies in each band.

## THE HITCHHIKER CCD CAMERA

Hitchhiker on the 4.2m William Herschel Telescope (WHT) images the off-axis field of view allowing it to be operated whenever small field of view or spectroscopic instruments are being used. It uses two CCDs to image one field in two broadband colours simultaneously. By parallel observing, we increase the cost-effectiveness and scientific output of the telescope. (See Driver *et al.* 1993.)

## OBSERVED NUMBER COUNTS

One constraint of any model that is consistent with the faint galaxy number counts is that of the upper limits set by observations of the EBL. Mattila (1989) gives an upper limit of 6 $S_{10,V}$ at 4000Å while Toller (1983) using the Pioneer 10 spacecraft gives EBL = 1.3 $S_{10,V}$ at 4000Å. At 5100Å Dube (1977) places a value of 1.0 $S_{10,V}$. Lower limits set by number counts include Tyson's (1988) 0.5 $S_{10,V}$ at 4500Å.

Below we present a model that successfully accounts for the faint galaxy number counts and is also consistent with the above observations of the EBL. Figure 1 shows number counts slopes for a high galactic latitude deep field with exposure times 2 hrs 15 mins for the B and R filters and 30 mins for V and I (Figs. 1a, 1b, 1c and 1d). The 30 min V band field provides the first deep V band number counts to be published. For the full reduction procedure and data analysis see Driver *et al.* (1993). It is important to remember that over 5 hours of integration time on a 4 metre telescope is shown here. This was obtained while the scheduled observer was also collecting data.

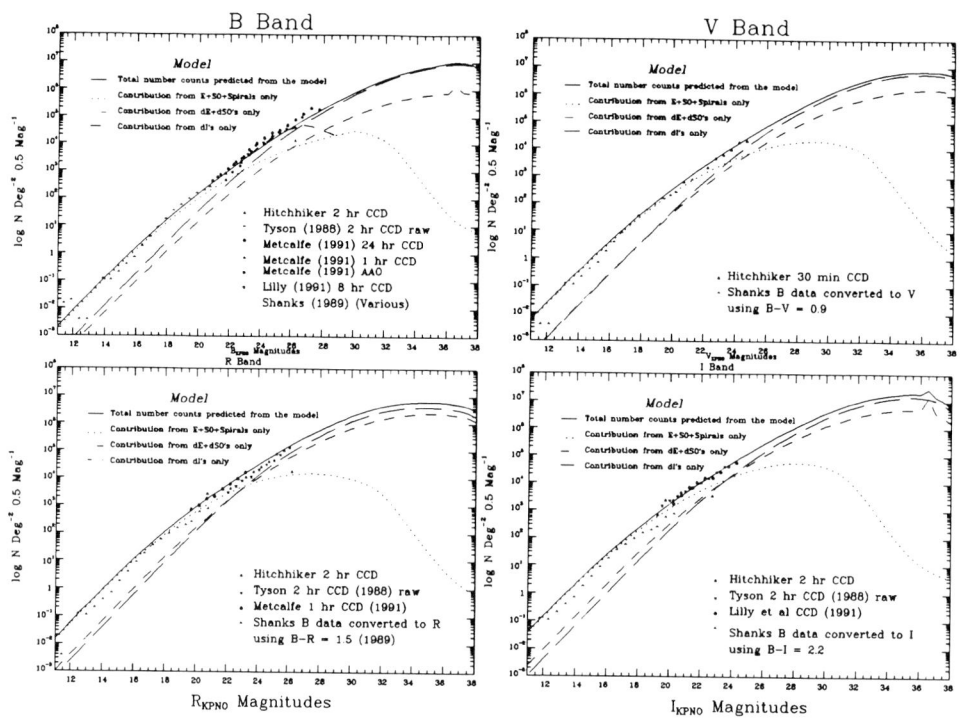

**Figure 1.** *Deep galaxy number count slopes for four filters B, V, R and I with the independent and total contributions for the 3 L.F.*

## A DWARF DOMINATED GALAXY LUMINOSITY FUNCTION

Since 1976, the Schechter luminosity function (LF) has dominated our view of the luminosity distribution of galaxies. More recently it has come under challenge (Binggeli 1986) as observations show that the Universe consists of many types of galaxy each with its own luminosity distribution. The sum total of these observed individual distributions forms an overall luminosity distribution which is inconsistent with a single Schechter function.

We therefore attempt to model the counts by using a new total LF made up of 3 Schechter functions, describing 3 broad morphological classes. The first represents normal giant galaxies (E+S0+spirals), the second early-type dwarfs (dE and dS0) and the third late-type dwarfs (dI+BCD). The steep faint end of the LF proposed by this model would imply that the universe is populated mainly by a dwarf galaxy population. The domination of numbers at faint magnitudes by dwarfs has recently been seen in nearby clusters like Virgo, (Binggeli, Sandage and Tammann 1985).

The model closely fits the number counts and slopes of the observations (see figures). Notice that using standard colour terms (*e.g.*, B-R=1.5 for giants, B-R=1.8 for dEs and B-R=0.6 for dIs) the model provides good fits for all four filters. In other words the observed number counts can be explained without the need to evoke any evolution, mergers or a past population of galaxies that has since faded.

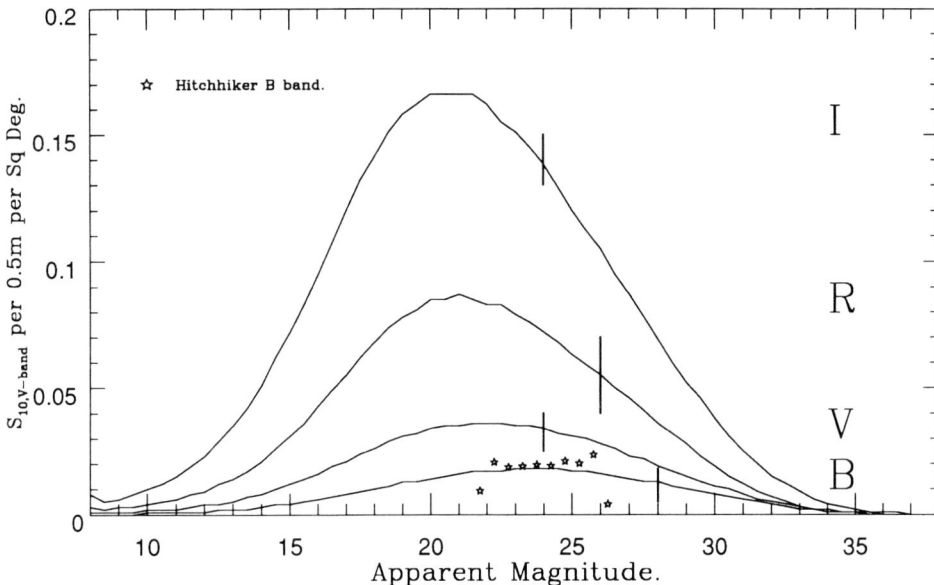

**Figure 2.** *EBL contributions for B, V, R and I filters in 0.5 magnitude bins for a dwarf dominated luminosity function.*

## THE DWARF'S CONTRIBUTION TO THE EBL

Using the number counts from the Hitchhiker camera and the simple model described above we can obtain estimates of the EBL. All observed values will be lower-limits as they are only calculated for discrete galactic sources and do not assume any diffuse background component. We adopt two approaches, firstly we use direct observations of deep number counts for the four broadband filters B, V, R, and I, and secondly the use of the dwarf dominated luminosity function discussed above. By extrapolation to fainter magnitudes this model can predict a rough value for the total EBL due to galaxies.

The values obtained for the limiting deep number counts are (observed magnitude range given in brackets):

B band $S_{10} = 0.18$ $(m_B = 21-26)$, V band $S_{10} = 0.18$ $(m_V = 22-24.5)$

R band $S_{10} = 0.79$ $(m_R = 20-26)$, I band $S_{10} = 1.43$ $(m_I = 19-22.5)$

For the model, by extrapolation down to a magnitude of $m_{Filter} = 38$ (see Fig. 1)

B band model $S_{10} = 0.45$, V band model $S_{10} = 0.90$

R band model $S_{10} = 2.02$, I band model $S_{10} = 4.09$

The contribution, to the EBL, for 0.5 magnitude bins between m=8 and m=38 for all four filters is shown in Fig. 2. The intensity is expressed in units of one $10^{th}$ magnitude A0 star per square degree for simplicity. The data points for the B band show the contribution to the EBL from galaxies detected by Hitchhiker in that magnitude bin. The vertical line marked on the plots show our present limit on deep observations.

## IMPLICATIONS

If the model is a reasonable representation of reality then in the blue band the main contribution to the EBL is due to dwarfs which (being blue) are seen at brighter magnitudes in the B band number counts. In the I band most of the EBL is due to giant galaxies at redshift of around 0.3. Notice also that the turn over is at brighter magnitudes for the I band than for the B band. This turnover occurs when the number count slope falls below 0.4; this is observed in all bands except the blue where the numbers are 'kept up' by a population of blue dwarf galaxies. If dwarfs do dominate the LF as we suggest then the EBL in the blue is dominated by the large numbers of dwarfs rather than evolved giants.

The values stated above are only a lower limit to the EBL as contributions due to diffuse light and fainter galaxies have been ignored. Encouragingly the values are similar to those quoted in the introduction. It is worth noting that most faint galaxy detection algorithms use a connected pixel technique, which bias towards the detection of compact high surface brightness objects at the limiting isophote. This strongly biases against the inclusion of very diffuse low surface brightness galaxies which would fail to meet this criteria. Any present, unobserved dwarf (or indeed giant) galaxies of central surface brightness $26B\mu$ or less would be missed and could also add a small component to the EBL.

## REFERENCES

Binggeli, B., Sandage, A., Tammann, E. A., 1985, *A.J.*, **90**, 1681

Binggeli, B., 1986, Santa Cruz Summer Workshop—Nearly Normal Galaxies, ed. Faber, p. 195

Driver, S. P., Phillipps, S., Davies, J. I., Morgan, I., Disney, M. J., 1992, *Multicolour faint galaxy number counts with the parallel CCD Hitchhiker camera*, MNRAS, in press.

Dube, R., 1977, *Ap.J.*, **215**, L51

Mattila, K., 1989, IAU Symposium No. 139, *Galactic and Extragalactic Light*

Toller, G., 1983, *Ap.J.*, **266**, L79

Tyson, A., 1988, *Ap.J.*, **96**

# ANTIPROTON LIFETIME LIMIT FROM OBSERVATION OF NUCLEON DECAY IN CLUSTERS OF GALAXIES

Daniel J. O'Connor
Istituto Nazionale di Fisica Nucleare
Via Livronese 582/A
56010 Pisa Italy

**Abstract.** The diffuse $\gamma$-ray background above 100 Mev is used to constrain antiproton lifetime limits in a universe which contains large domains of unmixed matter and antimatter on scales of cluster size and greater (*e.g.*, Steigman 1976). A limit on the antiproton lifetime times branching ratio, $\tau_{\bar{p}} \times Br$ is obtained in terms of $f_{gl}$, where $f_{gl}$ is the global antimatter:matter ratio. We use baryons in the hot gas of X-ray clusters of galaxies as a laboratory for the study of nucleon decay. For the antiproton decay mode, $\bar{p} \to e^- + \pi^\circ$, we are able calculate an expected absolute $\gamma$-ray flux as a function of $\tau_{\bar{p}} \times Br$ and the parameter $f_{gl}$. The predicted flux is compared to the measured diffuse $\gamma$-ray background above 100 Mev (Fichtel *et al.* 1978; Bignami *et al.* 1979). We obtain the limit $\tau_{\bar{p}} \times Br \geq f_{gl}\, 10^{24}$ (*sec*) (Fig. 1). The current sample of X-ray clusters contains $\sim 100$ object (Gioia *et al.* 1990; Edge *et al.* 1990). This situation will improve, with the maximum predicted number of clusters to be detected by ROSAT $\sim 10^4$ (Evrard & Henry 1991; Bohringer *et al.* 1992). Therefore for a universe with significant amounts of antimatter clusters, $10^{-4} \leq f_{gl} \leq 1.0$, we obtain an improvement in the antiproton lifetime limit from 6 to 10 orders of magnitude over current cosmic ray limits (Golden *et al.* 1979), and from 13 to 17 orders of magnitude over current laboratory limits (Bregman *et al.* 1978, Bell *et al.* 1979; Gabriclese *et al.* 1990).

## REFERENCES

Bell, M., *et al.*, 1979, *Antiproton Lifetime Measurement in the ICE Storage Ring Using a Counter Technique*, Physics Letters, **86B**, 215

Bohringer, H., *et al.*, 1992, *Clusters and Superclusters of Galaxies*, ed. A. C. Fabian, NATO ASI Series C, **366**, 71 (Kluwer Academic Press)

Bregman, M., 1978, *Measurement of Antiproton Lifetime Using the ICE Storage Ring*, Phys. Lett., **78B**, 174

Edge, A. C., *et al.*, 1990, *MNRAS*, **245**, 559

Evrard, A. E., Henry, J. P., 1991, *Expectations for X-Ray Cluster Observations by the ROSAT Satellite*, Ap.J., **383**, 95

Fabian, A. C., ed., 1992, *Clusters and Superclusters of Galaxies*, NATO ASI Series C, **366** (Kluwer Academic Press)

Fichtel, C. E., et al., 1978, *Diffuse Gamma Radiation*, Ap.J., **222**, 833
Gabrielse, G., et al., 1990, *Thousandfold Improvement in the Measured Antiproton Mass*, Phys. Rev. Lett., **65**, 1317
Gioia, I. M., et al., 1990, Ap.J., **356**, L35
Golden, R. L., et al., 1979, *Evidence for the Existence of Cosmic-Ray Antiprotons*, Phys. Rev. Lett., **43**, 1196
Steigman, G., 1976, *Observational Tests of Antimatter Cosmologies*, Ann. Rev. Astron. Astrophys.*, **14**, 339

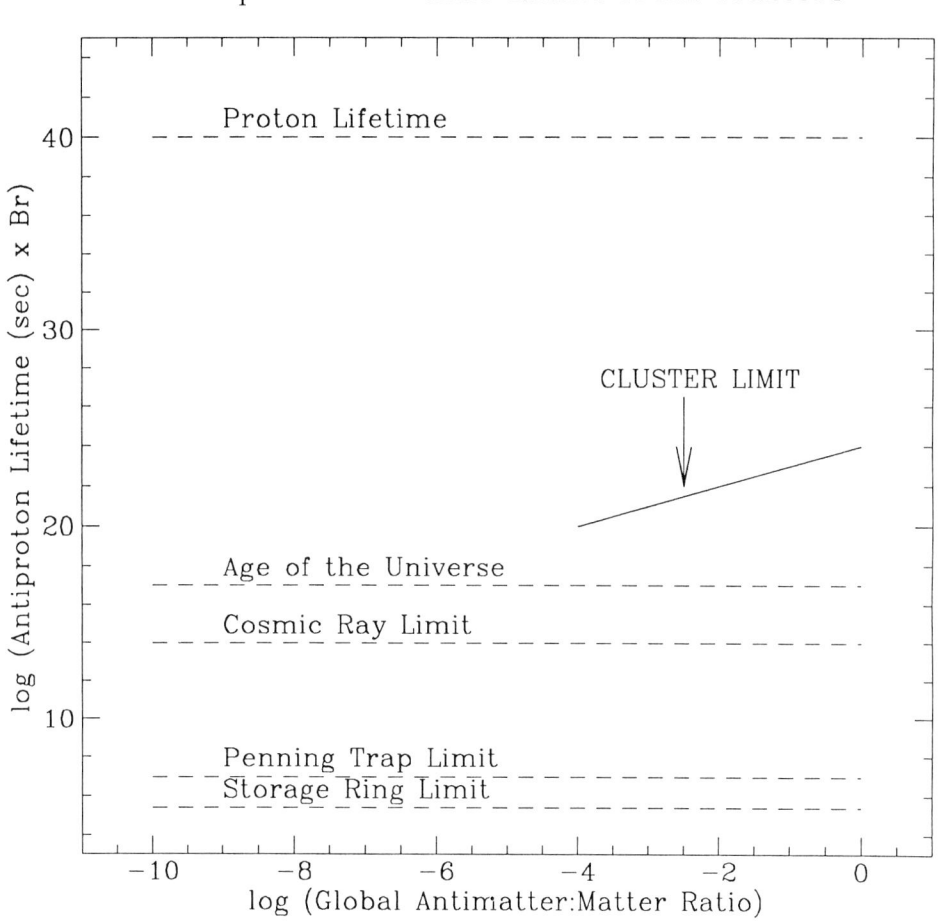

Figure 1.

# SECONDARY FLUCTUATIONS IN THE COSMIC MICROWAVE BACKGROUND

R. B. Partridge
Haverford College
Haverford, PA 19041
USA

**Abstract.** Most theoretical attention has focused on primary fluctuations in the cosmic microwave background radiation (CBR), those originating at the epoch of recombination, z ~ 1000. Secondary fluctuations originating at much lower redshifts are also possible, if the Universe is reionized. The clearest test of these possibilities is a measurement of CBR anisotropies on scales $\lesssim 5'$–$10'$. We report results of a VLA experiment at $\lambda = 3.6$ cm on scales $\lesssim 1'.5$ (Fomalont et al. 1993). The observations also constrain the evolution of the gas content of clusters of galaxies.

## 1. BACKGROUND

Observations of the isotropy of the CBR provide information about the distribution of matter on the surface of last scattering, conventionally taken to coincide with the epoch of recombination at z = 1000. CBR fluctuations arising from an inhomogeneous distribution of matter or from inhomogeneous velocities on this surface are called *primary*.

However, if the baryonic matter of the Universe is reionized at $z \ll 1000$, these primary fluctuations may be erased by the Thomson scattering of CBR photons by free electrons. The surface of last scattering moves to lower redshifts, for instance

$$z \sim 9 \quad \text{for } \Omega_b = \Omega_0 = 1, \; h = 1$$
$$\text{or } z \sim 60 \quad \text{for } \Omega_b = 0.06 \text{ and } \Omega = 1, \; h = 1.$$

As Ostriker and Vishniac (1986) pointed out, CBR fluctuations may arise on this surface as well—these are *secondary* fluctuations.

Primary fluctuations have a sharp cutoff at $\theta \sim 7'\Omega^{1/2}$ because of the non-zero thickness of the recombination surface. A similar cutoff at scales $< 10°$ would be true for secondary fluctuations as well if it were not for a variety of second order physical processes which greatly increase the amplitude of secondary fluctuations at small scales (*e.g.*, Vishniac 1987, and work in progress by Martinez-Gonzalez).

The angular resolution of the DMR instrument on COBE does not permit it to distinguish between primary and secondary fluctuations. The cleanest test would be provided by the detection of fluctuations on scales $\ll 7'\Omega^{1/2}$, which could only be secondary.

Another source of fluctuations on $\sim 1'$ scales is the inverse Compton scattering of CBR photons by hot gas in clusters of galaxies (Cole and Kaiser 1988; Schaeffer and Silk 1988; Markevitch et al. 1992). For the wavelengths of interest here, clusters would produce "dips" in the CBR temperature of order 10–100 $\mu$K.

## 2. OBSERVATIONS

We used the VLA at 8.44 GHz to search for CBR fluctuations in two deep $7' \times 7'$ sky images, both previously mapped at 5 GHz. "Bright" foreground radio sources (S > 14.5 $\mu$Jy) were removed from the map; the contribution of fainter sources was modeled by extrapolating our own (Windhorst et al. 1993) source counts to 4 $\mu$Jy using the relation $dN/dS = 4.6\ S^{-2.3}$ Jy$^{-1}$ ster$^{-1}$.

Instrument noise was determined by constructing images of one half the data subtracted from the other, and by comparing the center and the edge of the images. Unlike the case in earlier papers (Fomalont et al. 1988; Martin and Partridge 1988), we did not clean our images.

## 3. RESULTS

Upper limits on the residual variance in our images (after correction for instrument noise) are shown in the top row of the table. We find $\Delta T/T < 7.2 \times 10^{-5}$ at $\theta \sim 10''$ and at 95% confidence (Fomalont et al. 1993). Limits on fluctuations at larger scales were obtained by applying a Gaussian smoothing function (taper) to our images. At $\theta \sim 80''$, for instance, $\Delta T/T \leq 1.9 \times 10^{-5}$. Note that our sensitivity is close to the best filled-aperture results.

Our upper limits lie close to the predictions by Vishniac (1987). They also constrain the models of Markevitch et al. (1992). In these models, the number of clusters as a function of temperature decrement depends on $\Omega$ and especially $z_{\max}$, the "turn-on" epoch for cluster formation. They use earlier VLA results to show $z_{\max} < 5$. The stringent limit they obtain, however, is based on the incorrect assumption that the VLA response is equally sensitive across the entire field. I have corrected for this (approximately) in calculating our own, new, upper limits. As it happens, our results also require $z_{\max} < 5$.

**Limits to CBR Fluctuations (Both Fields)**

| Resolution | | Residual Variance $\sigma_r^2$ | | 95% Confidence Limits | |
|---|---|---|---|---|---|
| Nominal ('') | Effective ('') | Image $\mu$Jy$^2$ | Sky $\mu$Jy$^2$ | Sky $\mu$Jy$^2$ | $\Delta T_r/T_{cbr}$ $10^{-5}$ |
| 10 | 13.8 | 0.97 ± 0.87 | 2.43 ± 2.18 | < 6.03 | < 7.2 |
| 18 | 20.9 | 2.7 ± 3.4 | 6.8 ± 8.5 | < 20.8 | < 5.8 |
| 30 | 30.7 | 3.4 ± 9.0 | 8.5 ± 22.5 | < 45.6 | < 4.0 |
| 60 | 56.6 | −11 ± 41 | −28 ± 103 | < 170 | < 2.3 |
| 80 | 78.3 | −61 ± 126 | −153 ± 315 | < 520 | < 2.1 |
| 80* | 78.3 | 60 ± 73 | 150 ± 183 | < 452 | < 1.9 |

*Weak sources included.

## REFERENCES

Cole, S. and Kaiser, N. 1988, *M.N.R.A.S.*, **233**, 637.

Fomalont, E. B., Partridge, R. B., Lowenthal, J. D. and Windhorst, R. A. 1993, *Ap.J.*, **404**, 8.

Markevitch, M., Blumenthal, G. R., Forman, W., Jones, C. and Sunyaev, R. A. 1992, *Ap.J.*, **395**, 326.

Ostriker, J. P. and Vishniac, E. T. 1986, *Ap.J.(Letters)*, **306**, L51.

Schaeffer, R. and Silk, J. 1988, *Ap.J.*, **333**, 509.

Vishniac, E. T. 1987, *Ap.J.*, **322**, 597.

Windhorst, R. A., Fomalont, E. B., Partridge, R. B. and Lowenthal, J. D. 1993, *Ap.J.*, **405**, 498.